Lᵖ-Theory of Cylindrical Boundary Value Problems

Tobias Nau

L^p-Theory of Cylindrical Boundary Value Problems

An Operator-Valued Fourier Multiplier and Functional Calculus Approach

 Springer Spektrum

RESEARCH

Tobias Nau
Konstanz, Germany

Dissertation der Universität Konstanz
Tag der mündlichen Prüfung: 07.02.2012
1. Referent: Prof. Dr. Robert Denk
2. Referent: Prof. Dr. Jürgen Saal

ISBN 978-3-8348-2504-9   ISBN 978-3-8348-2505-6 (eBook)
DOI 10.1007/978-3-8348-2505-6

The Deutsche Nationalbibliothek lists this publication in the Deutsche Nationalbibliografie;
detailed bibliographic data are available in the Internet at http://dnb.d-nb.de.

Springer Spektrum
© Vieweg+Teubner Verlag | Springer Fachmedien Wiesbaden 2012

Cover design: KünkelLopka GmbH, Heidelberg

Printed on acid-free paper

Springer Spektrum is a brand of Springer DE. Springer DE is part of Springer Science+Business Media.
www.springer-spektrum.de

To Rebecca
and my parents
Gerhard & Verena Nau

Acknowledgments

This is to express my gratitude to several people who helped me during the time I worked on this thesis.

First of all, I am very grateful to my supervisor Prof. Dr. Robert Denk for his support when challenging questions had to be faced. I have highly appreciated the pleasant working atmosphere over the last three years and the possibility to take part in teaching and research activities both at the University of Konstanz and at other universities. Moreover, I would like to thank him very much for the way he encouraged me as a young researcher.

I express my gratitude to Prof. Dr. Jürgen Saal who always gave generously of his time when he was still part of the PDE research group in Konstanz. Furthermore, I am very grateful for his invitations to Darmstadt later on and for his great hospitality. Writing this thesis, I benefited a lot from what I have learned there.

I am also very grateful to Prof. Dr. Wolfgang Arendt for helpful suggestions as well as for interesting discussions during my stays at the University of Ulm.

Many thanks to the PDE research group in Konstanz. Special thanks to Mario Kaip, Johannes Schnur, and Tim Seger for their cooperativeness and the good times we had.

Thanks to Springer Spektrum, in particular to Mrs. Ute Wrasmann and Mrs. Anita Wilke for their kind support. Also many thanks to Thomas Melzer and Siegfried Maier for their helpful comments on the manuscript.

I thank my parents Gerhard and Verena Nau, who have continuously helped to make possible what I have enjoyed in my life so far. Special thanks also to my sister Dr. Stephanie Nau for good advice and assistance.

Above all, I thank my wife Rebecca for enduring with a smile the times when I was preoccupied with mathematics and for all the support she has been giving me throughout the years.

Tobias Nau

Zusammenfassung

Die vorliegende Doktorarbeit behandelt Anfangs-Randwert-Probleme der Form

$$(\text{ARP}) \quad \begin{cases} \partial_t u + \mathcal{A}(x, D)u &=& f & \text{in } \mathbb{R}_+ \times \Omega, \\ \mathcal{B}(x, D)u &=& 0 & \text{auf } \mathbb{R}_+ \times \partial\Omega, \\ u|_{t=0} &=& u_0 & \text{in } \Omega \end{cases}$$

in L^p-Räumen, wobei Ω ein zylinderförmiges Gebiet ist. Das heißt $\Omega = U \times V$ ist als das kartesische Produkt zweier Mengen gegeben. Dabei sind V und U Standardgebiete, was bedeutet, dass sich in diesen Gebieten Randwertprobleme mit bekannten Methoden der L^p-Theorie behandeln lassen (siehe z.B. [DHP03]). Ferner bezeichnet $\mathcal{A}(x, D)$ einen Differentialoperator in Ω, und $\mathcal{B}(x, D)$ repräsentiert je nach Problemstellung verschiedene Operatoren auf dem Rand $\partial\Omega$.

Ziel der Arbeit ist es, für den Operator $A_\mathcal{B}$ im zugeordneten Cauchy-Problem

$$(\text{CP}) \quad \begin{cases} \dot{u}(t) + A_\mathcal{B}u(t) &=& f(t), & t \in \mathbb{R}_+, \\ u(0) &=& u_0 \end{cases}$$

maximale (L^q-)Regularität in L^p-Räumen (siehe Definition 5.1) nachzuweisen. Dazu wird ein Resultat aus [Wei01b] verwendet, das maximale Regularität unter passenden Voraussetzungen durch $\mathcal{R}$-Sektorialität des entsprechenden Operators charakterisiert. Neben Aussagen zum Cauchy-Problem (CP) ermöglicht die $\mathcal{R}$-Sektorialität dank eines Resultats von Prüss zudem Aussagen zur maximalen Regularität des Operators $A_\mathcal{B}$ im Rahmen von Volterra-Integralgleichungen (siehe [Prü93]). Damit rückt die Behandlung des zu (CP) gehörenden Resolventenproblems in Form eines parameterabhängigen Randwertproblems in den Vordergrund.

Als zentrale Voraussetzung wird in dieser Arbeit die Anforderung gestellt, dass nicht nur Ω sondern das gesamte Anfangs-Randwert-Problem (ARP) zylinderförmig ist. Diese Anforderung beinhaltet zum einen, dass der Differentialoperator $\mathcal{A}$ in zwei Teile

$$\mathcal{A} = \mathcal{A}_1 + \mathcal{A}_2$$

zerfällt, wobei $\mathcal{A}_1$ sowohl in den Koeffizienten als auch in den Differentialausdrücken nur von U, und $\mathcal{A}_2$ entspechend nur von V abhängig ist. Andererseits soll sich auch der Randoperator $\mathcal{B}$ der Zylinderförmigkeit von Ω anpassen, indem er die Gestalt

$$\mathcal{B} = \begin{cases} \mathcal{B}_1, & \text{auf } \partial U \times V, \\ \mathcal{B}_2, & \text{auf } U \times \partial V \end{cases}$$

aufweist. An dieser Stelle sei darauf hingewiesen, dass diese Rahmenbedingungen auf zahlreiche Problemstellungen in der Anwendung zutreffen. So hat zum Beispiel die Wärmeleitungsgleichung mit Dirichlet- oder Neumann-Randbedingungen zylinderförmige Gestalt. Die geforderte Struktur erlaubt den Einsatz operatorwertiger Fouriermultiplikatoren und des operatorwertigen Dunford-Kalküls sektorieller Operatoren. Mithilfe dieser Methoden kann die $\mathcal{R}$-Sektorialität oder ein beschränkter $\mathcal{H}^\infty$-Kalkül von $A_\mathcal{B}$ auf die entsprechende Eigenschaft des induzierten Operators $A_{2\mathcal{B}_2}$ zurückgeführt werden. Dieser Zugang eröffnet einen eleganten Weg maximale Regularität des Operators $A_\mathcal{B}$ nachzuweisen.

Operatorwertige Fouriermultiplikatoren werden sowohl im Kontext vektorwertiger Fouriertransformation als auch im Kontext vektorwertiger Fourierreihen eingesetzt. Dies erfolgt zunächst auf abstrakter Ebene in den Kapiteln 6 und 7: Mit einem abgeschlossenen, linearen Operator A in einem Banachraum E werden Gleichungen der Form

$$\lambda u + P(D)u + Q(D)Au = f$$

in $L^p(U, E)$ behandelt. Dabei stellt U entweder den Ganzraum $\mathbb{R}^n$ oder den Würfel $(0, 2\pi)^n$ dar, wobei im letzteren Fall zudem verallgemeinerte periodische Randbedingungen gestellt werden. Außerdem bezeichnen $P(D)$ und $Q(D)$ Differentialoperatoren, die von unterschiedlicher Ordnung sein können, und λ ist ein komplexer Parameter.

Der Zusammenhang zu Resolventenproblemen zylinderförmiger Randwertprobleme wird nun hergestellt, indem der Operator A speziell als $A_{2\mathcal{B}_2}$ im Banachraum $L^p(V)$ gewählt wird. Unter geeigneten Voraussetzungen an die Koeffizienten ist $A_{2\mathcal{B}_2}$ ein $\mathcal{R}$-sektorieller Operator mit beschränktem $\mathcal{H}^\infty$-Kalkül. Diese Eigenschaft lässt sich über die gewonnenen Resultate auf den gesamten Operator $A_\mathcal{B}$ übertragen, sofern der Operator $\mathcal{A}_1(D)$ konstante Koeffizienten hat und eine passende Parameterelliptizität aufweist. Somit wird ein zylinderförmiges Modellproblem behandelt, in das $A_{2\mathcal{B}_2}$ in unveränderter Gestalt eingeht. Obwohl sich die Ausgangslage dadurch deutlich von herkömmlichen Modellproblemen im Ganzraum unterscheidet, können in ganz ähnlicher Weise nicht-konstante Koeffizienten von $\mathcal{A}_1$ durch Lokalisierung behandelt werden. Das Hauptresultat hierzu ist Theorem 8.10 in Kapitel 8.

Ist Ω als das kartesische Produkt mehrerer Standardgebiete gegeben, so wird anstelle der operatorwertigen Fouriermultiplikatoren der operatorwertige Dunford-Kalkül und ein Resultat von Kalton und Weis eingesetzt (siehe [KW01]). Dieser Zugang wird in Kapitel 10 verfolgt. Theorem 10.5 zeigt, dass sich so unter passenden Voraussetzungen die gleichen Resultate erzielen lassen, wie sie in [DHP03] und [DDH$^+$04] für Randwertprobleme in Standardgebieten bewiesen werden. Insbesondere lassen sich mit dieser Methode vektorwertige Randwertprobleme mit operatorwertigen Koeffizienten behandeln.

Andererseits können auch neue Aussagen und vereinfachte Beweise zu klassischen Problemstellungen wie etwa der Wärmeleitungsgleichung in zylinderförmigen

Gebieten gewonnen werden. Der verfolgte Lösungsansatz erlaubt es, gemischte Dirichlet-Neumann Randbedingungen auf unterschiedlichen Komponenten des Randes vorzugeben. Die entsprechenden Aussagen zum zugehörigen Resolventenproblem des Laplace-Operators werden in Theorem 8.22, Theorem 10.9 und Theorem 10.13 bewiesen.

Ein Anwendungsbereich dieser Ergebnisse ist die Kühlung rechtwinkliger oder zylinderförmiger elektronischer Bauteile über Kühlsysteme, die auf zwei gegenüberliegenden Seiten oder auch nur auf einer Seite des Bauteils angebracht sind. Des Weiteren sind Anwendungen im Bereich der Zellbiologie gegeben. Hier werden Diffusionsprozesse in sogenannten Beobachtungsfenstern modelliert. Dabei handelt es sich um kleine, würfelförmige Bereiche innerhalb einer Zelle, wobei davon ausgegangen wird, dass sich die Strukur der Zelle über angrenzende Zellbereiche hinweg periodisch fortsetzt. Dem wird durch periodische Randbedingungen Rechnung getragen. Da im Rahmen dieser Arbeit auch Mischungen aus periodischen und Dirichlet-Neumann Randbedingungen behandelt werden, lassen sich darüber hinaus auch membranöse Begrenzungen des Beobachtungsfensters modellieren.

Als weitere Anwendung wird in Kapitel 9 das Resolventenproblem zur Stokes-Gleichung in Schichten und in einer Klasse rechtwinkliger, zylinderförmiger Gebiete behandelt. Die Stokes-Gleichung geht durch Linearisierung aus der Navier-Stokes-Gleichung der Strömungsdynamik hervor, weshalb zylinderförmige Gebiete von besonderem Interesse sind. Zunächst werden periodische Randbedingungen behandelt. Insbesondere wird eine periodische Helmholtz-Zerlegung des zugehörigen L^p-Raums über Fouriermultiplikatoren hergeleitet. Durch den Einsatz von Reflektionstechniken wird schließlich die Existenz der klassischen Helmholtz-Zerlegung des zugehörigen L^p-Raums auf dieses periodische Analogon zurückgeführt. Die Hauptresultate dieses Kapitels sind in Theorem 9.15 und Theorem 9.17 zu finden.

Contents

List of Figures

1 Introduction and main results

The aim of this thesis is to develop a vector-valued L^p-approach to initial boundary value problems of the type

(1.1)
$$
\begin{aligned}
\partial_t u + \mathcal{A}(x, D)u &= f && \text{in } \mathbb{R}_+ \times \Omega, \\
\mathcal{B}(x, D)u &= 0 && \text{on } \mathbb{R}_+ \times \partial\Omega, \\
u|_{t=0} &= u_0 && \text{in } \Omega
\end{aligned}
$$

on cylindrical domains Ω. Here cylindrical means that Ω is of the form

(1.2)
$$
\Omega = U \times V,
$$

where V is a standard domain and U is given as a full space, a half space, a cube or a standard domain again. More than that, U itself can be given as the Cartesian product of finitely many domains of these types. For the sake of simplicity, throughout the introduction a standard domain is supposed to be a bounded, smooth domain. Furthermore, $\mathcal{A}(x, D)$ is a differential operator in Ω and $\mathcal{B}(x, D)$ is a multiple boundary operator, i.e. in general it represents several operators that act on the boundary $\partial\Omega$ of Ω.

Our main interest is to prove maximal $(L^q\text{-})$regularity on L^p-spaces (see Definition 5.1) of the operator $A_{\mathcal{B}}$ which appears in the corresponding Cauchy problem

(1.3)
$$
\begin{aligned}
\dot{u}(t) + A_{\mathcal{B}}u(t) &= f(t), && t \in \mathbb{R}_+, \\
u(0) &= u_0.
\end{aligned}
$$

Unless specified otherwise, we consider $1 < p < \infty$. To establish maximal regularity, we make use of the celebrated result of Weis in [Wei01b], roughly saying that $A_{\mathcal{B}}$ enjoys the property of maximal regularity if and only if it is $\mathcal{R}$-sectorial. Besides problem (1.3), integral equations of Volterra type involving $A_{\mathcal{B}}$ are considered. Due to a result of Prüss (see Theorem 5.8) questions on maximal regularity are answered in terms of $\mathcal{R}$-sectoriality also in this context.

Let the underlying cylindrical domain Ω in (1.1) be replaced by a full space, a half space or standard domain V for a moment. With $m \in \mathbb{N}$ consider a differential operator

(1.4)
$$
\mathcal{A}(x, D) = \sum_{|\alpha| \le 2m} a_\alpha(x) D^\alpha
$$

of order $2m$. Let $\mathcal{B}$ further represent m boundary operators

(1.5)
$$
B_j(x, D) = \sum_{|\beta| \le m_j} b_\beta(x) D^\beta, \quad m_j < 2m, \quad j = 1, \ldots, m.
$$

By means of results in [DHP03] and [DDH$^+$04], with a UMD space F, the corresponding $L^p(V, F)$-realization $A_{\mathcal{B}}^V$ with domain

$$D(A_{\mathcal{B}}^V) := \left\{ u \in W^{2m,p}(V, F);\ \mathcal{B}(\cdot, D)u = 0 \right\}$$

is known to enjoy the property of maximal regularity. In these works parameter-ellipticity and suitable assumptions on the coefficients are imposed to prove $\mathcal{R}$-sectoriality and boundedness of the $\mathcal{H}^\infty$-calculus (see Propositions 8.3 and 8.4). The proofs rely on a localization of the boundary which makes the class of standard domains advisable.

In this thesis we pursue a different strategy based on the restriction that not only Ω but the entire boundary value problem (1.1) is cylindrical. It implies that $\mathcal{A}$ resolves into two parts

$$\mathcal{A} = \mathcal{A}_1 + \mathcal{A}_2$$

such that $\mathcal{A}_1$ acts merely on U and $\mathcal{A}_2$ acts merely on V. Accordingly, $\mathcal{B}$ is assumed to resolve into two parts

$$\mathcal{B} = \begin{cases} \mathcal{B}_1, & \text{on } \partial U \times V, \\ \mathcal{B}_2, & \text{on } U \times \partial V \end{cases}$$

as well. Note that many standard systems such as the heat equation with Dirichlet or Neumann boundary conditions are of this form.

We essentially take advantage of this cylindrical structure of the entire problem in order to employ operator-valued multiplier theory and an operator-valued Dunford functional calculus to address problem (1.1). By means of these methods $\mathcal{R}$-sectoriality or boundedness of the $\mathcal{H}^\infty$-calculus of $A_{\mathcal{B}}$ in $L^p(\Omega, F)$ is reduced to the corresponding result on the induced operator $A_{2\mathcal{B}_2}$ in $L^p(V, F)$. This approach reveals a short and elegant way to prove maximal regularity for boundary value problems of type (1.1) on cylindrical domains of the form (1.2).

The operator-valued Fourier multiplier approach includes both multipliers in the context of vector-valued Fourier transform and Fourier series. In Chapters 6 and 7 it is introduced in the following rather abstract settings. Given a closed linear operator A in a Banach space E, we investigate equations of the form

$$(1.6) \qquad\qquad \lambda u + P(D)u + Q(D)Au = f$$

in $L^p(U, E)$, where U is given as the whole space $\mathbb{R}^n$ or the cube $(0, 2\pi)^n$. Here $P(D)$ and $Q(D)$ define partial differential operators of possibly different orders and λ is a complex parameter.

In the case $U = \mathbb{R}^n$ from (1.6) the operator-valued Fourier symbol of resolvent type

$$(1.7) \qquad\qquad m \colon \mathbb{R}^n \to \mathcal{L}(E);\ m(\xi) := (\lambda + P(\xi) + Q(\xi)A)^{-1}$$

occurs. Thus, we are faced with the question which conditions on m ensure that the associated Fourier operator $\mathcal{F}^{-1}m\mathcal{F}$ defines the solution operator to (1.6) in $L^p(\mathbb{R}^n, E)$. In Chapter 6 we give an answer to this question by specifying more abstract conditions from [Wei01b]; see also [ŠW07] and [HHN02].

In the case $U = (0, 2\pi)^n$ equation (1.6) is supplemented with generalized periodic boundary conditions and Fourier transform is replaced by Fourier series. Consequently, m turns into its discretization

$$M \colon \mathbb{Z}^n \to \mathcal{L}(E); \ M(k) := (\lambda + P(k) + Q(k)A)^{-1}$$

and the E-valued distributions under consideration change from $\mathcal{S}'(\mathbb{R}^n, E)$ to the space $\mathcal{D}'_{per}(\mathbb{R}^n, E)$ of periodic distributions. In contrast to extensive theory on E-valued distributions in general and E-valued tempered distributions in particular, the contribution in literature to E-valued periodic distributions is rather sparse. We fill this gap in Chapter 2 providing useful results on this class of E-valued distributions. The first result is Theorem 2.7, a representation theorem for E-valued periodic distributions by means of multiple, E-valued distributional Fourier series. The latter enables us to prove a powerful uniqueness theorem on Fourier series in Corollary 2.9. Besides that, it serves as a starting point to give a comprehensive characterization of functions belonging to periodic Sobolev spaces in Theorem 2.14. The representation theorem is extended to partially periodic E-valued distributions in Theorem 2.19. Later on, the construction of a Helmholtz projection in Chapter 9 relies on this result. Thus, apart from their own interest, the results achieved in this chapter are indispensable for later applications.

A first result on boundedness of the Fourier operator in $L^p((0, 2\pi)^n, E)$ associated with M was given by Arendt and Bu in [AB02]; see also [BK04]. As pointed out by the authors their results can as well be deduced from a more involved multiplier theorem in [ŠW07]. In so doing, we observe that the assumptions imposed in [AB02] and [BK04] can be relaxed. This result is made available in Theorem 3.24. The less restrictive assumptions of Theorem 3.24 are far from being artificial. In fact, this very improvement allows for a successful construction of the Helmholtz projection as indicated above.

The conditions on m and M which all operator-valued multiplier theorems mentioned have in common involve $\mathcal{R}$-boundedness of $\{m(\xi); \ \xi \in \mathbb{R}^n \setminus \{0\}\}$ and $\{M(k); \ k \in \mathbb{Z}^n\}$, respectively. In view of (1.7), $\mathcal{R}$-sectoriality of A thus provides a starting point to approach equation (1.6). Already in this rather abstract setting the idea of transference becomes apparent. Under suitable assumptions on the Banach space E, let $\mathbb{A}$ define an appropriate realization of (1.6) in $L^p(U, E)$. If A is $\mathcal{R}$-sectorial in E, with the aid of the operator-valued multiplier theorems the same property is inferred for the operator $\mathbb{A}$ in $L^p(U, E)$. Here the differential operators P and Q are assumed to be parameter-elliptic, where a Dore-Venni-type condition for the corresponding angles of parameter-ellipticity and the angle of $\mathcal{R}$-sectoriality of A has to be satisfied.

For the application we have in mind, A is specified to be given as $A_{2\mathcal{B}_2}$ in the Banach space $E := L^p(V)$. This leads to maximal regularity on $L^p(\Omega)$ for Cauchy problems associated with cylindrical boundary value problems that are partially endowed with generalized periodic boundary conditions. These types of boundary conditions appear, for instance, in the study of keratin network growth in biological cells (see [ABF$^+$08]). Note that $A_{2\mathcal{B}_2}$ is treated in full generality at once so that we are by no means restricted to constant coefficients in $\mathcal{A}_2$. In order to deal with non-constant coefficients also in the first part $\mathcal{A}_1$, a localization procedure is carried out. In contrast to existing literature, however, no whole space or half space but rather a cylindrical domain serves as a model problem here. The main result in this context is Theorem 8.10. One interesting outcome is that despite the splitting property of the differential operator $\mathcal{A}$ mixed derivatives of the solution with respect to all spacial directions belong to the underlying L^p-space.

As mentioned at the beginning, we can also achieve results on maximal regularity if $\Omega := \prod_{i=1}^N V_i$ is given as the Cartesian product of finitely many but more than just two or three domains. Then differential operators

$$\mathcal{A} = \mathcal{A}_1 + \ldots + \mathcal{A}_N$$

and multiple boundary operators of the form

$$\mathcal{B} = \{\mathcal{B}_i \text{ on } V_1 \times \ldots \times V_{i-1} \times \partial V_i \times V_{i+1} \times \ldots \times V_N\}_{i=1,\ldots,N}$$

are considered (cf. Definition 10.3). Here each part $\mathcal{A}_i$ is given as a differential operator of order $2m_i$ as introduced in (1.4) and each part $\mathcal{B}_i$ is a multiple boundary operator that represents m_i boundary operators of type (1.5). Again each pair $(\mathcal{A}_i, \mathcal{B}_i)$ acts on V_i, respectively on its boundary, only.

To investigate such type problems, we adopt the operator-valued Dunford functional calculus approach. It is based on the Kalton-Weis-Theorem (see [KW01]) which combines boundedness of the scalar $\mathcal{H}^\infty$-calculus with $\mathcal{R}$-boundedness conditions on operator-valued holomorphic functions. Theorem 10.5 on problems of this type basically says that the operator-valued Dunford functional calculus approach yields exactly the same results as they are known for problems on standard domains, provided the induced boundary value problems on the different cross-sections V_i match the requirements of [DHP03] and [DDH$^+$04]. In particular, parameter-ellipticity is not imposed for the entire problem but rather for each induced problem on the single cross-sections V_i. The regularity of the solution on the other hand is not subordinate to the splitting property. For instance, if all orders of the differential operators coincide, i.e. $m_1 = \ldots = m_N =: m$, maximal regularity results on the $L^p(\Omega, F)$-realization $A_\mathcal{B}$ can be derived where

$$(1.8) \qquad D(A_\mathcal{B}) := \left\{ u \in W^{2m,p}(\Omega, F); \ \mathcal{B}(\cdot, D)u = 0 \right\}.$$

Since non-smooth and non-convex domains Ω with non-compact boundary are covered, additional importance is attached to (1.8) and Theorem 10.5.

By virtue of this approach, for a class of Banach spaces F, we are thus able to consider F-valued solutions and to allow the coefficients of each part $\mathcal{A}_i$ and $\mathcal{B}_i$ to be $\mathcal{L}(F)$-valued. Applications for equations with $\mathcal{L}(F)$-valued coefficients are, for instance, given by coagulation-fragmentation systems (cf. [AW05b]) or spectral problems of parametrized differential operators in hydrodynamics (cf. [DMT02]).

Besides boundary value problems involving $\mathcal{L}(F)$-valued coefficients, we are able to establish new results and shorter proofs in the context of classical problems such as the scalar heat equation subject to Dirichlet, Neumann or mixed Dirichlet-Neumann boundary conditions $\mathcal{B} = \mathcal{B}(D)$. We focus on these problems which are covered choosing $A_{\mathcal{B}} := -\Delta_{\mathcal{B}}$ in Sections 8.3 and 10.3. Keep in mind that V being a standard domain by (1.8) always ensures that

$$D(-\Delta_{\mathcal{B}}) = \left\{ u \in W^{2,p}(\Omega);\ \mathcal{B}(D)u = 0 \right\}.$$

Furthermore, we can handle the heat equation with mixed Dirichlet-Neumann conditions also in cylindrical domains built up by rough cross-sections V_i. To be more precise, the cross-sections V_i are allowed to be bounded Lipschitz domains. In that case, we apply results from [Woo07] to the operators acting on the single cross-sections to deduce $\mathcal{R}$-sectoriality or even an $\mathcal{R}$-bounded $\mathcal{H}^{\infty}$-calculus for the entire problem in $L^p(\Omega)$. As it is well-known, in the context of rough domains the values of p are severely limited. Moreover, knowledge on the regularity of solutions decreases in general. Consequently, a weak formulation of the negative Laplacian with

$$D(-\Delta_{\mathcal{B}}) \subset \left\{ u \in W^{1,p}(\Omega);\ \Delta u \in L^p(\Omega) \right\}$$

is used. With regard on existing literature it is interesting to note that a large class of possibly unbounded Lipschitz domains Ω and simultaneously mixed Dirichlet-Neumann type boundary conditions are covered. Moreover, new results on the range of p are established such that the Dirichlet Laplacian admits an $\mathcal{R}$-bounded $\mathcal{H}^{\infty}$-calculus on $L^p(\Omega)$.

In Chapter 9 we apply the Fourier multiplier techniques to the Stokes equation in Ω given as infinite layer or infinite rectangular cylinder. The Stokes equation arises by linearization from the Navier-Stokes equation which describes fluid dynamics. This makes these special types of cylindrical domains particularly rewarding. To solve the Stokes resolvent problem in Ω both Fourier transform and Fourier series are used. In order to define a Stokes operator in the space $L^p_\sigma(\Omega)$ of solenoidal fields, we construct the Helmholtz projection $\mathbb{P} \in \mathcal{L}(L^p(\Omega))$. To this end, we first establish a rather non-physical, periodic Helmholtz projection $\mathbb{P}_{per}$ as a Fourier multiplier operator in $\mathcal{L}(L^p(\Omega_2))$. Here the thickness of Ω_2 compared to Ω is doubled. With suitable extension and restriction operators $\mathfrak{E}$ and $\mathfrak{R}$, in Theorem 9.15 we prove the relation

$$\mathbb{P} = \mathfrak{R}\mathbb{P}_{per}\mathfrak{E}.$$

Hence, besides existence of $\mathbb{P}$ a most elegant representation formula is obtained at the same time. For the special case of an infinite layer this gives an alternative description of $\mathbb{P}$ to the ones in [Far03] and [Abe05a]/[Abe05b]. For infinite rectangular domains, as far as the author knows, no according result on the Helmholtz projection has been available up to now.

This thesis is structured as follows.

In Chapter 2 we recall Fourier transform and Fourier series in a vector-valued and distributional framework. Here periodic and partially periodic vector-valued distributions are treated extensively. Having L^p-applications in mind, we introduce ν-periodic L^p-Sobolev spaces. In Theorem 2.14 numerous characterizations for functions u of this space are deduced. Finally, we develop the concept of partial Fourier coefficients of a partially periodic distribution and prove a representation theorem in Theorem 2.19.

In Chapter 3 we introduce continuous and discrete Fourier multipliers and establish regularity results on associated operators in Lemma 3.7 and Lemma 3.11. In order to present multiplier theorems, we first recall the notions of $\mathcal{R}$-bounded operator families, Banach spaces of class $\mathcal{HT}$, and property (α). With these concepts at hand, we state two operator-valued Michlin multiplier results from literature as Theorems 3.17 and 3.19. By means of a strong multiplier result due to Štrkalj and Weis we are able to weaken the rather strict $\mathcal{R}$-boundedness condition of Theorem 3.19 which is done in Theorem 3.24. Moreover, we discuss an intermediate condition that combines flexibility and strength in a way most suitable for later purposes.

In Chapter 4 classes of operators in Banach spaces are recalled. In particular, classes of pseudo-sectorial and sectorial operators are introduced. The latter allow for a Dunford calculus which is discussed rather extensively. We define boundedness of the $\mathcal{H}^\infty$-calculus and briefly comment on the class BIP of operators that admit bounded imaginary powers. By means of $\mathcal{R}$-boundedness introduced in Chapter 3 the classes of (pseudo-)$\mathcal{R}$-sectorial operators and of operators that admit an $\mathcal{R}$-bounded $\mathcal{H}^\infty$-calculus are defined. In the sequel well-known relations of these classes are recalled. Once again the notions from Banach space geometry introduced in Chapter 3 are needed. At the end of Chapter 4 the Dore-Venni-Theorem and a more recent theorem due to Kalton and Weis on the sum of two closed operators are presented.

Chapter 5 collects definitions and results on parabolic problems. We briefly comment on maximal regularity in the context of Cauchy problems and state the mentioned result of Weis on an equivalent description of maximal regularity by means of $\mathcal{R}$-sectoriality. Sufficient conditions for maximal regularity of a generator of an analytic semigroup are recalled for later use. The notion of maximal regularity is further defined for Volterra integral equations and sufficient conditions for this property in terms of $\mathcal{R}$-boundedness due to Prüss are presented.

Given a closed operator A in a Banach space X, Chapters 6 and 7 carry out the Fourier multiplier approach to A-dependent partial differential equations in the whole space $\mathbb{R}^n$ and in the cube $(0, 2\pi)^n$. In the latter case, ν-periodic boundary conditions are imposed. In both cases, properties like $\mathcal{R}$-sectoriality and a bounded $\mathcal{H}^\infty$-calculus are transferred from A to the corresponding realizations of these problems in the respective L^p-spaces. This is done in Section 6.2 for the case of the whole space, and in Section 7.3 for the case of the cube, respectively. In Section 7.2, moreover, a characterization of unique solvability of A-dependent problems in $(0, 2\pi)^n$ is derived (see Theorem 7.15). This result implies sufficient conditions for unique solvability of the corresponding Dirichlet-Neumann problem, provided all differential operators under consideration admit an appropriate structure. Employing reflection arguments, this is proved in Proposition 7.16. All of these results rely on verified multiplier conditions, that is, on verified $\mathcal{R}$-boundedness conditions. This is carried out in Sections 6.1 and 7.1, respectively. The crucial tool to estimate $\mathcal{R}$-bounds is known as the contraction principle of Kahane (see Lemma 3.2). It is combined with parameter-ellipticity assumptions on the differential operators and properties of A. Therefore, estimates for parameter-elliptic polynomials inferred from homogeneity arguments are summarized in Lemma 6.5. In addition, various representation formulas for continuous and discrete derivatives of operator-valued functions are needed. The corresponding results are Lemma 6.1 and Lemma 7.1. A proof of the latter includes tedious technical calculations, for one has to keep track of numerous shifts in $\mathbb{Z}^n$ at the same time. Thereby, a wide class of resolvent-type Fourier multipliers is deduced in Propositions 6.8 and 7.8.

In Chapter 8 these results are applied to parameter-elliptic, cylindrical boundary value problems in general and the Laplacian on Lipschitz cylinders in particular. In Section 8.1 we first collect known results from literature on parameter-elliptic boundary value problems in standard domains V. The Laplacian is also considered on bounded domains V of Lipschitz type. In the sequel L^p-realizations of these problems replace the abstract operator A considered in Chapter 6 and Chapter 7. This allows to treat problems with constant coefficients in unbounded cylindrical domains $\Omega := \mathbb{R}^{n_1} \times (0, 2\pi)^{n_2} \times V$. Non-constant coefficients are treated later on with the aid of a localization procedure. Recall that in contrast to existing literature no full space or half space but a cylindrical domain serves as a model problem. In Section 8.3 we focus on the Laplacian in cylindrical domains augmented with mixed ν-periodic and Dirichlet-Neumann boundary conditions. Furthermore, the modeling of keratin network growth in biological cells by the Laplacian with pure periodic as well as mixed periodic and Dirichlet-Neumann boundary conditions is discussed. The main results of this chapter are Theorem 8.10 and Theorem 8.22.

Chapter 9 considers an application to the Stokes problem in infinite layers and infinite rectangular domains Ω. First we investigate the Stokes resolvent problem subject to ν-periodic boundary conditions. Here partial Fourier series with respect to bounded coordinate directions and Fourier transform with respect to unbounded coordinate directions of the domain are employed. We recourse to the results from the previous chapters to prove unique solvability in $L^p(\Omega)$ for

$1 < p < \infty$. The precise spaces for velocity field and pressure in the ν-periodic setting are formulated in Theorem 9.5. A deeper investigation of the Stokes resolvent problem yields a ν-periodic analogue of the well-known Helmholtz projection in $L^p(\Omega)$ in terms of a Fourier multiplier operator. This rather non-physical projection turns out to be closely related to the standard Helmholtz projection in that space. A representation formula which reveals the connection of both projections is established in Theorem 9.15. Again appropriate reflection techniques come into play. The Helmholtz projection allows for a definition of the Stokes operator in the space of solenoidal fields $L_\sigma^p(\Omega)$. Supplemented with pure-slip boundary conditions, $\mathcal{R}$-boundedness of the $\mathcal{H}^\infty$-calculus for the Stokes operator is proved in Theorem 9.17.

In Chapter 10 we employ the operator-valued Dunford calculus to investigate cylindrical boundary value problems in cylindrical domains built up by finitely many standard domains. Here operator-valued Fourier multipliers are replaced by the Kalton-Weis-Theorem. An essential assumption of this theorem is that extensions of operators, first defined on the single cross-sections, are resolvent commuting. This to prove is the main task in the application to cylindrical boundary value problems in Section 10.2. The main result is Theorem 10.5. The Laplacian in Lipschitz cylinders is brought into focus once more in Section 10.3. Using weaker commutator conditions the approach outlined above is also applied to a class of operators whose resolvents do not commute. This can be used to investigate heat conduction in a Lipschitz cylinder with either in longitudinal directions or in cross-sections non constant heat conductivity coefficient.

Basic notation used in this thesis as well as definitions and facts on vector-valued function spaces and vector-valued distributions are collected in Appendix A. In Appendix B we finally comment on topics in the literature which are related to parts of this thesis.

2 Vector-valued Fourier transform and Fourier series

We start this chapter by stating the main definitions and results on the vector-valued Fourier transform as presented e.g. in [Ama95] and [ABHN01]. For a comprehensive introduction to the topic we refer to the mentioned monographs and the references therein. Later on, we establish the according results for vector-valued Fourier series.

Let E denote an arbitrary Banach space and let $f \in L^1(\mathbb{R}^n, E)$. Then the E-valued Fourier transform of f is the function $\mathcal{F}f \in L^\infty(\mathbb{R}^n, E)$ defined by

$$(2.1) \qquad \mathcal{F}f(\xi) := \frac{1}{(2\pi)^{\frac{n}{2}}} \int_{\mathbb{R}^n} e^{-ix\xi} f(x)dx.$$

To be more precise, formula (2.1) induces a mapping in $\mathcal{L}\big(L^1(\mathbb{R}^n, E), C_\infty(\mathbb{R}^n, E)\big)$ where the space $C_\infty(\mathbb{R}^n, E)$ consists of the continuous functions vanishing at infinity. In (2.1) we have used the abbreviation

$$x\xi := \langle x, \xi \rangle := \sum_{j=1}^{n} x_j \xi_j$$

for the standard scalar product in $\mathbb{R}^n$. Restricted to $\mathcal{S}(\mathbb{R}^n, E)$, the Fourier transform defines an isomorphism in this space whose inverse is given by

$$(2.2) \qquad \mathcal{F}^{-1}f(\xi) := \frac{1}{(2\pi)^{\frac{n}{2}}} \int_{\mathbb{R}^n} e^{ix\xi} f(x)dx.$$

This property extends to the space $\mathcal{S}'(\mathbb{R}^n, E)$ of E-valued tempered distributions, where the extension of the Fourier transform is defined via duality, i.e.

$$(2.3) \qquad \mathcal{F}u(f) := u(\mathcal{F}f) \quad (u \in \mathcal{S}'(\mathbb{R}^n, E), f \in \mathcal{S}(\mathbb{R}^n)).$$

For $u \in L^1(\mathbb{R}^n, E)$ understood as a tempered distribution it is easily seen that this definition coincides with the one given in (2.1). The following result is of technical interest in view of coefficients in differential expressions which involve closed operators.

Lemma 2.1. *Let A be a closed operator in E. For $u \in L^p(\mathbb{R}^n, D(A))$ it holds that $\mathcal{F}u \in \mathcal{S}'(\mathbb{R}^n, D(A))$ and $\mathcal{F}Au = A\mathcal{F}u$.*

Proof. Closedness of A yields

$$A \int_{\mathbb{R}^n} u(x)\mathcal{F}f(x)dx = \int_{\mathbb{R}^n} Au(x)\mathcal{F}f(x)dx \quad (f \in \mathcal{S}(\mathbb{R}^n))$$

(see e.g. [ABHN01, Proposition 1.1.7]). Since $Au \in L^p(\mathbb{R}^n, E)$ defines a tempered distribution in $\mathcal{S}'(\mathbb{R}^n, E)$, the claim follows from the definition in (2.3). $\qquad\square$

Remark 2.2. In Lemma 2.1 we can obviously replace $\mathcal{F}$ by $\mathcal{F}^{-1}$.

When dealing with differential operators and their resolvents, it is important to be aware of a relation between the Fourier transform of a Sobolev space function and the Fourier transforms of its partial derivatives. Even for tempered distributions $u \in \mathcal{S}'(\mathbb{R}^n, E)$ this relation is given by the fundamental property

$$(2.4) \qquad \mathcal{F}(D^\alpha u)(\xi) = \xi^\alpha \mathcal{F}u(\xi) \quad (u \in \mathcal{S}'(\mathbb{R}^n, E),\ \alpha \in \mathbb{N}_0^n),$$

where $D^\alpha := D_1^{\alpha_1} \ldots D_n^{\alpha_n}$ and $D_j := -i\partial_j$. This is obvious for $u \in \mathcal{S}(\mathbb{R}^n, E)$ due to the integration by parts formula and extends to $u \in \mathcal{S}'(\mathbb{R}^n, E)$ via (2.3).

We turn our attention to E-valued Fourier series, starting with the well-known definition of E-valued Fourier coefficients as presented in [ABHN01, Section 4.2.G] for continuous E-valued functions. This definition copies verbatim to the context of functions in $L^p(\mathcal{Q}_n, E)$, where $\mathcal{Q}_n := (0, 2\pi)^n$ (cf. [AB02]). More precisely, for $f \in L^p(\mathcal{Q}_n, E)$ and $k \in \mathbb{Z}^n$

$$\hat{f}(k) := \frac{1}{(2\pi)^n} \int_{\mathcal{Q}_n} e^{-ikx} f(x) dx$$

is a well-defined element of E which is called the k-th Fourier coefficient of f. Without further calculations, the respective analogue of Lemma 2.1 follows.

Lemma 2.3. *Let A be a closed operator in E. For $f \in L^p(\mathcal{Q}_n, D(A))$ it holds that $\hat{f}(k) \in D(A)$ and $A\hat{f}(k) = (Af)\hat{\ }(k)$ for all $k \in \mathbb{Z}^n$.*

Proof. As above, closedness of A yields

$$A \int_{\mathcal{Q}_n} e^{-ikx} f(x) dx = \int_{\mathcal{Q}_n} e^{-ikx} Af(x) dx \quad (k \in \mathbb{Z}^n).$$

$\square$

For $k \in \mathbb{Z}^n$ and $\eta \in E$ set $e_k(x) := e^{ikx}$ and $(\eta e^{ik\cdot})(x) := (\eta \otimes e_k)(x) := \eta e_k(x)$ for $x \in \mathbb{R}^n$. Given arbitrary multi-indices $\alpha, \beta \in \mathbb{Z}^n$, we define $\alpha \le \beta$ by $\alpha_j \le \beta_j$ for all $j = 1, \ldots, n$. With $[\alpha, \beta] := \{k \in \mathbb{Z}^n;\ \alpha \le k \le \beta\}$ and $\eta_k \in E$,

$$f = \sum_{k \in [\alpha, \beta]} \eta_k e^{ik\cdot} = \sum_{k \in [\alpha, \beta]} \eta_k \otimes e_k$$

defines the *trigonometric polynomial* given by

$$f(x) = \sum_{k \in [\alpha, \beta]} \eta_k e^{ikx} \quad (x \in \mathbb{R}^n).$$

It satisfies $\hat{f}(k) = 0$ if $k \notin [\alpha, \beta]$ and $\hat{f}(k) = \eta_k$ else. We denote the class of E-valued trigonometric polynomials on $\mathcal{Q}_n$ by $\mathbb{T}(\mathcal{Q}_n, E)$. It is well-known that the multiple series of Fourier coefficients $(\hat{f}(k))_{k \in \mathbb{Z}^n}$ of a function $f \in L^p(\mathcal{Q}_n, E)$ does not converge to f in $L^p(\mathcal{Q}_n, E)$ in general. However, two strategies seem to be natural in order to overcome this problem.

The first one is to change the series under consideration. This is done by summing up the Cesaro means of the Fourier coefficients instead of the Fourier coefficients themselves. For $m \in \mathbb{N}_0$ and $N \in \mathbb{N}$ recall the one dimensional Dirichlet and Fejer kernel

$$D_m(t) := \sum_{k=-m}^{m} e^{ikt}, \quad F_N(t) := \frac{1}{N+1} \sum_{m \leq N} D_m(t) \quad (t \in \mathbb{R}).$$

Furthermore, recall the multiple Fejer kernel

$$F_{n,N}(x) := \prod_{j=1}^{n} F_N(x_j) \quad (x \in \mathbb{R}^n).$$

The proof of the following result is well-known for scalar-valued functions and copies verbatim to E-valued functions.

Proposition 2.4. *Let* $f \in L^p(\mathcal{Q}_n, E)$. *Then* $F_{n,N} * f \in \mathbb{T}(\mathcal{Q}_n, E)$ *for each* $N \in \mathbb{N}$ *and* $F_{n,N} * f \to f$ *in* $L^p(\mathcal{Q}_n, E)$ *for* $N \to \infty$. *In particular, the space of trigonometric polynomials* $\mathbb{T}(\mathcal{Q}_n, E)$ *is dense in* $L^p(\mathcal{Q}_n, E)$.

The second strategy is closely related to the extension of the Fourier transform from $\mathcal{S}(\mathbb{R}^n, E)$ to $\mathcal{S}'(\mathbb{R}^n, E)$ as the idea is once more to extend the definition of Fourier coefficients to a certain subspace of $\mathcal{D}'(\mathbb{R}^n, E)$. In the present case, this subspace is given by the periodic distributions in $\mathcal{D}'(\mathbb{R}^n, E)$ which includes $L^p(\mathcal{Q}_n, E)$ in some appropriate sense. Replacing L^p-convergence by convergence in $\mathcal{D}'(\mathbb{R}^n, E)$, we will see that each periodic distribution has a representation as the limit of its Fourier series. Since contributions in the literature to this topic in an E-valued setting seem to be rather sparse, we give a more detailed presentation of this theory. The results are well-known in scalar valued context (see e.g. [Wal94]) and thanks to the extensive E-valued distribution theory in [Ama03] the proofs can be copied along the lines.

A distribution $T \in \mathcal{D}'(\mathbb{R}^n, E)$ is called 2π-periodic, or merely periodic, if

$$T(\varphi) = T(\tau_{2\pi k}\varphi) \quad (\varphi \in C_0^\infty(\mathbb{R}^n), \ k \in \mathbb{Z}^n),$$

where τ denotes the translation operator from Appendix A. We denote the subspace of all periodic distributions by $\mathcal{D}'_{per}(\mathbb{R}^n, E)$. Given $g \in L^p(\mathcal{Q}_n, E)$, the 2π-periodic extension g_{per} to the whole space $\mathbb{R}^n$ is locally integrable. For the sake of simplicity we write $g = g_{per}$ whenever we do not want to point out periodicity explicitly. Thus, g defines a regular, periodic distribution $T_g \in \mathcal{D}'_{per}(\mathbb{R}^n, E)$. In that sense, the class

$$C_{per}^\infty(\mathcal{Q}_n, E) := \{u|_{\mathcal{Q}_n}; \ u \in C^\infty(\mathbb{R}^n, E) : u(x) = u(x + 2\pi k) \ (k \in \mathbb{Z}^n)\}$$

is dense in $\mathcal{D}'_{per}(\mathbb{R}^n, E)$. Basically, this is due to the well-known fact that there exist $(\varphi_\varepsilon)_\varepsilon \subset C_0^\infty(\mathbb{R}^n)$ such that $T * \varphi_\varepsilon \to T * \delta = T$ in $\mathcal{D}'(\mathbb{R}^n, E)$ for any distribution T. Indeed, given $T \in \mathcal{D}'_{per}(\mathbb{R}^n, E)$ and $\varphi \in C_0^\infty(\mathbb{R}^n)$, $T * \varphi \in C^\infty(\mathbb{R}^n, E)$ is periodic again. This is clear since

$$(T * \varphi)(x) = T(\tau_x \check{\varphi}) = T(\tau_{x+2\pi k}\check{\varphi}) = (T * \varphi)(x + 2\pi k) \quad (k \in \mathbb{Z}^n).$$

The most important observation towards the definition of Fourier coefficients for periodic distributions is the existence of a function $\alpha \in C_0^\infty(\mathbb{R}^n)$ such that $\alpha^* := \sum_{k \in \mathbb{Z}^n} \tau_{2\pi k}\alpha = 1$, i.e.

$$(2.5) \qquad\qquad \alpha^*(x) := \sum_{k \in \mathbb{Z}^n} \alpha(x + 2\pi k) = 1 \quad (x \in \mathbb{R}^n).$$

Note that the sum is finite for each $x \in \mathbb{R}^n$. In case $n = 1$ consider $\varphi \in C_0^\infty(\mathbb{R})$ non-negative, such that $\varphi(x) \geq c > 0$ for $|x| \leq \pi$. Then $\varphi^*(x) > 0$ for all $x \in \mathbb{R}$ and $\beta := \varphi/\varphi^*$ enjoys property (2.5), i.e. $\beta^* = 1$. For arbitrary $n \in \mathbb{N}$ we simply set $\alpha(x) := \beta(x_1) \cdot \ldots \cdot \beta(x_n)$. If we additionally let $\operatorname{supp} \varphi \subset (-2\pi, 2\pi)$, we can achieve $\beta(0) = \alpha(0) = 1$.

In virtue of (2.5), the values of T_g for $g \in L^p(\mathcal{Q}_n, E)$ are explicitly given by

$$(g, \varphi) = \int_{\mathbb{R}^n} g(x)\varphi(x)dx = \int_{\mathcal{Q}_n} g(x)\varphi^*(x)dx.$$

Note that

$$\sum_{k \in \mathbb{Z}^n} \int_{2\pi k + \mathcal{Q}_n} g(x)\varphi(x)dx = \int_{\mathcal{Q}_n} \sum_{k \in \mathbb{Z}^n} g(x)\varphi(x + 2\pi k)dx$$

as all sums are finite. In particular, for fixed $k \in \mathbb{Z}^n$ and $f \in L^p(\mathcal{Q}_n, E)$ we have

$$\hat{f}(k) = \frac{1}{(2\pi)^n} \int_{\mathbb{R}^n} e^{-ikx} f(x)\alpha(x)dx = \frac{1}{(2\pi)^n}(f, \alpha e^{-ik\cdot}).$$

This gives rise to the definition of Fourier coefficients of $T \in \mathcal{D}'_{per}(\mathbb{R}^n, E)$ by means of

$$(2.6) \qquad\qquad \hat{T}(k) := \frac{1}{(2\pi)^n} T(\alpha e^{-ik\cdot}).$$

For instance, consider the periodic Dirac distribution $\delta_{2\pi} \in \mathcal{D}'_{per}(\mathbb{R}^n, E)$ defined by $\delta_{2\pi}(\varphi) = \sum_{q \in \mathbb{Z}^n} \varphi(2\pi q)$. Then $(2\pi)^n \hat{\delta}_{2\pi}(k) = \delta_{2\pi}(\alpha e^{-ik\cdot}) = \sum_{q \in \mathbb{Z}^n} \alpha(2\pi q) = 1$, that is,

$$\hat{\delta}_{2\pi}(k) = \frac{1}{(2\pi)^n} \quad (k \in \mathbb{Z}^n).$$

By density of $C_{per}^\infty(\mathcal{Q}_n, E)$ independence of $\hat{T}(k)$ of the particular choice of α can easily be deduced from the fact that $\hat{f}(k)$ does not depend on α at all.

Remark 2.5. Since independence can as well be shown in a direct and, in view of the definition in (2.5), a very representative manner, we carry it out briefly. Let $\alpha, \beta \in C_0^\infty(\mathbb{R}^n)$ both fulfill (2.5). Then $\beta = \alpha^* \beta = \sum_{q \in \mathbb{Z}^n} \alpha(\cdot + 2\pi q)\beta \in C_0^\infty(\mathbb{R}^n)$ and the sum is de facto finite. This allows for the calculation

$$T(\beta e^{-ik\cdot}) = T(\sum_{q \in \mathbb{Z}^n} \alpha(\cdot + 2\pi q)\beta e^{-ik\cdot}) = \sum_{q \in \mathbb{Z}^n} T(\alpha(\cdot + 2\pi q)\beta e^{-ik\cdot}).$$

Due to periodicity of T we have $T(\alpha(\cdot + 2\pi q)\beta e^{-ik\cdot}) = T(\beta(\cdot - 2\pi q)\alpha e^{-ik\cdot})$ and therefore

$$T(\beta e^{-ik\cdot}) = \sum_{q \in \mathbb{Z}^n} T(\beta(\cdot + 2\pi q)\alpha e^{-ik\cdot}) = T(\beta^* \alpha e^{-ik\cdot}) = T(\alpha e^{-ik\cdot}).$$

The fact that the Fourier coefficients of $f \in L^p(\mathcal{Q}_n, E)$ are uniformly bounded for $k \in \mathbb{Z}^n$ persists in the weaker form of polynomial boundedness for general $T \in \mathcal{D}'_{per}(\mathbb{R}^n, E)$. This is an easy observation in view of (A.1) since for arbitrary $K \subset \mathbb{R}^n$ with supp $\alpha \subset K$ there exists $m \in \mathbb{N}_0$ such that

$$(2.7) \qquad (2\pi)^n \|\hat{T}(k)\|_E = \|T(\alpha e^{-ik\cdot})\|_E \leq C p_{(m,K)}(\alpha e^{-ik\cdot}) \leq C|k|^m$$

for $k \neq 0$.

Lemma 2.6. *Let $(\eta_k)_{k \in \mathbb{Z}^n} \subset E$. Let $C > 0$ and $m \in \mathbb{N}_0$ such that*

$$\|\eta_k\|_E \leq C|k|^m \quad (k \in \mathbb{Z}^n \setminus \{0\}).$$

Then $T := \sum_{k \in \mathbb{Z}^n} \eta_k e^{ik\cdot}$ converges in $\mathcal{D}'(\mathbb{R}^n, E)$, i.e. $T \in \mathcal{D}'_{per}(\mathbb{R}^n, E)$.

Proof. With $|k| = |k|_2 = \sqrt{k_1^2 + \ldots + k_n^2}$ and $N \in \mathbb{N}$ we consider

$$f(x) = \sum_{k \in \mathbb{Z}^n \setminus \{0\}} \frac{\eta_k}{|k|^{2Nm}} e^{ikx}.$$

Then $|k|^{2Nm} = (k_1^2 + \ldots + k_n^2)^{Nm}$ and

$$\frac{\|\eta_k\|_E}{|k|^{2Nm}} \leq C \frac{1}{|k|^{2N}}.$$

Thus, the sum on the right-hand side is uniformly convergent, provided N is chosen to be large enough. Hence, f is continuous and convergence is also given in $\mathcal{D}'(\mathbb{R}^n, E)$. Moreover, there exists a differential operator $L = L(\partial)$ such that

$$LT_f(\varphi) = \sum_{k \in \mathbb{Z}^n \setminus \{0\}} \frac{\eta_k}{|k|^{2Nm}} |k|^{2Nm} (e^{ik\cdot}, \varphi) = \sum_{k \in \mathbb{Z}^n \setminus \{0\}} \eta_k(e^{ik\cdot}, \varphi).$$

This shows existence of $\eta_0 \in E$ such that $T = \eta_0 + LT_f \in \mathcal{D}'(\mathbb{R}^n, E)$. Since periodicity is obvious, the proof is finished. $\square$

With the help of Lemma 2.6 it is no longer difficult to prove the mentioned *representation theorem*.

Theorem 2.7. *For every* $T \in \mathcal{D}'_{per}(\mathbb{R}^n, E)$ *we have* $T = \sum_{k \in \mathbb{Z}^n} \hat{T}(k) e^{ik \cdot}$ *in* $\mathcal{D}'(\mathbb{R}^n, E)$.

Proof. Due to the estimate given in (2.7) and Lemma 2.6 the series on the right-hand side exists in $\mathcal{D}'(\mathbb{R}^n, E)$ and it remains to show equality. To this end, we find

$$T * \alpha e^{ik \cdot}(x) = T(\tau_x(\alpha e^{ik \cdot})^{\check{}}) = T((\tau_x \check{\alpha}) e^{ikx} e^{-ik \cdot}) = e^{ikx} (2\pi)^n \hat{T}(k)$$

since $\tau_x \check{\alpha}$ for fixed $x \in \mathbb{R}^n$ fulfills (2.5) again. This gives

$$\sum_{k \in \mathbb{Z}^n} \hat{T}(k) e^{ik \cdot} = \sum_{k \in \mathbb{Z}^n} \frac{1}{(2\pi)^n} T * \alpha e^{ik \cdot} = T * \sum_{k \in \mathbb{Z}^n} \frac{1}{(2\pi)^n} \alpha e^{ik \cdot} = T * \alpha \delta_{2\pi}.$$

Note that the convolutions are well-defined because of $\operatorname{supp} \alpha e^{ik \cdot} \subset \operatorname{supp} \alpha$ for all $k \in \mathbb{Z}^n$. Since $\alpha \in C_0^\infty(\mathbb{R}^n)$ can be chosen such that $\operatorname{supp} \alpha \subset (-2\pi, 2\pi)^n$ and $\alpha(0) = 1$, it follows that $T * \alpha \delta_{2\pi} = T * \delta = T$ and the proof is complete. $\qquad\square$

Remark 2.8. In the last proof we took the liberty to choose α in a suitable manner. In fact, any α subject to (2.5) fulfills $T * \alpha \delta_{2\pi} = T$ for every $T \in \mathcal{D}'_{per}(\mathbb{R}^n, E)$. To see this, let $\varphi \in \mathcal{D}(\mathbb{R}^n)$ be arbitrary and set

$$(\alpha \delta_{2\pi} * \varphi)(x) = \delta_{2\pi}(\alpha(\cdot)\varphi(x - \cdot)) = \sum_{k \in \mathbb{Z}^n} \alpha(2\pi k) \varphi(x - 2\pi k) =: \psi(x).$$

Then

$$(T * \psi)(y) = T(\sum_{k \in \mathbb{Z}^n} \alpha(2\pi k)\varphi(y + 2\pi k - \cdot)) = \sum_{k \in \mathbb{Z}^n} \alpha(2\pi k) T(\varphi(y + 2\pi k - \cdot))$$

which, by periodicity of T, is equal to

$$\sum_{k \in \mathbb{Z}^n} \alpha(2\pi k) T(\varphi(y - \cdot)) = T(\varphi(y - \cdot)) \sum_{k \in \mathbb{Z}^n} \alpha(2\pi k) = T(\varphi(y - \cdot)) \alpha^*(0) = T * \varphi.$$

Hence, $(T * \alpha \delta_{2\pi}) * \varphi = T * (\alpha \delta_{2\pi} * \varphi) = T * \varphi$ for arbitrary $\varphi \in \mathcal{D}(\mathbb{R}^n)$ which shows the claim.

An immediate consequence is an important uniqueness result for Fourier series. Namely, if $T = \sum_{q \in \mathbb{Z}^n} \eta_q e^{iq \cdot}$ with $(\eta_q)_{q \in \mathbb{Z}^n} \subset E$, then

$$T(\alpha e^{-ik \cdot}) = \sum_{q \in \mathbb{Z}^n} \eta_q (e^{iq \cdot}, \alpha e^{-ik \cdot}) = \sum_{q \in \mathbb{Z}^n} \eta_q \int_{\mathcal{Q}_n} e^{iqx} e^{-ikx} dx = (2\pi)^n \eta_k.$$

Hence, $\hat{T}(k) = \eta_k$ for all $k \in \mathbb{Z}^n$. On the other hand, from equality of Fourier coefficients of periodic distributions T and F on certain subsets of $\mathbb{Z}^n$ important relations between T and F themselves can be deduced.

Corollary 2.9. *Let* $T, F \in \mathcal{D}'_{per}(\mathbb{R}^n, E)$.

(i) *If* $\hat{T}(k) = \hat{F}(k)$ *for all* $k \in \mathbb{Z}^n$, *then* $T = F$ *in* $\mathcal{D}'_{per}(\mathbb{R}^n, E)$.

(ii) *If* $\hat{T}(k) = \hat{F}(k)$ *for all* $k \in \mathbb{Z}^n \setminus \{0\}$, *then* $T = F + \eta$ *with* $\eta := \hat{T}(0) - \hat{F}(0)$.

(iii) *If* $\hat{T}(k) = \hat{F}(k)$ *for all* $k \in \mathbb{Z}^n$ *such that* $k_j \neq 0$ *for some* $j \in \{1, \ldots, n\}$, *then there exists* $G \in \mathcal{D}'_{per}(\mathbb{R}^n, E)$ *independent of* x_j *such that* $T = F + G$.

Proof. Due to the representation theorem we have

$$T - F = \sum_{k \in \mathbb{Z}^n} \left(\hat{T}(k) - \hat{F}(k) \right) e^{ik\cdot}$$

which implies (i) and (ii) immediately. If $\hat{T}(k) = \hat{F}(k)$ for all $k \in \mathbb{Z}^n$ such that $k_j \neq 0$ as demanded in (iii), we get

$$\sum_{k \in \mathbb{Z}^n} \left(\hat{T}(k) - \hat{F}(k) \right) e^{ik\cdot} = \sum_{k' \in \mathbb{Z}^{n-1}} \left(\hat{T}((k', 0)) - \hat{F}((k', 0)) \right) e^{i(k', 0)\cdot}.$$

Hence, $T - F$ is independent of x_j for $e^{i(k', 0)\cdot}$ enjoys this property. $\qquad\square$

As a further consequence the converse assertion of Lemma 2.3 follows.

Corollary 2.10. *Let* A *be a closed operator in* E. *Let* $(\eta_k)_{k \in \mathbb{Z}^n} \subset D(A)$, $C > 0$, *and* $m \in \mathbb{N}_0$ *such that*

$$\|\eta_k\|_{(D(A))} \leq C |k|^m \quad (k \in \mathbb{Z}^n \setminus \{0\}).$$

Then $T := \sum_{k \in \mathbb{Z}^n} \eta_k e^{ik\cdot}$ *and* $F := \sum_{k \in \mathbb{Z}^n} A\eta_k e^{ik\cdot}$ *fulfill* $AT = F$ *in* $\mathcal{D}'_{per}(\mathbb{R}^n, E)$.

Proof. Let $\varphi \in \mathcal{D}(\mathbb{R}^n)$ be arbitrary. Then there exist $\eta_T, \eta_F \in E$ such that

$$\int_{\mathbb{R}^n} \sum_{k \in [-N, N]^n} \eta_k e^{ikx} \varphi(x) dx \to \eta_T$$

and, by closedness of A,

$$A \int_{\mathbb{R}^n} \sum_{k \in [-N, N]^n} \eta_k e^{ikx} \varphi(x) dx = \int_{\mathbb{R}^n} \sum_{k \in [-N, N]^n} A\eta_k e^{ikx} \varphi(x) dx \to \eta_F$$

in E for $N \to \infty$ due to convergence in $\mathcal{D}'(\mathbb{R}^n, E)$. Using the closedness of A another time yields $\eta_T \in D(A)$ and $A\eta_T = \eta_F$ which proves the claim. $\qquad\square$

In case of $L^p(\mathcal{Q}_n, E)$-functions this assertion can also be proved by means of Fejer's theorem (cf. [AB02, Lemma 3.1]). Indeed, let $f, g \in L^p(\mathcal{Q}_n, E)$ such that $\hat{f}(k) \in D(A)$ and $A\hat{f}(k) = \hat{g}(k)$ for all $k \in \mathbb{Z}^n$. Then $f \in L^p(\mathcal{Q}_n, D(A))$ and $Af = g$ in $L^p(\mathcal{Q}_n, E)$.

Finally, the representation theorem implies the crucial relation most similar to (2.4) between Fourier coefficients of distributional derivatives of $T \in \mathcal{D}'_{per}(\mathbb{R}^n, E)$ and Fourier coefficients of T itself. Switching again to $D^\alpha := (-i)^{|\alpha|}\partial^\alpha$, it reads as

$$(2.8) \qquad (D^\alpha T)\hat{}(k) = k^\alpha \hat{T}(k) \quad (T \in \mathcal{D}'_{per}(\mathbb{R}^n, E),\ k \in \mathbb{Z}^n,\ \alpha \in \mathbb{N}_0^n).$$

Besides uniqueness of Fourier coefficients we have used the fact that derivation acts as a continuous operator in $\mathcal{D}'(\mathbb{R}^n, E)$, i.e.

$$D^\alpha \sum_{k \in \mathbb{Z}^n} \hat{T}(k) e^{ik\cdot} = \sum_{k \in \mathbb{Z}^n} \hat{T}(k) D^\alpha e^{ik\cdot}.$$

Equation (2.8) will help to characterize periodic Sobolev spaces later on.

Definition 2.11. Let $m \in \mathbb{N}_0$. The E-valued periodic Sobolev space $W_{per}^{m,p}(\mathcal{Q}_n, E)$ of order m consists of all $u \in W^{m,p}(\mathcal{Q}_n, E)$ such that

$$\partial_j^\ell u|_{x_j=0} = \partial_j^\ell u|_{x_j=2\pi} \quad (j = 1, \ldots, n;\ 0 \le \ell < m).$$

Note that $W_{per}^{0,p}(\mathcal{Q}_n, E) = L^p(\mathcal{Q}_n, E)$ and for $m \in \mathbb{N}$

$$W^{m,p}(\mathcal{Q}_n, E) \hookrightarrow L^p(\mathcal{Q}_{n-1}, W^{m,p}((0, 2\pi), E)) \hookrightarrow L^p(\mathcal{Q}_{n-1}, C^{m-1}([0, 2\pi], E)).$$

Hence, all traces in the definition of $W_{per}^{m,p}(\mathcal{Q}_n, E)$ are well-defined by continuity.

Remark 2.12. Obviously, for all $m \in \mathbb{N}_0$ the space of trigonometric polynomials is a subset of $W_{per}^{m,p}(\mathcal{Q}_n, E)$.

Lemma 2.13. *Let $g \in L^p(\mathbb{R}^n, E)$, $j \in \{1, \ldots, n\}$, and $f(x) := \int_0^{x_j} g(x', s)ds$. Then for $r > 0$ we have $f|_{\mathbb{R}^{n-1} \times (-r,r)} \in L^p(\mathbb{R}^{n-1}, C((-r, r), E))$ and $\partial_j T_f = T_g$ in $\mathcal{D}'(\mathbb{R}^n, E)$.*

Proof. Let $r > 0$, and $\omega := f|_{\mathbb{R}^{n-1} \times (-r,r)}$, i.e.

$$\omega_{x'}(t) := \omega(x', t) := \int_0^t g(x', s)ds \quad (t \in (-r, r)).$$

Then Hölder's inequality gives

$$\|\omega_{x'}(t)\|_E \le \|g(x', \cdot)\|_{1,(-r,r),E} \le |t|^{\frac{p}{p-1}} \|g(x', \cdot)\|_{p,(-r,r),E} \le |t|^{\frac{p}{p-1}} \|g(x', \cdot)\|_{p,\mathbb{R},E}$$

and therefore $\omega \in L^p(\mathbb{R}^{n-1}, C((-r, r), E))$. An easy application of Fubini's theorem proves $\partial_j T_f = T_g$ in $\mathcal{D}'(\mathbb{R}^n, E)$. $\qquad\square$

Theorem 2.14. *Let $m \in \mathbb{N}$, $u \in W^{m,p}(\mathcal{Q}_n, E)$, and let u_{per} denote its periodic extension to $L^1_{loc}(\mathbb{R}^n, E)$. Then the following assertions are equivalent:*

(i) $u \in W^{m,p}_{per}(\mathcal{Q}_n, E)$.

(ii) $u_{per} \in W^{m,p}_{loc}(\mathbb{R}^n, E)$.

(iii) $(D^\alpha u_{per})\hat{}(k) = k^\alpha \hat{u}_{per}(k)$ *for $|\alpha| \leq m$ and $k \in \mathbb{Z}^n$.*

(iv) *For each $0 < |\alpha| \leq m$ and each $j = 1, \ldots, n$ such that $\alpha_j \neq 0$ there exists $\omega \in L^p(\mathcal{Q}_n, E)$ which is independent of x_j such that $D^{\alpha - e_j} u = \omega + V$, where $V \in L^p(\mathcal{Q}_n, E)$ defined by $V(x', x_j) := \int_0^{x_j} D^\alpha u(x', s) ds$ fulfills $V(\cdot, 2\pi) \equiv 0$ in $L^p(\mathcal{Q}_{n-1}, E)$.*

(v) $D^\alpha u|_{x_j = 0} = D^\alpha u|_{x_j = 2\pi}$ *for $j = 1, \ldots, n$ and $|\alpha| < m$.*

Proof. (i) $\Rightarrow$ (ii): Let $j \in \{1, \ldots, n\}$, $x = (x', x_j)$, and $u \in W^{m,p}(\mathcal{Q}_n, E)$. Then

$$u_{per}(x', \cdot) = g_{x'} + a_{x'} \sum_{k \in \mathbb{Z}} H(\cdot + 2\pi k)$$

x'-almost everywhere, where H denotes the Heaviside function and $g_{x'} \in C(\mathbb{R}, E)$ is absolutely continuous on $(2\pi k, 2\pi(k+1))$ for each $k \in \mathbb{Z}$. Furthermore, $a_{x'} \in E$ is precisely defined by $a_{x'} := \lim_{t \to 0} u(x', t) - \lim_{t \to 2\pi} u(x', t)$. Evidently, the condition $u(\cdot, 0) = u(\cdot, 2\pi)$ yields $a_{x'} = 0$, hence $\partial_j u_{per}(x', \cdot) = g'_{x'}$. The mentioned properties of $g_{x'}$ allow for the integration by parts formula to find

$$\int_{\mathbb{R}} g_{x'}(t) \varphi'(t) dt = - \int_{\mathbb{R}} g'_{x'}(t) \varphi(t) dt \quad (\varphi \in C_0^\infty(\mathbb{R})).$$

This shows $g'_{x'} \in L^1_{loc}(\mathbb{R}, E)$. Since j was arbitrary, $u_{per} \varphi \in W^{1,p}(\mathbb{R}^n, E)$ for arbitrary $\varphi \in C_0^\infty(\mathbb{R}^n)$, i.e. $u_{per} \in W^{1,p}_{loc}(\mathbb{R}^n, E)$ follows. These arguments applied to derivatives up to order m prove (ii).

(ii) $\Rightarrow$ (iii): $u_{per} \in W^{m,p}_{loc}(\mathbb{R}^n, E)$ implies $D^\alpha T_{u_{per}} = T_{D^\alpha u_{per}}$ for all $|\alpha| \leq m$. Now (iii) follows from (2.8).

(iii) $\Rightarrow$ (iv): Let $0 < |\alpha| \leq m$ be arbitrary and $j \in \{1, \ldots, n\}$ such that $\alpha_j \neq 0$. First note that $V \in L^p(\mathcal{Q}_n, E)$ by Lemma 2.13. Calculating Fourier coefficients of V yields

$$\hat{V}(k) = \frac{1}{k_j}(D^\alpha u)\hat{}(k) \quad (k \in \mathbb{Z}^n, \ k_j \neq 0),$$

which gives $\hat{V}(k) = (D^{\alpha - e_j} u)\hat{}(k)$ for all $k \in \mathbb{Z}^n$ such that $k_j \neq 0$ by (iii). Hence, there exists $G \in \mathcal{D}'_{per}(\mathbb{R}^n, E)$ independent of x_j such that $T_{D^{\alpha - e_j} u} = G + T_V$ by means of Corollary 2.9. In particular, since $D^{\alpha - e_j} u, V \in L^p(\mathcal{Q}_n, E)$, there exists $\omega \in L^p(\mathcal{Q}_n, E)$ independent of x_j such that $G = T_\omega$. Finally, $V(\cdot, 2\pi)$ is well-defined since $V \in L^p(\mathcal{Q}_{n-1}, C([0, 2\pi], E))$ again by Lemma 2.13. We calculate $(V(\cdot, 2\pi))\hat{}(k') = \hat{V}(k', 0) = 0$ for all $k' \in \mathbb{Z}^{n-1}$, that is, $V(\cdot, 2\pi) \equiv 0$ in

$L^p(Q_{n-1}, E)$ and (iv) is proven.

(iv) $\Rightarrow$ (v): $V(\cdot, 0) = V(\cdot, 2\pi) \equiv 0$ by (iv) implies $D^\alpha u|_{x_j=0} = \omega_{\alpha,j} = D^\alpha u|_{x_j=2\pi}$, where $\omega_{\alpha,j} \in L^p(Q_n, E)$ for $j = 1, \ldots, n$ and $|\alpha| < m$.

Finally, (v) $\Rightarrow$ (i) is obviously true and the proof is complete.   □

As seen in the previous proof, $u \in W_{per}^{m,p}(Q_n, E)$ implies $D^\alpha T_{u_{per}} = T_{D^\alpha u_{per}}$ for $|\alpha| \le m$. Thus, each derivative up to order m is given as a regular distribution. The following proposition proves the converse assertion to be valid. In fact, even more is true since each periodic distribution T such that all derivatives $D^\alpha T$ of highest order $|\alpha| = m$ define regular distributions is regular itself. Moreover, it is induced by a function $u \in W_{per}^{m,p}(Q_n, E)$. In that sense, periodic Sobolev spaces and *homogeneous* periodic Sobolev spaces coincide.

Proposition 2.15. *Let $m \in \mathbb{N}$ and $T \in \mathcal{D}'_{per}(\mathbb{R}^n, E)$. For each $\alpha \in \mathbb{N}^n$ subject to $|\alpha| = m$ let $v_\alpha \in L^p(Q_n, E)$ be given such that $\hat{v}_{\alpha,per}(k) = k^\alpha \hat{T}(k)$ for all $k \in \mathbb{Z}^n$. Then there exists $u \in W_{per}^{m,p}(Q_n, E)$ such that $D^\alpha u = v_\alpha$ for $|\alpha| = m$ and $T = T_u$.*

Proof. First note that $\hat{v}_{\alpha,per}(k) = k^\alpha \hat{T}(k) = (D^\alpha T)\hat{}(k)$ for all $k \in \mathbb{Z}^n$. This gives $D^\alpha T = T_{v_{\alpha,per}}$ for all $|\alpha| = m$ due to Corollary 2.9. Let β be given such that $|\beta| = m - 1$. Set $T_j := T_{v_j} := v_j := v_{\beta+e_j,per}$ and define $S := D^\beta T$. Then

$$D_j S = T_j \quad (j = 1, \ldots, n).$$

Recall that any $S \in \mathcal{D}'(\mathbb{R}^n, E)$ subject to these equations is uniquely determined except for a constant $\eta \in E$. We show that S is a regular distribution S_f, where $f|_{Q_n} \in L^p(Q_n, E)$ fulfills (iv) of Theorem 2.14. Obviously $S_1(\varphi) := -T_1(\psi_\varphi)$ as defined in (A.2) fulfills the equation for $j = 1$. The definition of ψ_φ shows

$$-T_1(\psi_\varphi) = T_V(\varphi) + T_\omega(\varphi),$$

where $V(x_1, x') := \int_0^{x_1} v_1(s, x') ds$. Here ω is independent of x_1 and both V and ω restricted to Q_n define functions in $L^p(Q_n, E)$. In particular S_1 is regular.

Let $1 < k < n$ and assume existence of a regular distribution $S_k \in \mathcal{D}'(\mathbb{R}^n, E)$ such that $D_j S_k = T_j$ holds true for all $j = 1, \ldots, k$. Following [Jan71, Satz 23.7], a regular distribution $S^* \in \mathcal{D}'(\mathbb{R}^n, E)$ can be found such that $S_{k+1} := S_k + S^*$ solves all equations with $j \le k+1$. Consequently, $S_{k+1} = S_{k+1,f}$ is regular as well. Moreover, it is independent of $x_1, \ldots, x_k$ and $f|_{Q_n} \in L^p(Q_n, E)$. By induction we find that $D^\beta T = S := S_{n,f}$ is a regular distribution for which $f|_{Q_n} \in L^p(Q_n, E)$ and (iv) of Theorem 2.14 is fulfilled at least for $j = 1$. Since we could have started with any coordinate $j \in \{1, \ldots, n\}$, by uniqueness of S except for a constant, the claim follows for $|\alpha| = 1$ and by iteration for $|\alpha| = m$.   □

More generally, with $\nu \in \mathbb{C}^n$ we will consider ν-periodic Sobolev spaces denoted by $W_{\nu,per}^{m,p}(Q_n, E)$. They represent the spaces of all $u \in W^{m,p}(Q_n, E)$ such that

$$(D^\alpha u)|_{x_j=2\pi} = e^{2\pi\nu_j}(D^\alpha u)|_{x_j=0} \quad (j = 1, \ldots, n, \ |\alpha| < m).$$

In particular, $W_{per}^{m,p}(\mathcal{Q}_n, E) = W_{0,per}^{m,p}(\mathcal{Q}_n, E)$. The following Lemma provides useful characterizations of ν-periodic Sobolev spaces $W_{\nu,per}^{m,p}(\mathcal{Q}_n, E)$.

Lemma 2.16. *The following assertions are equivalent:*

(i) $u \in W_{\nu,per}^{m,p}(\mathcal{Q}_n, E)$.

(ii) *There exists* $v \in W_{per}^{m,p}(\mathcal{Q}_n, E)$ *such that* $u = e^{\nu \cdot} v$.

(iii) $u \in W^{m,p}(\mathcal{Q}_n, E)$ *and for all* $|\alpha| \leq m$ *it holds that*

$$(e^{-\nu \cdot} D^\alpha u)\hat{}(k) = (k - i\nu)^\alpha (e^{-\nu \cdot} u)\hat{}(k) \quad (k \in \mathbb{Z}^n).$$

Proof. (i) $\Leftrightarrow$ (ii): This follows from the fact that the operator of multiplication with $e^{\pm \nu \cdot}$ is a bijection in $W^{m,p}(\mathcal{Q}_n, E)$.

(ii) $\Leftrightarrow$ (iii): Let $\nu \in \mathbb{C}^n$ and $m \in \mathbb{N}$ be arbitrary. Then $v := e^{-\nu \cdot} u \in W_{per}^{m,p}(Q_n, E)$ and Theorem 2.14 yields

$$(D^\alpha v)\hat{}(k) = k^\alpha \hat{v}(k) \quad (|\alpha| \leq m, \ k \in \mathbb{Z}^n).$$

By means of the Leibniz rule

$$D^\alpha(e^{\nu \cdot} v) = \sum_{\beta \leq \alpha} \binom{\alpha}{\beta} D^\beta e^{\nu \cdot} D^{\alpha-\beta} v = \sum_{\beta \leq \alpha} \binom{\alpha}{\beta} (-i)^\beta \nu^\beta e^{\nu \cdot} D^{\alpha-\beta} v$$

$$= e^{\nu \cdot} \sum_{\beta \leq \alpha} \binom{\alpha}{\beta} (-i\nu)^\beta D^{\alpha-\beta} v.$$

Thus,

$$\left(e^{-\nu \cdot} D^\alpha(e^{\nu \cdot} v)\right)\hat{}(k) = \sum_{\beta \leq \alpha} \binom{\alpha}{\beta} (-i\nu)^\beta (D^{\alpha-\beta} v)\hat{}(k)$$

$$= \sum_{\beta \leq \alpha} \binom{\alpha}{\beta} (-i\nu)^\beta k^{\alpha-\beta} \hat{v}(k) = (k - i\nu)^\alpha \hat{v}(k).$$

On the other hand, let $u \in W^{m,p}(\mathcal{Q}_n, E)$ be given such that

$$(e^{-\nu \cdot} D^\alpha u)\hat{}(k) = (k - i\nu)^\alpha (e^{-\nu \cdot} u)\hat{}(k) \quad (|\alpha| \leq m, \ k \in \mathbb{Z}^n)$$

holds true. Then $v := e^{-\nu \cdot} u$ fulfills

$$D^\alpha v = D^\alpha(e^{-\nu \cdot} u) = \sum_{\beta \leq \alpha} \binom{\alpha}{\beta} D^\beta e^{-\nu \cdot} D^{\alpha-\beta} u = \sum_{\beta \leq \alpha} \binom{\alpha}{\beta} (i\nu)^\beta e^{-\nu \cdot} D^{\alpha-\beta} u$$

which yields

$$(D^\alpha v)\hat{\ }(k) = \sum_{\beta \le \alpha} \binom{\alpha}{\beta} (i\nu)^\beta \big(e^{-\nu\cdot} D^{\alpha-\beta} u\big)\hat{\ }(k)$$

$$= \sum_{\beta \le \alpha} \binom{\alpha}{\beta} (i\nu)^\beta (k - i\nu)^{\alpha-\beta} (e^{-\nu\cdot} u)\hat{\ }(k) = k^\alpha \hat{v}(k).$$

Hence, $v \in W_{per}^{m,p}(Q_n, E)$ by Theorem 2.14 and the proof is complete. □

Accordingly, we define ν-periodic distributions $\mathcal{D}'_{\nu,per}(\mathbb{R}^n, E)$ by means of the condition

(2.9) $$T(e^{\nu\cdot}\varphi) = T(e^{\nu\cdot}\tau_{2\pi k}\varphi) \qquad (\varphi \in C_0^\infty(\mathbb{R}^n),\ k \in \mathbb{Z}^n).$$

Then, as indicated in the previous lemma, by definition of $e^{\nu\cdot}T$ we find

$$\mathcal{D}'_{\nu,per}(\mathbb{R}^n, E) = \big\{e^{\nu\cdot}T;\ T \in \mathcal{D}'_{per}(\mathbb{R}^n, E)\big\}$$

and $T \in \mathcal{D}'_{\nu,per}(\mathbb{R}^n, E)$ if and only if

$$(e^{-\nu\cdot} D^\alpha T)\hat{\ }(k) = (k - i\nu)^\alpha (e^{-\nu\cdot} T)\hat{\ }(k) \quad (k \in \mathbb{Z}^n).$$

For applications later on, in what follows we define partial Fourier series for a distribution $T \in \mathcal{D}'(\mathbb{R}^n \times \mathbb{R}^m, E)$ which is periodic with respect to $\mathbb{R}^n$ only. We write $\mathcal{D}'_{per,n}(\mathbb{R}^{n+m}, E)$ for this class of distributions to avoid any mix-up with $\mathcal{D}'_{per}(\mathbb{R}^{n+m}, E) = \mathcal{D}'_{per,n+m}(\mathbb{R}^{n+m}, E)$. Accordingly, we write $\mathcal{D}'_{\nu,per,n}(\mathbb{R}^{n+m}, E)$ for the class of partially ν-periodic distributions.

The definition

$$\hat{T}(k) := \frac{1}{(2\pi)^n} T(\alpha e^{-ik\cdot}) \quad (k \in \mathbb{Z}^n)$$

for $T \in \mathcal{D}'_{per}(\mathbb{R}^n, E)$ as given in (2.6) is extended to $T \in \mathcal{D}'_{per,n}(\mathbb{R}^{n+m}, E)$ by setting

$$\big(\hat{T}_{(k,y)}\big)(\varphi) := \frac{1}{(2\pi)^n} T_{(x,y)}(\alpha(x)e^{-ikx}\varphi(y)) \quad (\varphi \in C_0^\infty(\mathbb{R}^m),\ k \in \mathbb{Z}^n).$$

We write $T = T_{(x,y)}$ as well as $T(\varphi) = T(\varphi(x,y)) = T_{(x,y)}(\varphi(x,y))$ in the sequel to indicate the dependency on particular variables. Along the lines of Remark 2.5 we see that the definition of $\hat{T}_{(k,y)}$ is independent of the particular choice of α subject to equation (2.5).

In order to work with partial Fourier series efficiently, we have to establish a representation theorem extending Theorem 2.7. Throughout the remaining part of the chapter, let α satisfy equation (2.5). Given $w \in L^1_{loc}(\mathbb{R}^n, E)$ we frequently make use of the notation $[w] = [w]_x$ for the regular distribution $T_w \in \mathcal{D}'(\mathbb{R}^n, E)$. Finally, we set

$$C_{per,n}^\infty(\mathbb{R}^{n+m}, E) := \big\{u \in C^\infty(\mathbb{R}^{n+m}, E);\ u(x,y) = u(x + 2\pi k, y)\ (k \in \mathbb{Z}^n)\big\}.$$

Lemma 2.17. *Let* $T \in \mathcal{D}'_{per,n}(\mathbb{R}^{n+m}, E)$, $f \in \mathcal{D}(\mathbb{R}^n)$ *and* $g \in \mathcal{D}(\mathbb{R}^m)$. *Then for all* $k \in \mathbb{Z}^n$ *it holds that*

$$(2.10) \qquad T_{(u,v)}\left([\alpha e^{ik\cdot}]_x\big(f(x+u)\big)\cdot g(v)\right) = (2\pi)^n \hat{T}_{(k,y)}\big(g(y)\big) \cdot [e^{ik\cdot}]_x\big(f(x)\big).$$

Proof. By density it suffices to consider $T = T_\Psi$, where $\Psi \in C^\infty_{per,n}(\mathbb{R}^{n+m}, E)$. This allows for the calculation

$$T_{\Psi(u,v)}\left([\alpha e^{ik\cdot}]_x\big(f(x+u)\big)\cdot g(v)\right)$$

$$= \int_{\mathbb{R}^{n+m}} \Psi(u,v) \left(\int_{\mathbb{R}^n} \alpha(x)e^{ikx}f(x+u)dx\right) g(v)d(u,v)$$

$$= \int_{\mathbb{R}^{n+m}} \Psi(u,v) \left(\int_{\mathbb{R}^n} \alpha(x-u)e^{ik(x-u)}f(x)dx\right) g(v)d(u,v)$$

$$= \int_{\mathbb{R}^n} e^{ikx} \left(\int_{\mathbb{R}^{n+m}} \Psi(u,v)\alpha(x-u)e^{-iku}g(v)d(u,v)\right) f(x)dx.$$

Since $\tilde{\alpha}(u) = \tau_x \alpha(u)$ for fixed $x \in \mathbb{R}^n$ satisfies (2.5) again, we conclude

$$T_{(u,v)}\left([\alpha e^{ik\cdot}]_x\big(f(x+u)\big)\cdot g(v)\right) = \int_{\mathbb{R}^n} e^{ikx}T_{\Psi(u,v)}\big(\tilde{\alpha}(u)e^{-iku}g(v)\big)f(x)dx$$

$$= \int_{\mathbb{R}^n} e^{ikx}(2\pi)^n \hat{T}_{\Psi(k,v)}\big(g(v)\big)f(x)dx = (2\pi)^n \hat{T}_{\Psi(k,v)}\big(g(v)\big) \cdot [e^{ik\cdot}]_x\big(f(x)\big)$$

and the proof is complete. $\qquad\square$

Lemma 2.18. *For* $T \in \mathcal{D}'_{per,n}(\mathbb{R}^{n+m}, E)$ *and all* $k \in \mathbb{Z}^n$ *it holds that*

$$(2.11) \qquad\qquad T * \big([\alpha e^{ik\cdot}]_x \otimes \delta_y\big) = (2\pi)^n\big(\hat{T}_{(k,y)} \otimes [e^{ik\cdot}]_x\big).$$

Proof. It suffices to consider $\varphi \in \mathcal{D}(\mathcal{Q}_n \times \mathbb{R}^m)$ such that $\varphi(x,y) = f(x)g(y)$ with $f \in \mathcal{D}(\mathbb{R}^n)$ and $g \in \mathcal{D}(\mathbb{R}^m)$ due to density of $\mathcal{D}(\mathbb{R}^n) \times \mathcal{D}(\mathbb{R}^m)$ in $\mathcal{D}(\mathbb{R}^{n+m})$ (see e.g. [Ama03, Theorem 1.8.1]). In that case

$$\left(T * \big([\alpha e^{ik\cdot}]_x \otimes \delta_y\big)\right)\big(f(x)g(y)\big) = T\left(\big([\alpha e^{ik\cdot}]_x \otimes \delta_y\big) * \big(f(x)g(y)\big)\check{}\right)\check{}$$

$$= T\left(\big([\alpha e^{ik\cdot}]_x \otimes \delta_y\big) * \big(f(-x)g(-y)\big)\right)\check{}$$

$$= T_{(u,v)}\left(\big([\alpha e^{ik\cdot}]_x \otimes \delta_y\big)\big(f(x-u)g(y-v)\big)\right)\check{}$$

$$= T_{(u,v)}\left([\alpha e^{ik\cdot}]_x\big(f(x-u)\big)\cdot g(-v)\right)\check{} = T_{(u,v)}\left([\alpha e^{ik\cdot}]_x\big(f(x+u)\big)\cdot g(v)\right)$$

and Lemma 2.17 yields

$$\left(T*\left([\alpha e^{ik\cdot}]_x \otimes \delta_y\right)\right)\left(f(x)g(y)\right) = (2\pi)^n \hat{T}_{\Psi(k,v)}\left(g(v)\right) \cdot [e^{ik\cdot}]_x\left(f(x)\right)$$
$$= (2\pi)^n \left(\hat{T}_{(k,y)} \otimes [e^{ik\cdot}]_x\right)\left(f(x)g(y)\right).$$

$\square$

Now we are in the position to extend Theorem 2.7 to partially periodic distributions $T \in \mathcal{D}'_{per,n}(\mathbb{R}^{n+m}, E)$.

Theorem 2.19. *For* $T \in \mathcal{D}'_{per,n}(\mathbb{R}^{n+m}, E)$ *it holds that*

$$(2.12) \qquad\qquad T = \sum_{k\in\mathbb{Z}^n} \hat{T}_{(k,y)} \otimes [e^{ik\cdot}]_x.$$

Proof. First note that

$$T = T * \delta_{(x,y)} = T * (\delta_x \otimes \delta_y) = T * (\alpha\delta_{x,2\pi} \otimes \delta_y)$$

if $\alpha = \alpha(x)$ is chosen in such a way that $\operatorname{supp}\alpha \subset (-2\pi, 2\pi)^n$ and $\alpha(0) = 1$. By means of Lemma 2.18

$$\sum_{k\in\mathbb{Z}^n} \hat{T}_{(k,y)} \otimes [e^{ik\cdot}]_x = \sum_{k\in\mathbb{Z}^n} (2\pi)^{-n}\left(T * \left([\alpha e^{ik\cdot}]_x \otimes \delta_y\right)\right).$$

Recall $\delta_{x,2\pi} = \sum_{k\in\mathbb{Z}^n}(2\pi)^{-n}[e^{ik\cdot}]_x$. Because of separate continuity of both the convolution and the tensor product, the sum on the right-hand side converges in $\mathcal{D}'(\mathbb{R}^{n+m}, E)$ and

$$\sum_{k\in\mathbb{Z}^n} (2\pi)^{-n}\left(T * \left([\alpha e^{ik\cdot}]_x \otimes \delta_y\right)\right) = T * (\alpha\delta_{x,2\pi} \otimes \delta_y) = T.$$

$\square$

Lemma 2.20. *For* $T \in \mathcal{D}'_{per,n}(\mathbb{R}^{n+m}, E)$ *and all* $k \in \mathbb{Z}^n$ *it holds that*

$$\left(D_x^{\alpha^1} D_y^{\alpha^2} T\right)\hat{}_{(k,y)} = k^{\alpha^1} D^{\alpha^2}\left(\hat{T}_{(k,y)}\right).$$

Proof. This follows from the fact that

$$D_x^{\alpha^1} D_y^{\alpha^2} T = \sum_{k\in\mathbb{Z}^n} D_x^{\alpha^1} D_y^{\alpha^2}\left(\hat{T}_{(k,y)} \otimes [e^{ik\cdot}]_x\right)$$
$$= \sum_{k\in\mathbb{Z}^n} \left(D_y^{\alpha^2}\hat{T}_{(k,y)} \otimes D_x^{\alpha^1}[e^{ik\cdot}]_x\right) = \sum_{k\in\mathbb{Z}^n} \left(k^{\alpha^1} D_y^{\alpha^2}\hat{T}_{(k,y)} \otimes [e^{ik\cdot}]_x\right).$$

$\square$

In particular tangential derivation and calculation of partial Fourier coefficients commute. Finally, we extend the previous lemma to the context of ν-periodicity.

Lemma 2.21. *Let $\nu \in \mathbb{C}^n$. For $T \in \mathcal{D}'_{\nu,per,n}(\mathbb{R}^{n+m}, E)$ and all $k \in \mathbb{Z}^n$ it holds that*

$$\left(e^{-\nu \cdot} D_x^{\alpha^1} D_y^{\alpha^2} T\right)\hat{}_{(k,y)} = (k - i\nu)^{\alpha^1} D_y^{\alpha^2} \left((e^{-\nu \cdot} T_{(k,y)})\hat{}\right).$$

Proof. Let $\varphi \in \mathcal{D}(\mathbb{R}^{n+m})$. Then

$$\left(e^{-\nu \cdot} D_x^{\alpha^1} D_y^{\alpha^2} T\right)(\varphi) = D_x^{\alpha^1} D_y^{\alpha^2} T(e^{-\nu \cdot} \varphi) = (-1)^{|\alpha^1|} D_y^{\alpha^2} T\left(D_x^{\alpha^1}(e^{-\nu \cdot} \varphi)\right)$$

$$= (-1)^{|\alpha^1|} D_y^{\alpha^2} T\left(\sum_{\beta^1 \leq \alpha^1} \binom{\alpha^1}{\beta^1} (i\nu)^{\beta^1} e^{-\nu \cdot} D_x^{\alpha^1 - \beta^1} \varphi\right)$$

by the Leibniz rule. From

$$T(e^{-\nu \cdot} D_x^{\alpha^1 - \beta^1} \varphi) = e^{-\nu \cdot} T(D_x^{\alpha^1 - \beta^1} \varphi) = (-1)^{|\alpha^1 - \beta^1|} D_x^{\alpha^1 - \beta^1}(e^{-\nu \cdot} T)(\varphi)$$

we further deduce

$$\left(e^{-\nu \cdot} D_x^{\alpha^1} D_y^{\alpha^2} T\right)(\varphi) = (-1)^{|\beta^1|} D_y^{\alpha^2} \sum_{\beta^1 \leq \alpha^1} \binom{\alpha^1}{\beta^1} (i\nu)^{\beta^1} D_x^{\alpha^1 - \beta^1}(e^{-\nu \cdot} T)(\varphi)$$

$$= D_y^{\alpha^2} \sum_{\beta^1 \leq \alpha^1} \binom{\alpha^1}{\beta^1} (-i\nu)^{\beta^1} D_x^{\alpha^1 - \beta^1}(e^{-\nu \cdot} T)(\varphi).$$

By means of

$$D_x^{\alpha^1 - \beta^1}(e^{-\nu \cdot} T) = D_x^{\alpha^1 - \beta^1} \sum_{k \in \mathbb{Z}^n} (e^{-\nu \cdot} T)\hat{}_{(k,y)} \otimes [e^{ik \cdot}]_x$$

$$= \sum_{k \in \mathbb{Z}^n} k^{\alpha^1 - \beta^1} (e^{-\nu \cdot} T)\hat{}_{(k,y)} \otimes [e^{ik \cdot}]_x$$

we find

$$\left(e^{-\nu \cdot} D_x^{\alpha^1} D_y^{\alpha^2} T\right)(\varphi) = \sum_{k \in \mathbb{Z}^n} (k - i\nu)^{\alpha^1} D_y^{\alpha^2} (e^{-\nu \cdot} T)\hat{}_{(k,y)} \otimes [e^{ik \cdot}]_x(\varphi)$$

which proves the assertion. $\square$

3 $\mathcal{R}$-boundedness and operator-valued Fourier multiplier theorems

In this chapter we present results on operator-valued Fourier multipliers both in the context of Fourier transform and Fourier series. They employ the concept of $\mathcal{R}$-boundedness which we introduce next. With $\mathcal{R}$-boundedness at hand, conditions can be deduced which make sure that a function defines a Fourier multiplier. Besides that, $\mathcal{R}$-boundedness is as well involved in necessary conditions for Fourier multipliers.

We refer to [DHP03] and [KW04] for a comprehensive introduction to the notion of $\mathcal{R}$-bounded operator families and restrict ourselves in this thesis to the definition and some basic properties only.

Definition 3.1. A family $\mathcal{T} \subset \mathcal{L}(X, Y)$ is called $\mathcal{R}$-*bounded* if there exist a $C > 0$ and a $p \in [1, \infty)$ such that for all $N \in \mathbb{N}$, $T_j \in \mathcal{T}$, $x_j \in X$ and all independent symmetric $\{-1, 1\}$-valued random variables ε_j on a probability space $(G, \mathcal{M}, P)$ for $j = 1, \ldots, N$, we have that

$$(3.1) \qquad \|\sum_{j=1}^{N} \varepsilon_j T_j x_j\|_{L^p(G,Y)} \leq C \|\sum_{j=1}^{N} \varepsilon_j x_j\|_{L^p(G,X)}.$$

The smallest $C > 0$ such that (3.1) is satisfied is called $\mathcal{R}$-*bound* of $\mathcal{T}$ and is denoted by $\mathcal{R}_p(\mathcal{T})$.

For our purposes there is no need to distinguish the p-dependent $\mathcal{R}$-bounds. Hence, we omit the index p and merely write $\mathcal{R}(\mathcal{T})$. Observe that $\mathcal{R}$-boundedness implies uniform norm boundedness. In Hilbert spaces both concepts are equivalent (see e.g. [KW04]).

The following two results on $\mathcal{R}$-boundedness will be used frequently in subsequent proofs. The first one shows that $\mathcal{R}$-bounds behave like uniform bounds concerning sums and products. This follows as a direct consequence of the definition of $\mathcal{R}$-boundedness. The second one is known as the contraction principle of Kahane. A proof can be found in [KW04] or [DHP03]. The contraction principle of Kahane and $\mathcal{R}$-boundedness of the sum of $\mathcal{R}$-bounded families yield $\mathcal{R}$-boundedness also for the union of $\mathcal{R}$-bounded families.

Lemma 3.2. a) *Let X, Y, and Z be Banach spaces and let $\mathcal{T}, \mathcal{S} \subset \mathcal{L}(X, Y)$ as well as $\mathcal{U} \subset \mathcal{L}(Y, Z)$ be $\mathcal{R}$-bounded. Then $\mathcal{T} \cup \mathcal{S} \subset \mathcal{L}(X, Y)$, $\mathcal{T} + \mathcal{S} \subset \mathcal{L}(X, Y)$ and $\mathcal{U}\mathcal{T} \subset \mathcal{L}(X, Z)$ are $\mathcal{R}$-bounded as well. More precisely, we have*

$$\mathcal{R}(\mathcal{T} \cup \mathcal{S}), \ \mathcal{R}(\mathcal{T} + \mathcal{S}) \leq \mathcal{R}(\mathcal{S}) + \mathcal{R}(\mathcal{T}) \quad \text{and} \quad \mathcal{R}(\mathcal{U}\mathcal{T}) \leq \mathcal{R}(\mathcal{U})\mathcal{R}(\mathcal{T}).$$

Furthermore, if $\overline{\mathcal{T}}$ denotes the closure of $\mathcal{T}$ with respect to the strong operator topology, we have $\mathcal{R}(\overline{\mathcal{T}}) = \mathcal{R}(\mathcal{T})$.

b) [Contraction principle of Kahane] *Let $1 \leq p < \infty$. Then for all $N \in \mathbb{N}$, $x_j \in X$, and ε_j as above, and for all $a_j, b_j \in \mathbb{C}$ with $|a_j| \leq |b_j|$ for $j = 1, \ldots, N$ we have*

$$(3.2) \qquad \| \sum_{j=1}^{N} a_j \varepsilon_j x_j \|_{L^p(G,X)} \leq 2 \| \sum_{j=1}^{N} b_j \varepsilon_j x_j \|_{L^p(G,X)}.$$

We present two further results on $\mathcal{R}$-boundedness we will need in the sequel. The first one (see e.g. [KW04, Corollary 2.14]) is of importance for applications which involve $\mathcal{R}$-bounds of integral operator families. In particular, special families including Dunford integral representations are captured.

Lemma 3.3. *Let $\mathcal{T} \subset \mathcal{L}(X)$ be $\mathcal{R}$-bounded, let $G \subset \mathbb{R}^n$ be a domain and $C > 0$. Given $N \in L^\infty(G, \mathcal{L}(X))$ such that $N(G) \subset \mathcal{T}$ and $g \in L^1(G)$ we set*

$$T_{N,g} f := \int_G g(x) N(x) f dx \quad (f \in X).$$

Then $\{T_{N,g}; \ N, g \text{ as above}, \ \|g\|_1 \leq C\} \subset \mathcal{L}(X)$ is $\mathcal{R}$-bounded.

For the moment we denote by m_φ the operator of multiplication with a function φ in L^p-spaces. The upcoming result is of particular interest in view of localization procedures we will have to carry out while treating non-constant coefficients in boundary value problems. A proof can be found e.g. in [DHP03, Corollary 3.7].

Lemma 3.4. *Let $1 \leq p < \infty$, let E and F be Banach spaces and let $G \subset \mathbb{R}^n$ be a domain. Set $X := L^p(G, E)$ and $Y := L^p(G, F)$ and let $\mathcal{T} \subset \mathcal{L}(X, Y)$ be $\mathcal{R}$-bounded. Then for each $C > 0$ the set*

$$\{m_\varphi T m_\psi; \ T \in \mathcal{T}, \ \varphi, \psi \in L^\infty(G), \ \|\varphi\|_\infty, \|\psi\|_\infty \leq C\}$$

is $\mathcal{R}$-bounded.

With $\mathcal{R}$-boundedness in mind, we first turn to continuous Fourier multipliers, whereas discrete Fourier multipliers are considered later on. Let E and F denote arbitrary Banach spaces and let $m \in L^\infty(\mathbb{R}^n, \mathcal{L}(E, F))$. Then the operator

$$T_m f := \mathcal{F}^{-1} m \mathcal{F} f \quad (f \in \mathcal{S}(\mathbb{R}^n, E))$$

is a well-defined mapping from $\mathcal{S}(\mathbb{R}^n, E)$ to $\mathcal{S}'(\mathbb{R}^n, F)$ since $m \mathcal{F} f \in L^\infty(\mathbb{R}^n, F)$ defines a regular distribution in $\mathcal{S}'(\mathbb{R}^n, F)$.

Definition 3.5. Let $1 \leq p < \infty$. A function $m \in L^\infty(\mathbb{R}^n, \mathcal{L}(E, F))$ is called a *continuous, operator-valued, $(L^p\text{-})$Fourier multiplier*, if $T_m f \in L^p(\mathbb{R}^n, F)$ for all $f \in \mathcal{S}(\mathbb{R}^n, E)$ and if $C > 0$ exists such that

$$\|T_m f\|_{p,F} \leq C \|f\|_{p,E} \quad (f \in \mathcal{S}(\mathbb{R}^n, E)).$$

In that case $T_m \in \mathcal{L}(L^p(\mathbb{R}^n, E), L^p(\mathbb{R}^n, F))$ by density of $\mathcal{S}(\mathbb{R}^n, E) \subset L^p(\mathbb{R}^n, E)$. The operator T_m is called the *Fourier multiplier operator associated with m*.

Starting with $f \in \mathcal{F}^{-1}(\mathcal{D}(\mathbb{R}^n, E))$, the assumption $m \in L^\infty(\mathbb{R}^n, \mathcal{L}(E, F))$ can be replaced by the weaker condition $m \in L^1_{loc}(\mathbb{R}^n, \mathcal{L}(E, F))$ (cf. [DHP03]). However, we stick to L^∞-functions right from the definition since it is well-known that Fourier multipliers are necessarily essentially bounded. The corresponding proof carried out for $n = 1$ as well as comments on the extension to arbitrary $n \in \mathbb{N}$ can be found in [DHP03], [ŠW07], or [KW04].

Proposition 3.6. *Let E, F be Banach spaces, $m \in L^1_{loc}(\mathbb{R}^n, \mathcal{L}(E, F))$ and let $L_m \subset \mathbb{R}^n$ denote the set of Lebesgue points of m. If m defines a continuous Fourier multiplier, then $\{m(\xi); \, \xi \in L_m\}$ is $\mathcal{R}$-bounded.*

In applications often questions on regularity of $T_m f$ arise. The following lemma provides a condition on m which is equivalent for $W^{\ell, p}$-regularity of $T_m f$.

Lemma 3.7. *Let $1 \leq p < \infty$, $\ell \in \mathbb{N}_0$ and let $m \in L^\infty(\mathbb{R}^n \setminus \{0\}, \mathcal{L}(E, F))$. Then the following assertions are equivalent:*

(i) $T_m \in \mathcal{L}(L^p(\mathbb{R}^n, E), W^{\ell, p}(\mathbb{R}^n, F))$.

(ii) For each $|\alpha| \leq \ell$ the function $m_\alpha \colon \xi \mapsto \xi^\alpha m(\xi)$ defines a continuous Fourier multiplier.

Proof. (i) $\Rightarrow$ (ii): Set $\kappa_\alpha(\xi) := \xi^\alpha$. For arbitrary $f \in \mathcal{S}(\mathbb{R}^n, E)$ we have

$$\mathcal{F} D^\alpha T_m f = \kappa_\alpha \mathcal{F} \mathcal{F}^{-1} m \mathcal{F} f = m_\alpha \mathcal{F} f$$

in $\mathcal{S}'(\mathbb{R}^n, F)$. Hence, $\mathcal{F}^{-1} m_\alpha \mathcal{F} f = D^\alpha T_m f$ and there exists $C > 0$ such that

$$\|\mathcal{F}^{-1} m_\alpha \mathcal{F} f\|_{p, F} = \|D^\alpha T_m f\|_{p, F} \leq C \|f\|_{p, E}$$

for all $|\alpha| \leq \ell$ and all $f \in \mathcal{S}(\mathbb{R}^n, E)$.

(ii) $\Rightarrow$ (i): For arbitrary $f \in \mathcal{S}(\mathbb{R}^n, E)$ first note that $T_m f \in L^p(\mathbb{R}^n, F)$ and that

$$\mathcal{F} D^\alpha T_m f = \kappa_\alpha \mathcal{F} T_m f = m_\alpha \mathcal{F} f$$

holds true in $\mathcal{S}'(\mathbb{R}^n, F)$. Applying $\mathcal{F}^{-1}$ yields $D^\alpha T_m f = T_{m_\alpha} f$ and

$$\|D^\alpha T_m f\|_{p, F} = \|T_{m_\alpha} f\|_{p, F} \leq C \|f\|_{p, E}$$

for all $|\alpha| \leq \ell$ and all $f \in \mathcal{S}(\mathbb{R}^n, E)$. Hence, $D^\alpha T_m$ extends to a bounded operator in $\mathcal{L}(L^p(\mathbb{R}^n, E), L^p(\mathbb{R}^n, F))$. $\qquad\square$

In close analogy, the notion of a Fourier multiplier is defined in the context of Fourier series. We refer e.g. to [CW77] for useful results on how continuous and discrete multipliers are linked to each other. See also [ŠW07, Lemma 3.3] and the discussion on Theorem 3.23 below. Let $M \in \ell^\infty(\mathbb{Z}^n, \mathcal{L}(E, F))$, that is $M \colon \mathbb{Z}^n \to \mathcal{L}(E, F)$ uniformly bounded. Then the operator T_M defined by

$$(T_M f)\hat{}(k) = M(k) \hat{f}(k) \quad (k \in \mathbb{Z}^n)$$

represents a well-defined mapping from $L^p(\mathcal{Q}_n, E)$ to $\mathcal{D}'_{per}(\mathbb{R}^n, F)$. This mapping has to be understood in the sense that there exists $C > 0$ such that

$$\|M(k)\hat{f}(k)\|_F \leq C \quad (k \in \mathbb{Z}^n)$$

and therefore

$$T_M f := \sum_{k \in \mathbb{Z}^n} M(k)\hat{f}(k)e^{ik\cdot} \in \mathcal{D}'_{per}(\mathbb{R}^n, F)$$

by Lemma 2.6.

Definition 3.8. Let $1 \leq p < \infty$. A function $M \in \ell^\infty(\mathbb{Z}^n, \mathcal{L}(E, F))$ is called a *discrete, operator-valued, (L^p-)Fourier multiplier*, if $C > 0$ exists such that

$$\|T_M f\|_{p,F} \leq C\|f\|_{p,E} \quad (f \in \mathbb{T}(\mathcal{Q}_n, E)).$$

In that case $T_M \in \mathcal{L}(L^p(\mathcal{Q}_n, E), L^p(\mathcal{Q}_n, F))$ by density of $\mathbb{T}(\mathcal{Q}_n, E) \subset L^p(\mathcal{Q}_n, E)$. The operator T_M is called the *Fourier multiplier operator associated with M*.

In the definition of T_M, uniform boundedness of M can be relaxed to the assumption of M being polynomially bounded for Lemma 2.6 can be applied if

$$\|M(k)\hat{f}(k)\|_F \leq C|k|^m \quad (k \in \mathbb{Z}^n).$$

When restricted to the space of trigonometric polynomials $\mathbb{T}(\mathcal{Q}_n, E)$, the operator T_M maps to $\mathbb{T}(\mathcal{Q}_n, F)$ of course without any assumptions on $M: \mathbb{Z}^n \to \mathcal{L}(E, F)$.

However, the following result shows the assumption $M \in \ell^\infty(\mathbb{Z}^n, \mathcal{L}(E, F))$ to be a natural choice. A proof carried out for $n = 1$ can be found in [AB02]. Comments on the extension to arbitrary $n \in \mathbb{N}$ are given in [BK04] and [ŠW07].

Proposition 3.9. *Let E, F be Banach spaces and $M: \mathbb{Z}^n \to \mathcal{L}(E, F)$. If M defines a discrete Fourier multiplier, then $\{M(k); \ k \in \mathbb{Z}^n\}$ is $\mathcal{R}$-bounded.*

It is worth noting that the property of M being a discrete Fourier multiplier is characterized by the set inclusion $T_M(L^p(\mathcal{Q}_n, E)) \subset L^p(\mathcal{Q}_n, F)$ without any continuity condition. We make this result more precise in the following lemma.

Lemma 3.10. *For every function $M: \mathbb{Z}^n \to \mathcal{L}(E, F)$ the following two statements are equivalent:*

(i) M is a discrete L^p-multiplier.

(ii) For each $f \in L^p(\mathcal{Q}_n, E)$ there exists $g \in L^p(\mathcal{Q}_n, F)$ such that

$$\hat{g}(k) = M(k)\hat{f}(k) \quad (k \in \mathbb{Z}^n).$$

Proof. $(i) \Rightarrow (ii)$: As already mentioned in the definition, T_M first defined for trigonometric polynomials only extends uniquely to $T_M: L^p(\mathcal{Q}_n, E) \to L^p(\mathcal{Q}_n, F)$ by continuity.

$(ii) \Rightarrow (i)$: We define $T_M f = g$, where g fulfills $\hat{g}(k) = M(k)\hat{f}(k)$ for $k \in \mathbb{Z}^n$.

As $D(T_M) = L^p(\mathcal{Q}_n, E)$, the closed graph theorem yields continuity of T_M if we can show that T_M is closed. To this end, let $f_n \to f$ in $L^p(\mathcal{Q}_n, E)$ as well as $g_n := T_M f_n \to g$ in $L^p(\mathcal{Q}_n, F)$. Since $L^p(\mathcal{Q}_n, E) \subset L^1(\mathcal{Q}_n, E)$, Lebesgue's theorem of dominated convergence shows $\hat{f}_n(k) \to \hat{f}(k)$ and $\hat{g}_n(k) \to \hat{g}(k)$ for all $k \in \mathbb{Z}^n$. Since further $M(k) \in \mathcal{L}(E, F)$, we additionally have $\hat{g}_n(k) \to M(k)\hat{f}(k)$. Hence, $\hat{g}(k) = M(k)\hat{f}(k)$ for all $k \in \mathbb{Z}^n$ and Corollary 2.9 yields $g = T_M f$. $\quad\square$

With Lemma 3.10 at hand, questions on regularity of $T_M f$ can be answered more easily than those on regularity of $T_m f$ (cf. Lemma 3.7). This aspect is made available in the following lemma.

Lemma 3.11. *Let* $1 \le p < \infty$, $\ell \in \mathbb{N}_0$ *and let* $M \in \ell^\infty(\mathbb{Z}^n, \mathcal{L}(E, F))$. *Then the following assertions are equivalent:*

(i) $T_M \in \mathcal{L}(L^p(\mathcal{Q}_n, E), W_{per}^{\ell, p}(\mathcal{Q}_n, F))$.

(ii) $T_M \in \mathcal{L}(L^p(\mathcal{Q}_n, E), L^p(\mathcal{Q}_n, F))$ *maps* $L^p(\mathcal{Q}_n, E)$ *into* $W_{per}^{\ell, p}(\mathcal{Q}_n, F)$.

(iii) $M_\alpha \colon k \mapsto k^\alpha M(k)$ *defines a discrete Fourier multiplier for each* $|\alpha| \le \ell$.

(iv) $M_\alpha \colon k \mapsto k^\alpha M(k)$ *defines a discrete Fourier multiplier for each* $|\alpha| = \ell$.

Proof. (i) $\Rightarrow$ (ii): Obvious.
(ii) $\Rightarrow$ (iii): For arbitrary $f \in L^p(\mathcal{Q}_n, E)$ we have $T_M f \in W_{per}^{\ell, p}(\mathcal{Q}_n, F)$ and

$$(D^\alpha T_M f)\hat{}(k) = k^\alpha M(k)\hat{f}(k) = M_\alpha(k)\hat{f}(k)$$

by (2.8). Since $D^\alpha T_M f \in L^p(\mathcal{Q}_n, F)$ it follows that M_α defines a Fourier multiplier for all $|\alpha| \le \ell$ by Lemma 3.10.
(iii) $\Rightarrow$ (iv): Obvious.
(iv) $\Rightarrow$ (i): Let $f \in L^p(\mathcal{Q}_n, E)$ and α subject to $|\alpha| = \ell$ be arbitrary. By assumption we have $v_\alpha := T_{M_\alpha} f \in L^p(\mathcal{Q}_n, F)$. Thus there exists $C > 0$ such that $\|M_\alpha(k)\hat{f}(k)\|_F \le C$. Set

$$\tilde{M}^{(\alpha)}(k) := \begin{cases} 0, & k^\alpha = 0, \\ M(k), & k^\alpha \ne 0. \end{cases}$$

Since $|k^\alpha| \ge 1$ we deduce

$$\|\tilde{M}^{(\alpha)}(k)\hat{f}(k)\|_F \le \|k^\alpha M(k)\hat{f}(k)\|_F \le C \qquad (k \in \mathbb{Z}^n).$$

In particular, there exists $C > 0$ such that this estimate is valid for all $\alpha = \ell e_j$, where $j = 1, \ldots, n$. This shows $\|M(k)\hat{f}(k)\|_F \le C$ for all $k \in \mathbb{Z}^n \setminus \{0\}$ and further

$$\|M(k)\hat{f}(k)\|_F \le C \qquad (k \in \mathbb{Z}^n).$$

Lemma 2.6 yields $T := \sum\limits_{k \in \mathbb{Z}^n} M(k)\hat{f}(k)e^{ik\cdot} \in \mathcal{D}'_{per}(\mathbb{R}^n, F)$ and by (2.8) we get

$$D^\alpha T = \sum_{k \in \mathbb{Z}^n} M_\alpha(k)\hat{f}(k)e^{ik\cdot} = v_\alpha.$$

Proposition 2.15 therefore implies existence of $g \in W^{\ell,p}_{per}(\mathcal{Q}_n, F)$ such that $T = T_g$ and $(D^\alpha g)\hat{}(k) = M(k)\hat{f}(k)$ for all $|\alpha| \le \ell$. Lemma 3.10 now finishes the proof. $\square$

We head for multiplier results and return to continuous Fourier multipliers first. The question under what circumstances a function defines an operator-valued Fourier multiplier had been unanswered for a long time. A satisfying answer eventually was given by Weis in [Wei01b], who has been able to prove an operator-valued version of the Michlin multiplier theorem. Many other contributions, both in the context of Fourier transform and Fourier series, such as [AB02], [ŠW07], [HHN02], and [BK04] followed. In order to present these Fourier multiplier results, a few preparations are in order. The idea Weis pursued to prove the operator-valued version of the Michlin multiplier theorem in case $n = 1$ was to build up a class of Fourier multipliers out of one basic multiplier function, namely

$$m \colon \mathbb{R} \setminus \{0\} \to \mathcal{L}(E); \ m(\xi) := \xi|\xi|^{-1}\,\mathrm{id}_E\,.$$

However, there exist Banach spaces E such that this function fails to define a Fourier multiplier. Thus, the condition that it fulfills enters as a restriction on E in the multiplier theorem of Weis. A commonly used description of this property is given in terms of the Hilbert transform $H \colon \mathcal{S}(\mathbb{R}, E) \to \mathcal{S}'(\mathbb{R}, E)$ defined by $Hf := \mathcal{F}^{-1}m_i\mathcal{F}f$, where $m_i(\xi) := i\xi|\xi|^{-1}\,\mathrm{id}_E$.

Definition 3.12. A Banach space E is called *Banach space of class $\mathcal{HT}$* if there exists a $q \in (1, \infty)$ such that H extends to a bounded operator in $\mathcal{L}\big(L^q(\mathbb{R}, E)\big)$.

Remark 3.13. a) The property of a Banach space E to be of class $\mathcal{HT}$ is equivalent to E being a UMD space. This in turn is equivalent to the condition that E is ζ-convex. For more information on these equivalences and on ζ-convexity in particular we refer to [Dor93, Section 3] and the survey article [RdF86] referred to therein.
b) As a consequence of ζ-convexity, each Banach space of class $\mathcal{HT}$ is superreflexive and therefore reflexive ([RdF86, Proposition 2]).

We collect some more properties of the class $\mathcal{HT}$ from [Ama95, Theorem 4.5.2] which we will use frequently without further comments.

Lemma 3.14. a) *Hilbert spaces are spaces of class $\mathcal{HT}$.*
b) *Every Banach space isomorphic to a space of class $\mathcal{HT}$ is of class $\mathcal{HT}$.*
c) *Finite products of spaces of class $\mathcal{HT}$ are of class $\mathcal{HT}$.*
d) *The spaces $L^p(G, F)$ are of class $\mathcal{HT}$ for $1 < p < \infty$ if F is of class $\mathcal{HT}$.*
e) *Closed linear subspaces of spaces of class $\mathcal{HT}$ are of class $\mathcal{HT}$.*

The whole process of building larger classes of multiplier functions out of the basic multiplier function is carried out in detail in [KW04]. With a second Banach space F, functions of type

$$m\colon \mathbb{R} \setminus \{0\} \to \mathcal{L}(E,F); \ m(\xi) := \chi_{(a,b]}(\xi) \otimes B$$

are established as Fourier multipliers, where $\chi_{(a,b]}$ denotes the characteristic function of the interval $(a,b]$ and $B \in \mathcal{L}(E,F)$ is a fixed bounded linear operator. This allows for $\mathcal{L}(E,F)$-valued functions to be considered. It finally ends with the operator-valued Michlin multiplier result due to Weis which relies on $\mathcal{R}$-boundedness. To present the Fourier multiplier results in their full strength, in addition to class $\mathcal{HT}$, a further notion from Banach space geometry called property (α) is needed. One situation where it comes into play is when a whole family of multiplier functions is considered and if the family of associated operators is required to be $\mathcal{R}$-bounded again.

Definition 3.15. A Banach space X is said to have *property* (α) if there exists a $C > 0$ such that for all $n \in \mathbb{N}$, $\alpha_{ij} \in \mathbb{C}$ with $|\alpha_{ij}| \leq 1$, all $x_{ij} \in X$, and all independent symmetric $\{-1,1\}$-valued random variables ε_i^1 on a probability space $(G_1, \mathcal{M}_1, P_1)$ and ε_j^2 on a probability space $(G_2, \mathcal{M}_2, P_2)$ for $i,j = 1, \ldots, N$, we have

$$\int_{G_1} \int_{G_2} \| \sum_{i,j=1}^{N} \varepsilon_i^1(u)\varepsilon_j^2(v)\alpha_{ij}x_{ij} \|_X \, du \, dv \leq C \int_{G_1} \int_{G_2} \| \sum_{i,j=1}^{N} \varepsilon_i^1(u)\varepsilon_j^2(v)x_{ij} \|_X \, du \, dv.$$

The following facts on property (α) can be found in [KW04].

Lemma 3.16. *For $n \in \mathbb{N}$, $\mathbb{C}^n$ enjoys property (α) and, if E enjoys property (α), for $1 \leq p < \infty$ the spaces $L^p(G,E)$ enjoy property (α) as well.*

Now we are in the position to state the Michlin multiplier theorem in operator-valued context due to Weis which gives sufficient conditions on m to define a Fourier multiplier. See [Wei01b], [HHN02], and [ŠW07]. Recall that $0 \leq \gamma \leq 1$ is defined by means of the different components, i.e.

$$0 \leq \gamma \leq 1 \quad :\Longleftrightarrow \quad 0 \leq \gamma_i \leq 1 \quad (i = 1, \ldots, n).$$

Theorem 3.17. a) *Let $1 < p < \infty$, let E and F be Banach spaces of class $\mathcal{HT}$ and let $\mathcal{T} \subset \mathcal{L}(E,F)$ be $\mathcal{R}$-bounded. If $m \in C^n(\mathbb{R}^n \setminus \{0\}, \mathcal{L}(E,F))$ satisfies*

$$(3.3) \qquad \{ |\xi|^{|\gamma|} D^\gamma m(\xi); \ \xi \in \mathbb{R}^n \setminus \{0\}, \ 0 \leq \gamma \leq 1 \} \subset \mathcal{T},$$

then m defines a Fourier multiplier.
b) *If E and F additionally enjoy property (α), then*

$$(3.4) \qquad \{ \xi^\gamma D^\gamma m(\xi); \ \xi \in \mathbb{R}^n \setminus \{0\}, \ 0 \leq \gamma \leq 1 \} \subset \mathcal{T}$$

is sufficient. In this case, the set

$$\{T_m; \ m \ \textit{satisfies condition} \ (3.4)\} \subset \mathcal{L}\big(L^p(\mathbb{R}^n, E), L^p(\mathbb{R}^n, F)\big)$$

is $\mathcal{R}$-bounded again.

The additional result on $\mathcal{R}$-boundedness of operators associated with a family of multipliers in part b) is due to Girardi and Weis [GW03]. For further information on operator-valued Fourier multipliers we refer to [KW04] and [DHP03].

Remark 3.18. In case $n = 1$ the weights $|\xi|^{|\gamma|}$ in a) and $|\xi^\gamma|$ in b) are equal. For $n > 1$ the weight $|\xi|^{|\gamma|}$ is larger in the sense that for $|\cdot|_i \in \{|\cdot|_1, |\cdot|_2\}$ we have

$$|\xi|_i^{|\gamma|} \geq |\xi|_\infty^{|\gamma|} \geq \big(\max_{j, \ \gamma_j \neq 0} |\xi_j|\big)^{|\gamma|} \geq \prod_{j, \ \gamma_j \neq 0} |\xi_j| = |\xi^\gamma|.$$

The upcoming result is the analogue of Theorem 3.17 in the context of discrete Fourier multipliers. It was first proved by Arendt and Bu in [AB02] for $n = 1$. For arbitrary n it is due Bu and Kim ([BK04]). In the article [Bu06], Bu gives a shorter proof for arbitrary n by means of induction based on the result for $n = 1$. Instead of derivatives as above, in the context of Fourier series discrete derivatives are involved which we introduce next. Let E and F be Banach spaces and let $G \subset \mathbb{Z}^n$. For a function $M: \mathbb{Z}^n \to \mathcal{L}(E, F)$ the restriction of M to G is defined by

$$M_G(k) := \begin{cases} M(k), & k \in G, \\ 0, & k \notin G. \end{cases}$$

In particular $M_{\mathbb{Z}^n} = M$. Let $\alpha, \beta \in \big(\mathbb{Z} \cup \{-\infty, \infty\}\big)^n$ with $\alpha \leq \beta$, i.e. $\alpha_j \leq \beta_j$ for $j = 1, \ldots, n$ and let e_j denote the j-th unit vector in $\mathbb{Z}^n$. The difference operators Δ^{e_j} are defined as

$$\Delta^{e_j} M_{[\alpha,\beta]}(k) := \begin{cases} M_{[\alpha,\beta]}(k) - M_{[\alpha,\beta]}(k - e_j), & k_j \neq \alpha_j, \\ 0, & k_j = \alpha_j, \end{cases}$$

where $k \in [\alpha, \beta]$. For arbitrary $\gamma \in \{0, 1\}^n$ we set

(3.5) $\Delta^0 M_{[\alpha,\beta]} = 0, \quad \Delta^\gamma M_{[\alpha,\beta]} := \Delta^{\gamma_1 e_1} \ldots \Delta^{\gamma_n e_n} M_{[\alpha,\beta]}.$

We will sometimes refer to Δ^γ as the discrete derivative of order γ.

Theorem 3.19. a) *Let $1 < p < \infty$, let E and F be Banach spaces of class $\mathcal{HT}$ and let $\mathcal{T} \subset \mathcal{L}(E, F)$ be $\mathcal{R}$-bounded. If $M: \mathbb{Z}^n \to \mathcal{L}(E, F)$ satisfies*

(3.6) $\big\{|k|^{|\gamma|} \Delta^\gamma M(k); \ k \in \mathbb{Z}^n, \ 0 \leq \gamma \leq 1\big\} \subset \mathcal{T},$

then M defines a Fourier multiplier.
b) *If E and F additionally enjoy property (α), then*

(3.7) $\big\{k^\gamma \Delta^\gamma M(k); \ k \in \mathbb{Z}^n, \ 0 \leq \gamma \leq 1\big\} \subset \mathcal{T}$

is sufficient. In this case, the set

$$\{T_M; \ M \ satisfies \ condition \ (3.7)\} \subset \mathcal{L}\big(L^p(\mathcal{Q}_n, E), L^p(\mathcal{Q}_n, F)\big)$$

is $\mathcal{R}$-bounded again.

Remark 3.20. In [BK04] Theorem 3.19 is stated with discrete derivatives $\tilde{\Delta}$ defined in such a way that $\Delta^\gamma M(k + \gamma) = \tilde{\Delta}^\gamma M(k)$. However, as for fixed $\gamma \in \{0, 1\}^n$ there exist $c, C > 0$ such that $c|k - \gamma| \leq |k| \leq C|k - \gamma|$ for $k \in \mathbb{Z}^n \setminus \{0, 1\}^n$, the contraction principle of Kahane shows our formulation to be equivalent to the one given in [BK04] (cf. condition (3.12) below).

As mentioned by Arendt and Bu, their result can as well be deduced from a discrete multiplier theorem due to Štrkalj and Weis in [ŠW07], see Theorem 3.23 below. The latter also serves to prove the Michlin multiplier theorem in the multidimensional case. Thus, it can be used as baseline to prove both Theorem 3.17 and Theorem 3.19. We briefly sketch the expressions and ideas used in [ŠW07] in order to present this result. Based on this, we can quickly derive a version of the multiplier result of Bu and Kim with a slight but useful weakening of the assumptions on M.

Given a bounded subset $\mathcal{T} \subset \mathcal{L}(E, F)$, the Minkowski functional of the absolute convex hull

$$\mathrm{aco}(\mathcal{T}) := \Big\{ \sum_{j=1}^N \lambda_j T_j; \ N \in \mathbb{N}, \ T_j \in \mathcal{T}, \ \lambda_j \in \mathbb{C} \ \text{with} \ \sum_{j=1}^N |\lambda_j| = 1 \Big\}$$

is defined as $\| \cdot \|_{\mathcal{T}} \colon \mathcal{L}(E, F) \to [0, \infty]$; $T \mapsto \inf\{t > 0; \ T \in t\,\mathrm{aco}(\mathcal{T})\}$. Let $M \colon \mathbb{Z}^n \to \mathcal{L}(E, F)$ and $\mathcal{T} \subset \mathcal{L}(E, F)$ be bounded. Then the $\mathcal{T}$-variation of M in the interval $[\alpha, \beta]$ is defined as

$$\mathop{\mathrm{var}}_{[\alpha,\beta],\mathcal{T}} M_{[\alpha,\beta]} := \sum_{k \in [\alpha,\beta]} \|\Delta^{\gamma_k} M_{[\alpha,\beta]}(k)\|_{\mathcal{T}},$$

where $\gamma_k = (\gamma_{k_1}, \ldots, \gamma_{k_n})$ with

$$\gamma_{k_j} := \begin{cases} 1, & k_j \neq \alpha_j, \\ 0, & k_j = \alpha_j. \end{cases}$$

The following decompositions of $\mathbb{Z}^n$ are crucial.

Definition 3.21. a) $D_0 := \{0\} \subset \mathbb{Z}^n$ and for $d = nr + j, \ r \in \mathbb{N}_0, \ j \in \{1, \ldots, n\}$

$$D_d := \{k \in \mathbb{Z}^n; \ |k_1|, \ldots, |k_{j-1}| < 2^{r+1}, \ 2^r \leq |k_j| < 2^{r+1}, \ |k_{j+1}|, \ldots, |k_n| < 2^r\}$$

is called the *coarse decomposition* of $\mathbb{Z}^n$.
b) For $\nu = (\nu_1, \ldots, \nu_n) \in \mathbb{N}_0^n$, $I_r = \{\ell \in \mathbb{Z}; \ 2^{r-1} \leq |\ell| < 2^r\}$ for $r \in \mathbb{N}$, and $I_0 := \{0\}$

$$D_\nu := I_{\nu_1} \times \ldots \times I_{\nu_n}$$

is called the *fine decomposition* of $\mathbb{Z}^n$.

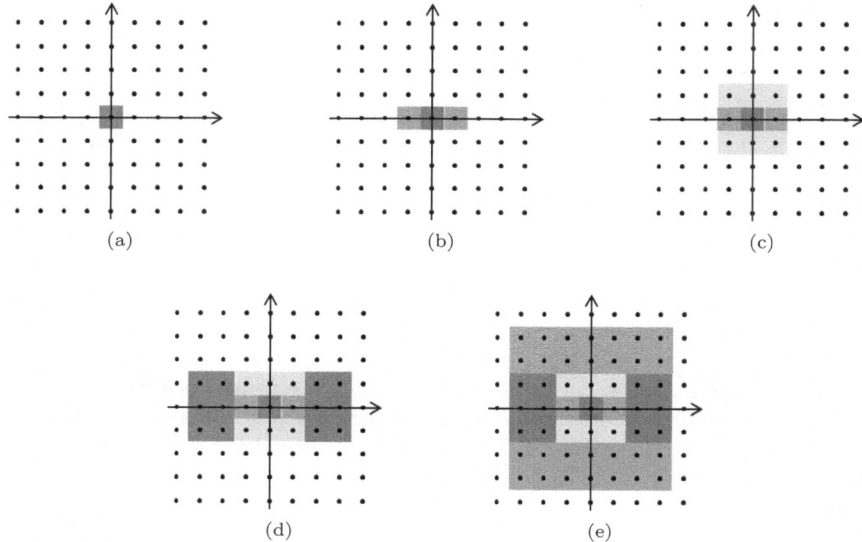

Figure 3.1: Steps in the coarse decomposition of $\mathbb{Z}^2$: (a): D_0 to (e): $\bigcup_{d=0}^{4} D_d$

Note that D_d is the union of $s_d \leq 2$ intervals $[\alpha_{d,i}, \beta_{d,i}]$. Respectively, D_ν is the union of $s_\nu \leq 2^n$ intervals $[\alpha_{\nu,i}, \beta_{\nu,i}]$. One thus defines the $\mathcal{T}$-*variation* of $M\colon \mathbb{Z}^n \to \mathcal{L}(E, F)$ with respect to the coarse decomposition by

$$\operatorname*{var}_{D_d, \mathcal{T}} M := \sum_{i=1}^{s_d} \operatorname*{var}_{[\alpha_{d,i}, \beta_{d,i}], \mathcal{T}} M_{[\alpha_{d,i}, \beta_{d,i}]}$$

and with respect to the fine decomposition by

$$\operatorname*{var}_{D_\nu, \mathcal{T}} M := \sum_{i=1}^{s_\nu} \operatorname*{var}_{[\alpha_{\nu,i}, \beta_{\nu,i}], \mathcal{T}} M_{[\alpha_{\nu,i}, \beta_{\nu,i}]}.$$

In case $n = 1$ we have $D_d = D_\nu$ for $d = \nu \in \mathbb{N}_0$. In case $n > 1$ for each $\nu \in \mathbb{N}_0^n$ there exists $d = d_\nu \in \mathbb{N}_0$ such that $D_\nu \subset D_d$. See Figure 3.1 for an illustration of the coarse decomposition and Figure 3.2 for an illustration of the fine decomposition of $\mathbb{Z}^2$.

Definition 3.22. M is said to be of *bounded $\mathcal{T}$-variation* with respect to the coarse decomposition D_d, respectively the fine decomposition D_ν, if there exists an $\mathcal{R}$-bounded subset $\mathcal{T} \subset \mathcal{L}(E, F)$ such that the condition

$$(3.8) \qquad\qquad\qquad\qquad \sup_{d \in \mathbb{N}_0} \operatorname*{var}_{D_d, \mathcal{T}} M < \infty,$$

respectively

(3.9) $$\sup_{\nu \in \mathbb{N}_0^n} \operatorname*{var}_{D_\nu, \mathcal{T}} M < \infty,$$

is fulfilled.

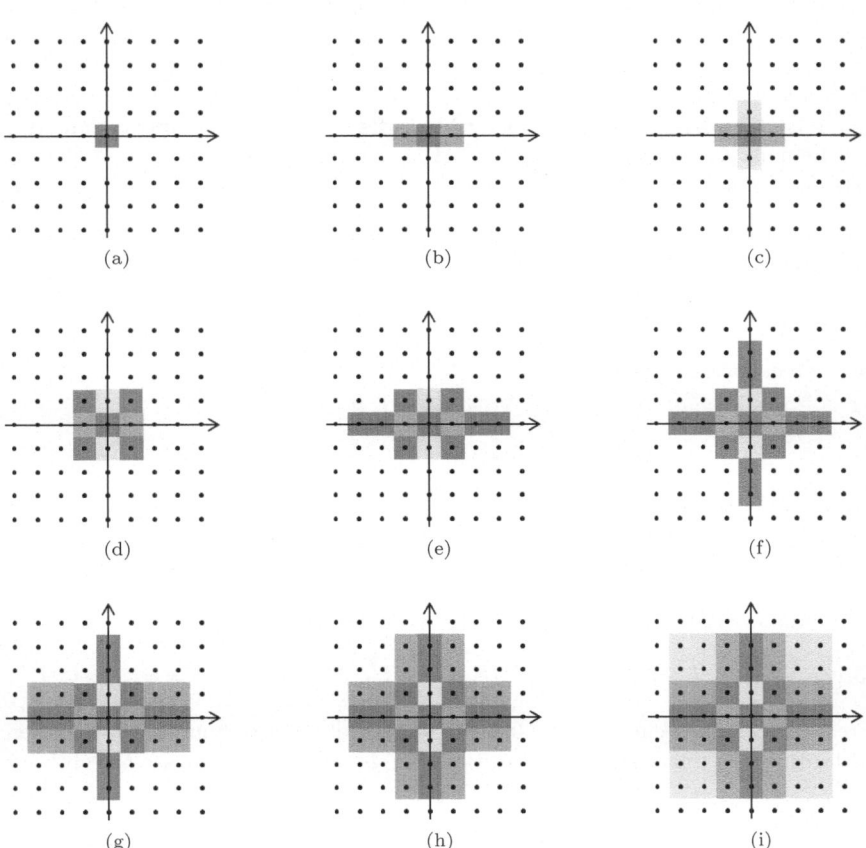

Figure 3.2: Steps in the fine decomposition of $\mathbb{Z}^2$: (a): $D_{(0,0)}$ to (i): $\bigcup_{\nu \leq (2,2)} D_\nu$

With this notation at hand, we can now state the discrete multiplier theorem due to Štrkalj and Weis from [ŠW07].

Theorem 3.23.   a) *Let* $1 < p < \infty$ *and let* E *and* F *be Banach spaces of class* $\mathcal{HT}$. *Let* $M : \mathbb{Z}^n \to \mathcal{L}(E, F)$ *be of bounded* $\mathcal{T}$-*variation with respect to the coarse decomposition. Then* M *defines a discrete* L^p-*multiplier.*

b) *Let E and F additionally enjoy property (α) and let $M\colon \mathbb{Z}^n \to \mathcal{L}(E,F)$ be of bounded $\mathcal{T}$-variation with respect to the fine decomposition. Then M defines a discrete L^p-multiplier.*

The authors of [BK04] also consider the coarse and fine decomposition of $\mathbb{Z}^n$. Therefore, it is not astonishing that their result can be deduced from Theorem 3.23. Doing that, it is obvious that the assumptions (3.6) and (3.7) on M can be relaxed. While the readability of the weaker condition decreases, the restrictions for M anyhow similarly do so. Due to the definition of the coarse and fine decomposition most of the benefit is gained close to zero. It will turn out that this relaxation helps to avoid undesirable shifts of operators in applications (see e.g. Lemma 9.4 and Proposition 7.25). For this reason we present the following version of Theorem 3.19.

Theorem 3.24. a) *Let $1 < p < \infty$, let E and F be be Banach spaces of class $\mathcal{HT}$ and let $\mathcal{T} \subset \mathcal{L}(E,F)$ be $\mathcal{R}$-bounded. If $M\colon \mathbb{Z}^n \to \mathcal{L}(E,F)$ satisfies*

$$(3.10) \qquad \{|k|^{|\gamma_k|} \Delta^{\gamma_k} M_{D_d}(k);\ d \in \mathbb{N}_0,\ k \in D_d\} \subset \mathcal{T},$$

then M defines an L^p-multiplier.
b) *If E and F additionally enjoy property (α), then*

$$(3.11) \qquad \{k^{\gamma_k} \Delta^{\gamma_k} M_{D_\nu}(k);\ \nu \in \mathbb{N}_0^n,\ k \in D_\nu\} \subset \mathcal{T}$$

is sufficient. In this case, the set

$$\{T_M;\ M \text{ satisfies condition } (3.11)\} \subset \mathcal{L}\big(L^p(\mathcal{Q}_n, E), L^p(\mathcal{Q}_n, F)\big)$$

is $\mathcal{R}$-bounded again. In particular, (3.11) is fulfilled if

$$(3.12) \quad \{M(k);\ k \in \mathbb{Z}^n\} \cup \{k^\gamma \Delta^\gamma M(k);\ k \in \mathbb{Z}^n \setminus \{-1,0,1\}^n,\ 0 \neq \gamma \leq 1\} \subset \mathcal{T}.$$

Proof. a) By Theorem 3.23 it suffices to show that M is of bounded $\mathcal{T}$-variation with respect to the coarse decomposition. Let $d \in \mathbb{N}_0$ with $d = rn+j$ be arbitrary and

$$D_d = [\alpha_{d,-}, \beta_{d,-}] \cup [\alpha_{d,+}, \beta_{d,+}],$$

where $k_j \leq 0$ for $k \in [\alpha_{d,-}, \beta_{d,-}]$ and $k_j \geq 0$ for $k \in [\alpha_{d,+}, \beta_{d,+}]$. Then

$$\|\Delta^{\gamma_k} M_{D_d}(k)\|_{\mathcal{T}} \leq \frac{1}{|k|^{|\gamma_k|}}$$

follows by assumption (3.10). The definition of $[\alpha_{d,\pm}, \beta_{d,\pm}]$ and γ_k shows

$$\sup_{d \in \mathbb{N}_0} \operatorname*{var}_{D_d, \mathcal{T}} \leq 2 \sup_{d \in \mathbb{N}_0} \sum_{k \in [\alpha_{d,-}, \beta_{d,-}]} \frac{1}{|k|^{|\gamma_k|}}.$$

We rearrange the sum and use the fact that $|\cdot| \geq |\cdot|_\infty$ for $|\cdot| \in \{|\cdot|_1, |\cdot|_2\}$ to get

$$\sum_{k \in [\alpha_{d,-},\beta_{d,-}]} \frac{1}{|k|^{|\gamma_k|}} = \sum_{\gamma \leq 1} \sum_{k;\gamma_k=\gamma} \frac{1}{|k|^{|\gamma|}} \leq \sum_{\gamma \leq 1} \sum_{k;\gamma_k=\gamma} \frac{1}{|k|_\infty^{|\gamma|}}.$$

Clearly, for $0 \leq \ell \leq n$ we have

$$\left|\{\gamma \in \mathbb{N}_0^n;\ 0 \leq \gamma \leq 1,\ |\gamma| = \ell\}\right| = \binom{n}{\ell}.$$

As the length of any edge of $[\alpha_{d,-},\beta_{d,-}]$ is smaller than 2^{r+2} and $|k|_\infty \geq 2^r$ for $d \geq 1$ we have

$$\sum_{\gamma \leq 1} \sum_{k;\gamma_k=\gamma} \frac{1}{|k|_\infty^{|\gamma|}} \leq \sum_{\ell=0}^n \binom{n}{\ell} \left(\frac{2^{r+2}}{2^r}\right)^l \leq \sum_{\ell=0}^n \binom{n}{\ell} 4^l \leq 2^{3n}.$$

Since this estimate is independent of d, the proof of part a) is complete.

b) Again by Theorem 3.23 it suffices to show that M is of bounded $\mathcal{T}$-variation with respect to the fine decomposition. Let $\nu = (\nu_1, \ldots, \nu_n) \in \mathbb{N}_0^n$ be arbitrary and

$$D_\nu = I_{\nu_1} \times \ldots \times I_{\nu_n} = \bigcup_{i=1}^{s_\nu} [\alpha_{\nu,i}, \beta_{\nu,i}].$$

By assumption (3.11)

$$\|\Delta^{\gamma_k} M_{D_\nu}(k)\|_{\mathcal{T}} \leq \frac{1}{|k^{\gamma_k}|}$$

holds true. Let $i_- \in \{1, \ldots, s_\nu\}$ such that $\alpha_{\nu,i_-} \leq 0$. By definition of γ_k and D_ν we then have

$$\sum_{k \in [\alpha_{\nu,i}, \beta_{\nu,i}]} \frac{1}{|k^{\gamma_k}|} \leq \sum_{k \in [\alpha_{\nu,i_-}, \beta_{\nu,i_-}]} \frac{1}{|k^{\gamma_k}|},$$

hence,

$$\sup_{\nu \in \mathbb{N}_0^n} \operatorname*{var}_{D_\nu,\mathcal{T}} = \sup_{\nu \in \mathbb{N}_0^n} \sum_{i=1}^{s_\nu} \sum_{k \in [\alpha_{\nu,i}, \beta_{\nu,i}]} \|(\Delta^{\gamma_k} M_{[\alpha_{\nu,i}, \beta_{\nu,i}]})(k)\|_{\mathcal{T}}$$

$$\leq \sup_{\nu \in \mathbb{N}_0^n} \sum_{i=1}^{s_\nu} \sum_{k \in [\alpha_{\nu,i}, \beta_{\nu,i}]} \frac{1}{|k^{\gamma_k}|} \leq 2^n \sup_{\nu \in \mathbb{N}_0^n} \sum_{k \in [\alpha_{\nu,i_-}, \beta_{\nu,i_-}]} \frac{1}{|k^{\gamma_k}|}.$$

Again we rearrange the sum to get

$$\sum_{k\in[\alpha_{\nu,i_-},\beta_{\nu,i_-}]}\frac{1}{|k^{\gamma_k}|} = \sum_{|\gamma|\leq 1}\sum_{k;\gamma_k=\gamma}\frac{1}{|k^{\gamma_k}|}$$

$$= 1 + \sum_{j=1}^{n}\sum_{k;\gamma_k=e_j}\frac{1}{|k^{e_j}|} + \sum_{j,i=1}^{n}\sum_{k;\gamma_k=e_j+e_i}\frac{1}{|k^{e_j+e_i}|} + \cdots$$

$$\cdots + \sum_{j=1}^{n}\sum_{k;\gamma_k=1-e_j}\frac{1}{|k^{1-e_j}|} + \sum_{k;\gamma_k=1}\frac{1}{|k^{1}|}.$$

For arbitrary $\ell\in\{1,\ldots,n\}$ we find

$$\sum_{k;\gamma_k=e_{j_1}+\ldots+e_{j_\ell}}\frac{1}{|k^{e_{j_1}+\ldots+e_{j_\ell}}|} = \sum_{k;\gamma_k=e_{j_1}+\ldots+e_{j_\ell}}\frac{1}{|k_{j_1}|}\cdot\ldots\cdot\frac{1}{|k_{j_\ell}|}$$

$$= \sum_{k';\gamma'_{k'}=e_{j_2}+\ldots+e_{j_\ell}}\left(\sum_{r=0}^{2^{\nu_{j_1}-1}-2}\frac{1}{2^{\nu_{j_1}-1}+r}\right)\frac{1}{|k_{j_2}|}\cdot\ldots\cdot\frac{1}{|k_{j_\ell}|}$$

$$= \prod_{i=1}^{\ell}\left(\sum_{r=0}^{2^{\nu_{j_i}-1}-2}\frac{1}{2^{\nu_{j_i}-1}+r}\right).$$

As $\sum_{r=0}^{2^{\nu_j-1}-2}\frac{1}{2^{\nu_j-1}+r}\leq 1$ for all $\nu\in\mathbb{N}_0^n$ and all $j=1,\ldots,n$, this gives

$$\sup_{\nu\in\mathbb{N}_0^n}\operatorname*{var}_{D_\nu,\mathcal{T}}\leq 2^n\sum_{j=0}^{n}\binom{n}{j} = 2^{2n}$$

independent of $\nu\in\mathbb{N}_0^n$.

The result on $\mathcal{R}$-boundedness of operators associated with a family of multipliers is now proved as in [GW03, Theorem 3.2]. $\square$

Remark 3.25. As long as we investigate finitely many functions with regard to being discrete Fourier multipliers, it suffices to check conditions (3.10) or (3.11) for large $k\in\mathbb{Z}^n$, i.e. for $k\in\mathbb{Z}^n\setminus G$ with a finite set $G\subset\mathbb{Z}^n$. This is due to the fact that a family of finitely many bounded linear operators as well as the union of finitely many $\mathcal{R}$-bounded families is $\mathcal{R}$-bounded.

By our choice of the *intermediate condition* (3.12) it is now apparent what 'benefit close to zero' exactly means. Within the cube $\{-1,0,1\}^n$ only the values of M itself enter into the $\mathcal{R}$-boundedness condition, whereas no values of the discrete derivatives $\Delta^\gamma M$ have to be considered. In particular, since $\gamma\leq 1$, the value $M(0)$ does not occur in any expression resulting from shifts of M as a consequence of discrete derivation of order γ. For an infinite family $\{M_\lambda;\ \lambda\in\Lambda\}$ of functions

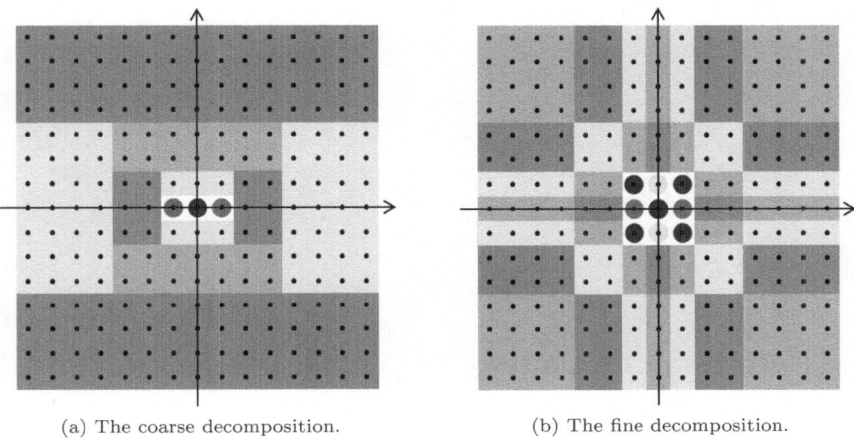

(a) The coarse decomposition.   (b) The fine decomposition.

Figure 3.3: The intermediate conditions.
The balls in the center indicate that here the intervals $[\alpha_{d,i}, \beta_{d,i}]$ and $[\alpha_{\nu,i}, \beta_{\nu,i}]$ do only contain one single point $k \in \mathbb{Z}^2$. Hence, $k_j = \alpha_j$ for $j = 1, 2$ and $\gamma_k = 0$. This implies that no discrete derivatives of order $\gamma \neq 0$ enter in the calculation of $\text{var}_{D_d, \mathcal{T}} M$ and $\text{var}_{D_\nu, \mathcal{T}} M$.

under consideration recall that the family $\{M_\lambda(k);\ k \in \{-1, 0, 1\}^n,\ \lambda \in \Lambda\}$ of operators cannot be neglected (see Proposition 3.9). Similarly, an *intermediate condition* can be defined for part a) of the multiplier theorem. However, we cannot exclude the whole cube $\{-1, 0, 1\}^n$ in case $\gamma \neq 0$. Still, the value $M(0)$ does not occur in any expression resulting from shifts of M as a consequence of discrete derivation of order γ (see Figure 3.3).

4 Classes of operators and Dunford functional calculus

This chapter provides important classes of linear operators in Banach spaces. Each class plays a crucial role in the investigation of parabolic problems. For later application we will distinguish between injective and non-injective operators. It starts with the classes of pseudo-sectorial and sectorial operators which allow for a Dunford functional calculus. With $\mathcal{R}$-boundedness from Chapter 3 at hand, the class of operators admitting an $\mathcal{R}$-bounded $\mathcal{H}^\infty$-calculus is discussed. This class is of particular interest in view of our purposes later on. We refer to [DHP03], [KW04], and [Haa06] for a deeper investigation of the different classes.

Definition 4.1. A closed, densely defined linear operator A in a Banach space X is called *pseudo-sectorial* if $(-\infty, 0) \subset \rho(A)$ and if there exists $C > 0$ such that

$$(4.1) \qquad \|t(t+A)^{-1}\|_{\mathcal{L}(X)} \leq C \quad (t > 0).$$

In this case, it is well-known that there exists a $\phi \in (0, \pi)$ such that the uniform estimate extends to all

$$\lambda \in \Sigma_{\pi-\phi} := \{z \in \mathbb{C} \backslash \{0\}; \ |\arg(z)| < \pi - \phi\},$$

see e.g. [DHP03]. The number

$$\phi_A := \inf \left\{ \phi; \ \rho(-A) \supset \Sigma_{\pi-\phi}, \ \sup_{\lambda \in \Sigma_{\pi-\phi}} \|\lambda(\lambda+A)^{-1}\|_{\mathcal{L}(X)} < \infty \right\}$$

is called *spectral angle* of A. The class of pseudo-sectorial operators is denoted by $\Psi S(X)$. A pseudo-sectorial operator A is called *sectorial* if additionally $R(A)$ is dense in X and $N(A) = \{0\}$. The class of sectorial operators is denoted by $S(X)$.

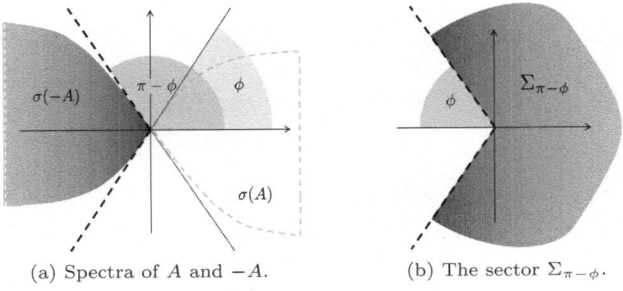

(a) Spectra of A and $-A$.   (b) The sector $\Sigma_{\pi-\phi}$.

Figure 4.1: Relevant sets in the complex plane.

For $A \in \Psi S(X)$ observe that $\sigma(A) \subset \overline{\Sigma}_{\phi_A}$. It is well-known that $-A$ generates a bounded analytic C_0-semigroup on X in case that $\phi_A < \frac{\pi}{2}$.

The definition of sectoriality as presented above follows [DHP03]. It collects all additional properties of an operator A subject to condition (4.1) which allow for an intensive functional calculus. Pseudo-sectoriality, however, is an advisable weakening that is motivated, for instance, by the Neumann Laplacian on bounded domains which obviously fails to be injective. Besides pseudo-sectoriality and sectoriality, a third notion called quasi-sectoriality is used in literature. It defines operators which are sectorial after a suitable shift (cf. [Haa06]). Hence, pseudo-sectorial operators are a special class of quasi-sectorial operators. For the sake of convenience we distinguish between pseudo-sectorial and sectorial operators only. Note that injectivity of A could have been canceled in the definition of sectoriality for it is implied by the assumption $\overline{R(A)} = X$. Moreover, if X is known to be reflexive, most additional properties besides (4.1) come for free. The following proposition comments on these relations briefly. We refer to [Haa06] and [DHP03] for proofs and comprehensive introductions to (pseudo-)sectorial operators.

Proposition 4.2. *Let A define a closed operator subject to condition* (4.1).
a) *It holds that $N(A) \cap \overline{R(A)} = \{0\}$, hence, density of $R(A)$ implies $N(A) = \{0\}$.*
b) *If X is reflexive, we have $\overline{D(A)} = X$ and $X = N(A) \oplus \overline{R(A)}$. Consequently, $A \in \Psi S(X)$. Moreover, density of $R(A)$ and $N(A) = \{0\}$ are equivalent. Hence, each of the latter conditions implies $A \in S(X)$.*

Next we introduce a holomorphic functional calculus for sectorial operators. This will lead to the notion of an $\mathcal{H}^\infty$-calculus and a first important subclass of $S(X)$. For a comprehensive introduction to this concept we refer to [CDMY96], [KW01], [DHP03], [KW04], [DV05], and [Haa06].

For $\sigma \in (0, \pi]$ we denote by $\mathcal{H}^\infty(\Sigma_\sigma)$ the commutative algebra of bounded, holomorphic functions on Σ_σ, that is,

$$\mathcal{H}^\infty(\Sigma_\sigma) := \{f \colon \Sigma_\sigma \to \mathbb{C}; \ f \text{ is holomorphic}, \ |f|_\infty^\sigma < \infty\},$$

where

$$|f|_\infty^\sigma := \sup\{|f(z)|; \ z \in \Sigma_\sigma\}.$$

Using $\rho(z) := \frac{z}{(1+z)^2}$ we define the subalgebra

$$\mathcal{H}_0^\infty(\Sigma_\sigma) := \{f \in \mathcal{H}^\infty(\Sigma_\sigma); \ \text{there exist } C, \varepsilon > 0 \text{ such that}$$
$$|f(z)| \le C|\rho(z)|^\varepsilon \text{ for all } z \in \Sigma_\sigma\}.$$

Let A be a pseudo-sectorial operator in X. Pick $\sigma \in (\phi_A, \pi]$ and $\psi \in (\phi_A, \sigma)$. The path

$$\Gamma := (\infty, 0]e^{i\psi} \cup [0, \infty)e^{-i\psi}$$

oriented counterclockwise, i.e. the positive real axis $\mathbb{R}_+$ lies to the left, stays with the only possible exception at zero in the resolvent of A. Hence, by Cauchy's integral formula and the pseudo-sectoriality of A, for every $f \in \mathcal{H}_0^\infty(\Sigma_\sigma)$ the Bochner integral

$$f(A) := \frac{1}{2\pi i} \int_\Gamma f(\mu)(\mu - A)^{-1} d\mu$$

represents a well-defined element in $\mathcal{L}(X)$. Moreover, the above formula defines an algebra homomorphism

(4.2) $$\Phi_A \colon \mathcal{H}_0^\infty(\Sigma_\sigma) \to \mathcal{L}(X); \quad f \mapsto f(A)$$

known as the Dunford calculus.

Example 4.3. It holds that

$$\rho(A) := \frac{1}{2\pi i} \int_\Gamma \rho(\mu)(\mu - A)^{-1} d\mu = A(1 + A)^{-2}.$$

If A is sectorial, one can extend the Dunford calculus to arbitrary $f \in \mathcal{H}^\infty(\Sigma_\sigma)$ if one accepts that $f(A)$ is possibly unbounded. To see this, we make use of the fact that $\rho f \in \mathcal{H}_0^\infty(\Sigma_\sigma)$ for $f \in \mathcal{H}^\infty(\Sigma_\sigma)$. Furthermore, since $A \in S(X)$ is injective, the same is true for $\rho(A)^{-1}$. This is sufficient to define

$$f(A) := \rho(A)^{-1}(\rho f)(A), \quad D(f(A)) := \{x \in X; \ (\rho f)(A)x \in D(A) \cap R(A)\},$$

which gives rise to a closed, densely defined operator in X. By Cauchy's theorem this definition is consistent with the former one if $f \in \mathcal{H}_0^\infty(\Sigma_\sigma)$.

Moreover, if A is sectorial, functions f of polynomial growth at 0 and at infinity can be handled. In that case the above definition has to be modified. More precisely, there exists $k \in \mathbb{N}$ such that

$$f(A) := \rho(A)^{-k}(\rho^k f)(A), \quad D(f(A)) := \{x \in X; \ (\rho^k f)(A)x \in D(A^k) \cap R(A^k)\},$$

gives rise to a closed, densely defined operator in X (see [DHP03, Theorem 2.1] or [Haa06, Proposition 2.3.13]). For $\mu \in \mathbb{C}$ the complex powers $f_\mu(z) := z^\mu$ are holomorphic in the sliced complex plane Σ_π and admit polynomial behavior at 0 and at infinity. Thus, A^μ is well-defined as a closed, densely defined operator in X. In particular, fractional powers A^α, $0 < \alpha < 1$ are captured.

It is worthwhile to mention that pseudo-sectoriality only is not sufficient to apply these kinds of extension procedures for the injectivity of A cannot be dropped. Indeed, if A is not injective and if $g \in \mathcal{H}_0^\infty(\Sigma_\sigma)$ exists such that $gf \in \mathcal{H}_0^\infty(\Sigma_\sigma)$ and $g(A)^{-1}$ is injective, then f has to show a certain limit behavior at 0. In fact, $\lambda \in \mathbb{C}$ and $s > 0$ have to exist, such that $f(z) - \lambda$ tends to zero by the rate of $|z|^s$ as $z \to 0$ (see [Haa06, Lemma 2.3.8]).

This, however, still allows for the consideration of the subalgebra

$$\mathcal{E}(\Sigma_\sigma) := \mathcal{H}_0^\infty(\Sigma_\sigma) \oplus \langle (1 + z)^{-1} \rangle \oplus \langle 1 \rangle$$
$$:= \{g(z) + c(1 + z)^{-1} + d; \quad g \in \mathcal{H}_0^\infty(\Sigma_\sigma), \ c, d \in \mathbb{C}\}$$

which weakens the condition of decay at infinity. For $f \in \mathcal{E}(\Sigma_\sigma)$ one sets

$$f(A) := g(A) + c(1 + A)^{-1} + d$$

whenever $f(z) = g(z) + c(1 + z)^{-1} + d$ with $g \in \mathcal{H}_0^\infty(\Sigma_\sigma)$ and $c, d \in \mathbb{C}$. Given a pseudo-sectorial operator A, the Dunford calculus extends to an algebra homomorphism

$$\Phi_A : \mathcal{E}(\Sigma_\sigma) \to \mathcal{L}(X); \quad f \mapsto f(A)$$

(see [Haa06, Theorem 2.3.3]). For many other special classes of functions f a definition of $f(A)$ is meaningful in case of a pseudo-sectorial operator A; for instance, there exists $g \in \mathcal{E}(\Sigma_\sigma)$ such that the definition $f(A) := g(A)^{-1}(gf)(A)$ extends the Dunford calculus to holomorphic f of polynomial growth at infinity which fulfill the necessary limit behavior at 0 ([Haa06, Proposition 2.3.11]).

Definition 4.4. Let $A \in \Psi S(X)$ and let $\mathcal{F} \in \{\mathcal{H}_0^\infty, \mathcal{E}, \mathcal{H}^\infty\}$. We say that the $\mathcal{F}$-*calculus of A is bounded* if there exists $\sigma > \phi_A$ such that $f(A) \in \mathcal{L}(X)$ for all $f \in \mathcal{F}(\Sigma_\sigma)$ and

$$\Phi_A : \mathcal{F}(\Sigma_\sigma) \to \mathcal{L}(X); \quad f \mapsto f(A)$$

is bounded with respect to the topologies on $\mathcal{H}^\infty(\Sigma_\sigma)$ and $\mathcal{L}(X)$, that is, if there exists $C > 0$ such that

(4.3) $$\|f(A)\|_{\mathcal{L}(X)} \leq C_\sigma |f|_\infty^\sigma \quad (f \in \mathcal{F}(\Sigma_\sigma)).$$

The bound for Φ_A in general depends on σ. The infimum over all $\sigma > \phi_A$ such that this bound remains finite is called $\mathcal{F}$-*angle of A* and is denoted by $\phi_A^\mathcal{F}$.
We denote the class of operators $A \in \Psi S(X)$ which admit a bounded $\mathcal{H}_0^\infty$-calculus on X by $\Psi \mathcal{H}^\infty(X)$ and the class of operators $A \in S(X)$ which admit a bounded $\mathcal{H}^\infty$-calculus on X by $\mathcal{H}^\infty(X)$. In both cases, we write as well ϕ_A^∞ instead of $\phi_A^\mathcal{F}$.

Remark 4.5. Instead of (4.3) we can equivalently require

(4.4) $$\|f(A)\|_{\mathcal{L}(X)} \leq C_\sigma \quad (f \in \mathcal{F}(\Sigma_\sigma), \ |f|_\infty^\sigma \leq 1).$$

Moreover, the $\mathcal{E}$-calculus of an operator $A \in \Psi S(X)$ is bounded if and only if the $\mathcal{H}_0^\infty$-calculus is bounded. To see this, consider for a moment all $f \in \mathcal{E}(\Sigma_\sigma)$ such that $|f|_\infty^\sigma \leq 1$, say $f(z) = g_f(z) + c_f(1 + z)^{-1} + d_f$ with $g_f \in \mathcal{H}_0^\infty(\Sigma_\sigma)$ and $c_f, d_f \in \mathbb{C}$. Then $c = c_f$, $d = d_f$ and $g = g_f$ are bounded independently of f. On the one hand, since $g(z) \to 0$ and $c(1 + z)^{-1} + d \to c + d$ for $z \to 0$ it follows that $|c + d| \leq 1$. On the other hand, since $g(z) + c(1 + z)^{-1} \to 0$ for $|z| \to \infty$ it follows that $|d| \leq 1$, hence $|c| \leq 2$. Altogether $|g|_\infty^\sigma \leq 4$ follows and this estimate does not depend on f.

The following result known as the convergence lemma ([CDMY96, Lemma 2.1], see also [Haa06, Proposition 5.1.4]) is the crucial tool to get a grip on $f(A)$ in case $f \notin \mathcal{E}(\Sigma_\sigma)$.

Lemma 4.6. *Let $A \in \Psi S(X)$, $\sigma \in (\phi_A, \pi)$ and let $(f_n)_{n \in \mathbb{N}} \subset \mathcal{H}^\infty(\Sigma_\sigma)$ such that $f_n(A)$ is well-defined for $n \in \mathbb{N}$. Suppose that $\sup_{n \in \mathbb{N}} \|f_n\|_\infty < \infty$, let the limit $f(z) := \lim_{n \to \infty} f_n(z)$ exists pointwise on Σ_σ, and assume that $f(A)$ is well-defined. Then $f_n(A)x \to f(A)x$ for all $x \in D(A) \cap R(A)$.*
If additionally $A \in S(X)$, $f_n(A) \in \mathcal{L}(X)$, and $\sup_{n \in \mathbb{N}} \|f_n(A)\|_{\mathcal{L}(X)} < \infty$, then $f(A) \in \mathcal{L}(X)$ and $f_n(A) \to f(A)$ strongly.

In case $A \in S(X) \cap \Psi\mathcal{H}^\infty(X)$, thanks to the convergence lemma and the following result (see e.g. [CDMY96, Corollary 2.2]), Φ_A extends boundedly from $\mathcal{H}_0^\infty(\Sigma_\sigma)$ to $\mathcal{H}^\infty(\Sigma_\sigma)$.

Lemma 4.7. *Let $f \in \mathcal{H}^\infty(\Sigma_\sigma)$ and let $\rho_n \in \mathcal{H}_0^\infty(\Sigma_\sigma)$ be defined by*

$$\rho_n(z) := \frac{1}{1 + z/n} - \frac{1}{1 + zn} = \frac{n^2 z - z}{(1 + nz)(n + z)}.$$

Then $f_n := \rho_n f$ and f fulfill the assumptions of the convergence lemma.

Corollary 4.8. *For $A \in S(X)$ we additionally have $A \in \Psi\mathcal{H}^\infty(X)$ if and only if $A \in \mathcal{H}^\infty(X)$.*

An important subclass of $\mathcal{H}^\infty(X)$ is given by the class of sectorial operators which admit bounded imaginary powers.

Definition 4.9. Let $A \in S(X)$. Then A is said to admit *bounded imaginary powers* if $A^{is} \in \mathcal{L}(X)$ for all $s \in \mathbb{R}$ and if there is a constant $C > 0$ such that

$$\|A^{is}\|_{\mathcal{L}(X)} \leq C \quad (|s| \leq 1).$$

The class of such operators is denoted by $\mathrm{BIP}(X)$. Moreover, the quantity

$$\theta_A := \varlimsup_{|s| \to \infty} \frac{1}{|s|} \log \|A^{is}\|_{\mathcal{L}(X)}$$

is called *power angle of $A \in \mathrm{BIP}(X)$.*

Equivalently, $A \in \mathrm{BIP}(X)$ if there exists $M \geq 1$ and θ such that

$$(4.5) \qquad\qquad \|A^{is}\|_{\mathcal{L}(X)} \leq M e^{\theta|s|} \quad (s \in \mathbb{R}).$$

Moreover, $\theta_A = \inf\{\theta;\ (4.5)\text{ is valid}\}$ (see [DHP03], [KW04], or [Haa06]).

The non-trivial implication follows from the group property of imaginary powers A^{is} and the convergence lemma. Let $s \in \mathbb{R}$ and $\sigma \in (0, \pi)$. For the imaginary power function $f \colon \Sigma_\sigma \to \mathbb{C}$; $f(z) := z^{is}$ we then have

$$|f(z)| = |z^{is}| = |e^{is \ln z}| = e^{|s| \arg(z)} \leq e^{\sigma|s|} \quad (z \in \Sigma_\sigma).$$

Thus $|f|_\infty^\sigma \leq e^{\sigma|s|}$ which particularly shows $\theta_A \leq \phi_A^\infty$. Altogether we have

$$\mathcal{H}^\infty(X) \subset \mathrm{BIP}(X) \subset S(X)$$

and
$$\phi_A \leq \theta_A \leq \phi_A^\infty.$$

One remarkable property of operators which belong to the class $BIP(X)$ is the following result on fractional powers (see e.g. [Tri78]). It describes the spaces

$$X_\alpha := X_{A^\alpha} := \left(D(A^\alpha), \|\cdot\|_\alpha\right), \quad \|x\|_\alpha := \|x\|_X + \|A^\alpha x\|_X \quad (0 < \alpha < 1)$$

by means of complex interpolation. Here the embeddings

$$D(A) \subset X_\alpha \subset X \quad (0 < \alpha < 1)$$

are valid.

Proposition 4.10. *Suppose $A \in BIP(X)$. Then for $0 < \alpha < 1$ the space X_α is isomorphic to $[X, D(A)]_\alpha$.*

So far uniform norm boundedness of different families of operators has been considered. With the stronger concept of $\mathcal{R}$-boundedness at hand (see Chapter 3), the above definition of sectoriality can be adjusted to $\mathcal{R}$-sectoriality.

Definition 4.11. A pseudo-sectorial operator $A \in \Psi S(X)$ is called *pseudo-$\mathcal{R}$-sectorial* if there exist an angle $\phi \in (0, \pi)$ and a constant $C_\phi > 0$ such that

$$(4.6) \qquad\qquad \mathcal{R}(\{\lambda(\lambda + A)^{-1}; \ \lambda \in \Sigma_{\pi - \phi}\}) \leq C_\phi.$$

The class of pseudo-$\mathcal{R}$-sectorial operators is denoted by $\Psi\mathcal{R}S(X)$ and we call $\phi_A^{\mathcal{R}S}$ given as the infimum over all angles ϕ such that (4.6) holds the $\mathcal{R}$-*angle of A*. If in addition $A \in S(X)$, then A is called $\mathcal{R}$-*sectorial* and we write $A \in \mathcal{R}S(X)$.

As $\mathcal{R}$-boundedness is stronger than the uniform boundedness with respect to operator norm in general, $\mathcal{R}$-sectoriality always implies the sectoriality of an operator A and we have

$$\phi_A \leq \phi_A^{\mathcal{R}S}.$$

Accordingly, we can strengthen the condition of a bounded $\mathcal{H}^\infty$-calculus by means of $\mathcal{R}$-boundedness.

Definition 4.12. Let $A \in \Psi S(X)$ and let $\mathcal{F} \in \{\mathcal{H}_0^\infty, \mathcal{E}, \mathcal{H}^\infty\}$. We say that the $\mathcal{F}$-*calculus of A is $\mathcal{R}$-bounded* if there exists $\sigma > \phi_A$ such that $f(A) \in \mathcal{L}(X)$ for all $f \in \mathcal{F}(\Sigma_\sigma)$ and if there exist a $\sigma > \phi_A$ and a constant $C_\sigma > 0$ such that

$$(4.7) \qquad\qquad \mathcal{R}(\{f(A); \ f \in \mathcal{F}(\Sigma_\sigma), \ |f|_\infty^\sigma \leq 1\}) \leq C_\sigma.$$

The infimum over all $\sigma > \phi_A$ such that (4.7) holds true with some $C_\sigma > 0$ is called the $\mathcal{R}\mathcal{F}$-*angle of A* and is denoted by $\phi_A^{\mathcal{R}\mathcal{F}}$.
We denote the class of operators $A \in \Psi S(X)$ which admit an $\mathcal{R}$-bounded $\mathcal{H}_0^\infty$-calculus on X by $\Psi\mathcal{R}\mathcal{H}^\infty(X)$ and the class of operators $A \in S(X)$ which admit an $\mathcal{R}$-bounded $\mathcal{H}^\infty$-calculus on X by $\mathcal{R}\mathcal{H}^\infty(X)$. In both cases, we write as well $\phi_A^{\mathcal{R}\infty}$ instead of $\phi_A^{\mathcal{R}\mathcal{F}}$.

Remark 4.13. Observe that the $\mathcal{E}$-calculus of an operator $A \in \Psi S(X)$ is $\mathcal{R}$-bounded if and only if $A \in \Psi \mathcal{R}\mathcal{H}^\infty(X)$. In particular, $A \in \Psi \mathcal{R}\mathcal{H}^\infty(X)$ implies $A \in \Psi \mathcal{R}S(X)$ (cf. Remark 4.5).

In order to compare all classes of operators introduced above, the notions of class $\mathcal{H}\mathcal{T}$ and property (α) are employed. For instance, if X is of class $\mathcal{H}\mathcal{T}$, then it is shown in [CP01] that bounded imaginary powers imply $\mathcal{R}$-sectoriality.

Proposition 4.14. *Let X be a Banach space of class $\mathcal{H}\mathcal{T}$ and let $A \in BIP(X)$ with power angle θ_A. Then $A \in \mathcal{R}S(X)$ and $\phi_A^{\mathcal{R}} \leq \theta_A$.*

To sum it up, in a Banach space X of class $\mathcal{H}\mathcal{T}$ the inclusions

$$\mathcal{R}\mathcal{H}^\infty(X) \subset \mathcal{H}^\infty(X) \subset BIP(X) \subset \mathcal{R}S(X) \subset S(X)$$

are valid and the corresponding angles fulfill

$$\phi_A \leq \phi_A^{\mathcal{R}} \leq \theta_A \leq \phi_A^\infty \leq \phi_A^{\mathcal{R}\infty}.$$

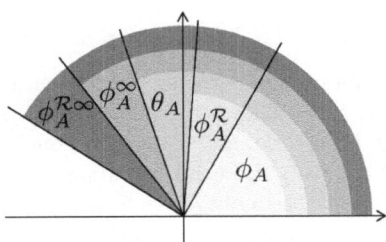

Figure 4.2: Relation of angles in a Banach space of class $\mathcal{H}\mathcal{T}$.

In a Banach space enjoying property (α) more can be derived. Indeed, boundedness and $\mathcal{R}$-boundedness of the $\mathcal{H}_0^\infty$-calculus coincide. Equality of boundedness and $\mathcal{R}$-boundedness of the $\mathcal{H}^\infty$-calculus can be found in [KW01, Theorem 5.3]. In fact, all steps of the proof carry over to the case of an $\mathcal{H}_0^\infty$-calculus.

Proposition 4.15. *Let X have property (α). Then $A \in \Psi\mathcal{H}^\infty(X)$ if and only if $A \in \Psi \mathcal{R}\mathcal{H}^\infty(X)$. Accordingly, $A \in \mathcal{H}^\infty(X)$ if and only if $A \in \mathcal{R}\mathcal{H}^\infty(X)$. In these cases $\phi_A^\infty = \phi_A^{\mathcal{R}\infty}$.*

Furthermore, if A is known to be $\mathcal{R}$-sectorial and to admit a bounded $\mathcal{H}^\infty$-calculus, equality of according angles follows ([KW01, Proposition 5.1]). In fact, the proof applies to the according assertion for pseudo-sectorial operators. Here no restriction on the underlying Banach space is needed.

Proposition 4.16. *Let X be a Banach space. If $A \in \mathcal{H}^\infty(X) \cap \mathcal{R}S(X)$, respectively $A \in \Psi\mathcal{H}^\infty(X) \cap \Psi\mathcal{R}S(X)$, then $\phi_A^\infty = \phi_A^{\mathcal{R}}$.*

For a long time it remained an open question under what circumstances in a Banach space X the sum of two closed operators A and B is closed again. A first suitable answer was given by the celebrated result of Dore and Venni in [DV87], roughly saying that the assertion is valid if both operators have bounded imaginary powers and if their resolvents commute, i.e. if

$$(\lambda - A)^{-1}(\mu - B)^{-1} = (\mu - B)^{-1}(\lambda - A)^{-1} \quad (\lambda \in \rho(A),\ \mu \in \rho(B)).$$

In fact, even the property of having bounded imaginary powers is inherited by $A + B$. In their paper, Dore and Venni imposed $0 \in \rho(B) \cap \rho(A)$. This restriction could be removed in [PS90, Theorem 4].

Theorem 4.17. *Let X be a Banach space of class $\mathcal{HT}$. Let $A, B \in BIP(X)$ be resolvent commuting and assume $\theta_A + \theta_B < \pi$. Then $A + B$ with domain $D(A + B) := D(A) \cap D(B)$ is closed and sectorial. Moreover, there exists $C > 0$ such that*

$$\|Ax\|_X + \|Bx\|_X \leq C\|(A + B)x\|_X \quad (x \in D(A + B))$$

is satisfied and $0 \in \rho(A) \cup \rho(B)$ implies $0 \in \rho(A + B)$.
Let $\theta_A \neq \theta_B$. Then $A + B \in BIP(X)$ and $\theta_{A+B} \leq \max\{\theta_A, \theta_B\}$.

Another answer to the question under what conditions $A+B$ is closed was given by Kalton and Weis in [KW01]. They were able to establish closedness of $A + B$ if one of the operators is $\mathcal{R}$-sectorial and if the other one admits a bounded $\mathcal{H}^\infty$-calculus ([KW01, Theorem 6.3]). Here no condition on the underlying Banach space X is needed. Besides closedness, $\mathcal{R}$-sectoriality is established for $A + B$ provided the underlying Banach space enjoys property (α).

Theorem 4.18. *Let X be a Banach space and let $A \in \mathcal{H}^\infty(X)$ and $B \in \mathcal{RS}(X)$ be two resolvent commuting operators such that $\phi_A^\infty + \phi_B^\mathcal{R} < \pi$. Then $A + B$ with domain $D(A + B) := D(A) \cap D(B)$ is closed and sectorial. Moreover, there exists $C > 0$ such that*

$$\|Ax\|_X + \|Bx\|_X \leq C\|(A + B)x\|_X \quad (x \in D(A + B))$$

is satisfied and $0 \in \rho(A) \cup \rho(B)$ implies $0 \in \rho(A + B)$.
If X enjoys property (α), then $A + B \in \mathcal{RS}(X)$ and $\phi_{A+B}^\mathcal{R} \leq \max\{\phi_A^\infty, \phi_B^\mathcal{R}\}$.

This result is particularly important for the situation of Cauchy problems, i.e. $B = \partial_t$, since then the assumption for A of having bounded imaginary powers is reduced to the weaker property of $\mathcal{R}$-sectoriality. In many other situations such as in the investigation of Volterra integral equations the latter can be employed successfully.

In view of our applications, we want sums $A + B$ or as well products AB not only to be $\mathcal{R}$-sectorial again, but even to admit an $\mathcal{H}^\infty$-calculus. This leads to the following result which is obtained in the same way as Theorem 4.18 (see [NS11a]).

Proposition 4.19. a) *Let X be a Banach space of class $\mathcal{HT}$ with property (α). Let $A, B \in \mathcal{H}^\infty(X)$ be two resolvent commuting operators such that $\phi_A^\infty + \phi_B^\infty < \pi$. Then $A + B \in \mathcal{RH}^\infty(X)$ and $\phi_{A+B}^{\mathcal{R},\infty} \leq \max\{\phi_A^\infty, \phi_B^\infty\}$.*
b) *If additionally $0 \in \rho(A)$, then $AB \in \mathcal{RH}^\infty(X)$ and $\phi_{AB}^{\mathcal{R},\infty} \leq \phi_A^\infty + \phi_B^\infty$.*

Remark 4.20. By iteration it readily follows that the assertions remain true for finite sums (respectively finite products) as long as in each step the condition for the $\mathcal{H}^\infty$-angles and commutativity of the resolvents is satisfied. Accordingly, Proposition 4.18 applies in iteration to $B \in \mathcal{RS}(X)$ and $A_1, \ldots, A_N \in \mathcal{H}^\infty(X)$.

In application, we also employ a corresponding result to Proposition 4.19a) for the non-commuting case from [PS07]. Indeed, the same assertion holds if the following so-called *Labbas-Terreni condition* is satisfied.

$$(4.8) \quad \begin{cases} \text{Let } 0 \in \rho(A) \text{ and let there exist constants } c > 0, \ 0 \leq \alpha < \beta < 1, \\ \psi_A > \phi_A, \psi_B > \phi_B, \psi_A + \psi_B < \pi, \\ \text{such that for all } \lambda \in \Sigma_{\pi - \psi_A}, \ \mu \in \Sigma_{\pi - \psi_B} \text{it holds that} \\ \left\| A(\lambda + A)^{-1}[A^{-1}, (\mu + B)^{-1}] \right\| \leq c/(1 + |\lambda|)^{1-\alpha} |\mu|^{1+\beta}. \end{cases}$$

Here $[S, T] = ST - TS$. The result given in [PS07] then reads as follows.

Proposition 4.21. *Let X be a Banach space of class $\mathcal{HT}$ with property (α) and let $A, B \in \mathcal{H}^\infty(X)$. Suppose that (4.8) holds for some angles $\psi_A > \phi_A^\infty$, $\psi_B > \phi_B^\infty$ with $\psi_A + \psi_B < \pi$. Then there exists $\delta \geq 0$ such that $A + B + \delta$ is invertible and such that $A + B + \delta \in \mathcal{RH}^\infty(X)$ and $\phi_{A+B+\delta}^\infty \leq \max\{\psi_A, \psi_B\}$. In case that the resolvents commute or if c in (4.8) is small enough, we can take $\delta = 0$.*

Remark 4.22. Again iteration is possible, provided the angle and commutator conditions are satisfied in every step.

Remark 4.23. Both Proposition 4.19 and Proposition 4.21 exist in slightly different versions if X is an arbitrary Banach space, (cf. [KW01] and [PS07]). In our applications, however, the assumptions that the underlying Banach space is of class $\mathcal{HT}$ and that it enjoys property (α) are always satisfied.

5 Parabolic problems and maximal regularity

This chapter recalls the notion of maximal regularity both in the context of Cauchy problems (see e.g. [Dor93] and [KW04]) and Volterra integral equations (see e.g. [Prü93]). In both cases, sufficient conditions for maximal regularity in terms of properties of linear operators from the previous chapter are derived.

First we turn our attention to Cauchy problems

$$
\begin{aligned}
\dot{u} + Au &= f \quad \text{in } (0, T), \\
u(0) &= 0,
\end{aligned}
$$

where $T \in (0, \infty]$. Here $A \colon X \supset D(A) \to X$ is supposed to be a closed and densely defined operator.

Recall that in case $A \in \Psi S(X)$, $\phi_A < \frac{\pi}{2}$, the operator $-A$ generates a bounded analytic C_0-semigroup on X. For a suitable treatment of related nonlinear problems, however, the generation of an analytic semigroup might not be enough. Then the stronger property of maximal regularity is required which is defined as follows.

Definition 5.1. Let $1 \leq q \leq \infty$, let X be a Banach space, and let A be closed and densely defined. Then A is said to have *maximal L^q-regularity* on X if for each $f \in L^q(\mathbb{R}_+, X)$ there is a unique solution $u \colon \mathbb{R}_+ \to D(A)$ of the Cauchy problem

$$
\begin{aligned}
(5.1) \qquad \dot{u} + Au &= f \quad \text{in } \mathbb{R}_+, \\
u(0) &= 0,
\end{aligned}
$$

satisfying the estimate

$$
\|\dot{u}\|_{L^q(\mathbb{R}_+, X)} + \|Au\|_{L^q(\mathbb{R}_+, X)} \leq C\|f\|_{L^q(\mathbb{R}_+, X)}
$$

with a $C > 0$ independent of $f \in L^q(\mathbb{R}_+, X)$.

Due to a result of Sobolevskii ([Sob64], see also [Dor93, Theorem 4.2]), the class of operators having L^q-maximal regularity does not depend on q. Indeed, if A as L^q-maximal regularity for one $q \in (1, \infty)$, then A as L^q-maximal regularity for all $q \in (1, \infty)$. We therefore drop the indication and agree to speak of maximal regularity of A, only. Furthermore, if the operator A has the property of maximal regularity, then $-A$ is known to be the generator of an analytic semigroup (see e.g. [Dor93, Theorem 2.1] and the references given there).

Remark 5.2. Note that maximal regularity does not imply $u \in W^{1,q}(\mathbb{R}_+, X)$ for the solution of the Cauchy problem. In fact, in case A has maximal regularity, the solution u of (5.1) fulfills $u \in W^{1,q}(\mathbb{R}_+, X)$ and

$$
\|u\|_{W^{1,q}(\mathbb{R}_+, X)} + \|Au\|_{L^q(\mathbb{R}_+, X)} \leq C\|f\|_{L^q(\mathbb{R}_+, X)}
$$

if and only if the condition $0 \in \rho(A)$ is added (see e.g. [KW04]). Since there is no common differentiation in terminology, we agree to speak of A having *strong* maximal regularity if A has maximal regularity and if $0 \in \rho(A)$. Moreover, if A has the property of (strong) maximal regularity, A has also the property of (strong) maximal regularity if $\mathbb{R}_+$ in (5.1) is replaced by any finite interval $(0, T)$, $T < \infty$ (see e.g. [KW04]). Note that maximal regularity and strong maximal regularity coincide if $\mathbb{R}_+$ is replaced by any finite interval $(0, T)$, $T < \infty$. In applications, sometimes maximal regularity is established for $A + \delta$, $\delta > 0$ only but not for A itself. Restricted to finite intervals $(0, T)$, $T < \infty$ this still yields maximal regularity of A substituting f by $e^{-\delta \cdot} f$ and u by $e^{\delta \cdot} u \in W^{1,p}((0, T), X)$.

The following result due to Weis ([Wei01b, Theorem 4.2]) characterizes the property of maximal regularity on Banach spaces of class $\mathcal{HT}$ by means of pseudo-$\mathcal{R}$-sectoriality of the operator under consideration.

Theorem 5.3. *Let X be a Banach space of class $\mathcal{HT}$ and let A generate a bounded analytic semigroup on X. Then A has the property of maximal regularity if and only if $A \in \Psi \mathcal{RS}(X)$ and $\phi_A^{\mathcal{R}} < \frac{\pi}{2}$.*

As a consequence of Theorem 5.3 we can use pseudo-$\mathcal{R}$-sectoriality to reformulate the result given in [Wei01a, Corollary 4d)].

Corollary 5.4. *Let $1 < p < \infty$ and let $G \subset \mathbb{R}^n$ be an arbitrary domain. If $-A$ is the generator of a positive analytic semigroup of contractions on $L^p(G)$, then $A \in \Psi \mathcal{RS}(L^p(G))$ and $\phi_A^{\mathcal{R}} < \frac{\pi}{2}$.*

By definition $A \in \mathcal{RS}(L^p(G))$ follows, provided A is injective. In order to deduce a bounded $\mathcal{H}^\infty$-calculus for A, the condition of injectivity has to be added necessarily. In fact, it is the only additional condition which has to be imposed. This result relies on Corollary 5.4, Proposition 4.16 and a result of Duong ([Duo90]) and Hieber and Prüss ([HP98]) based on the transference principle. In the absence of injectivity the proof in [Duo90] still applies. In that case, of course only an $\mathcal{H}_0^\infty$-calculus for A can be deduced.

Corollary 5.5. *Let $1 < p < \infty$ and let $G \subset \mathbb{R}^n$ be an arbitrary domain. If $-A$ is the generator of a positive analytic semigroup of contractions on $L^p(G)$, then $A \in \Psi \mathcal{H}^\infty(L^p(G))$ and $\phi_A^\infty < \frac{\pi}{2}$. If in addition A is injective, it holds that $A \in \mathcal{H}^\infty(L^p(G))$ and $\phi_A^\infty < \frac{\pi}{2}$.*

As a second application of $\mathcal{R}$-sectoriality in the context of parabolic equations, we consider the abstract Volterra equation

$$(5.2) \qquad u(t) + \int_0^t b(t - s) A u(s) ds = f(t) \quad (t \in J)$$

in a Banach space X. Here $A \colon X \supset D(A) \to X$ is again a closed and densely defined operator, $J = [0, T]$ with $T \in (0, \infty)$, and $b \in L_{loc}^1(\mathbb{R}_+)$ is a scalar-valued kernel.

This equation is often referred to as the strong Volterra equation. The mild formulation of equation (5.2) reads as

$$(5.3) \qquad u(t) + A \int_0^t b(t-s)u(s)ds = f(t) \quad (t \in J).$$

In order to discuss maximal L^q-regularity for abstract Volterra equations, we introduce the spaces $H_q^\alpha(\mathbb{R}, X)$ by means of Fourier transform. For $\alpha \in \mathbb{R}_+$ we define (see [Tri78, Definition 4.2.1])

$$H_q^\alpha(\mathbb{R}, X) := \{f \in \mathcal{S}'(\mathbb{R}, X); \, \exists f_\alpha \in L^q(\mathbb{R}, X) \colon \mathcal{F}f_\alpha(\xi) = (1 + |\xi|^2)^{\frac{\alpha}{2}} \mathcal{F}f(\xi)\}$$

and

$$\|f\|_{\alpha,q} := \|f\|_{\alpha,q,\mathbb{R}} := \|f_\alpha\|_q.$$

For $J = [0, T]$ with $T \in (0, \infty)$ we set

$$H_q^\alpha(J, X) := \{f|_J; \, f \in H_q^\alpha(\mathbb{R}, X)\}$$

and

$$\|f\|_{\alpha,q} := \|f\|_{\alpha,q,J} := \inf_{g \colon g|_J = f, \, g \in H_q^\alpha(\mathbb{R},X)} \|g_\alpha\|_q.$$

Definition 5.6. Let $1 < q < \infty$, $J = [0, T]$ with $T \in (0, \infty)$, and $b \in L_{loc}^1(\mathbb{R}_+)$. We say that equations (5.2) and (5.3) have *maximal L^q-regularity* if there exists $\alpha \geq 0$ such that

(i) for each $f = b * g$ where $g \in L^q(J, X)$ and b as above there is a unique solution $u \in H_q^\alpha(J, X) \cap L^q(J, D(A))$ of (5.2), and there is a constant $C(T) > 0$ such that

$$\|u\|_{\alpha,q} + \|Au\|_q \leq C(T)\|g\|_q,$$

(ii) for each $f \in L^q(J, D(A))$ there is a unique solution $u \in L^q(J, D(A))$ of (5.2) with $u - f \in H_q^\alpha(J, X)$, and there is a constant $C(T) > 0$ such that

$$\|u\|_q + \|u - f\|_{\alpha,q} + \|Au\|_q \leq C(T)\big(\|f\|_q + \|Af\|_q\big),$$

(iii) for each $f \in L^q(J, X)$ there is a unique solution $u \in L^q(J, X)$ of (5.3) with $b * u \in H_q^\alpha(J, X) \cap L^q(J, D(A))$, and there is a constant $C(T) > 0$ such that

$$\|u\|_q + \|b * u\|_{\alpha,q} + \|Ab * u\|_q \leq C(T)\|f\|_q.$$

To present a result on maximal L^q-regularity of equations (5.2) and (5.3) based on $\mathcal{R}$-sectoriality of A, some definitions of useful properties of b are in order. First recall that a function $b \in L_{loc}^1(\mathbb{R}_+)$ is of *subexponential growth* if for each $\varepsilon > 0$

$$\int_0^\infty e^{-\varepsilon t}|b(t)|dt < \infty.$$

If b is of subexponential growth, the Laplace transform $\mathfrak{L}b$ of b is defined to be

$$\mathfrak{L}b(z) := \int_0^\infty e^{-zt}b(t)dt \quad (\mathrm{Re}\, z > 0).$$

Definition 5.7. Let $b \in L^1_{loc}(\mathbb{R}_+)$ be of subexponential growth.

(i) Let $k \in \mathbb{N}$. Then b is called *k-regular* if there is a constant $c > 0$ such that

$$|z^n \mathfrak{L}b^{(n)}(z)| \le c|\mathfrak{L}b(z)| \quad (\text{Re } z > 0, \ 0 \le n \le k).$$

(ii) Let $\mathfrak{L}b(z) \ne 0$ for all Re $z > 0$. Then b is called *sectorial with angle ϕ_b* if

$$\phi_b := \sup\{|\arg(\mathfrak{L}b(z))|; \text{ Re } z > 0\} < \pi.$$

If $b \in L^1_{loc}(\mathbb{R}_+)$ is of subexponential growth and 1-regular, for each $\xi \in \mathbb{R} \setminus \{0\}$ the limit

$$\mathfrak{L}b(i\xi) := \lim_{\text{Re}\lambda > 0, \ \lambda \to i\xi} \mathfrak{L}b(\lambda)$$

exists ([Prü93, Lemma 8.1]). Now we are in the position to state the result on maximal L^q-regularity that we already mentioned.

Theorem 5.8. *Let X be a Banach space of class $\mathcal{HT}$. Let $b \in L^1_{loc}(\mathbb{R}_+)$ be of subexponential growth, 1-regular, sectorial, and let $\overline{\lim}_{s \to \infty}|\mathfrak{L}b(s)|s^\alpha < \infty$ for some $\alpha \ge 0$. Let $A \in \mathcal{RS}(X)$ such that the parabolicity condition*

$$\phi_b + \phi_A^{\mathcal{R}} < \pi$$

is fulfilled. In addition, let either $0 \in \rho(A)$ or $b \in L^1(\mathbb{R}_+)$. Then for each $1 < q < \infty$ equations (5.2) and (5.3) have maximal L^q-regularity.

This result due to Prüss is basically taken from [Prü93, Theorem 8.7] where it is proved for $A \in \text{BIP}(X)$ and $\phi_A^{\mathcal{R}}$ replaced by θ_A. This is due to the fact that the proof there is based on Theorem 4.17 by Prüss and Sohr. Hence, it is proved in [Prü93, Theorem 8.6] that the multiplier operator B associated with

$$m(\xi) := \frac{1}{\mathfrak{L}b(i\xi)}$$

admits bounded imaginary powers with $\theta_B = \phi_b$. Since furthermore $B \in \mathcal{H}^\infty(X)$ and $\phi_B^\infty = \phi_b$, the result in the version as stated can now be deduced from the more recent Theorem 4.18.

In the subsequent chapters we establish pseudo-$\mathcal{R}$-sectoriality for numerous operators under consideration. If possible, we always emphasize that the $\mathcal{R}$-angle is less than $\frac{\pi}{2}$. However, we do not formulate the implications on the parabolic problems presented above each time.

6 Fourier transform approach to operator-dependent problems

In this chapter we employ the continuous Fourier multiplier result Theorem 3.17 to investigate partial differential equations in the whole space $\mathbb{R}^n$. These equations are allowed to depend on a closed linear operator in a suitable way. Apart from its own interest, this chapter provides the first part of an abstract background for the treatment of cylindrical boundary value problems later on.

6.1 Preliminaries

The multiplier conditions of Theorem 3.17 rely on derivatives of operator-valued functions. Hence, we start this section with appropriate representation formulas.

Given $\alpha \in \mathbb{N}_0^n \setminus \{0\}$, let

$$(6.1) \qquad \mathcal{Z}_\alpha := \left\{ \mathcal{W} = (\omega^1, \dots, \omega^r); \ 1 \leq r \leq |\alpha|, \ 0 < \omega^j \leq \alpha, \ \sum_{j=1}^r \omega^j = \alpha \right\}$$

denote the set of all additive decompositions of α into $r = r_{\mathcal{W}}$ multi-indices. For the sake of consistence we set $\mathcal{Z}_0 := \{\emptyset\}$ and $r_\emptyset := 0$.

The following lemma collects well-known representation formulas for derivatives of operator-valued smooth functions, where $m, r \in \mathbb{N}$ and X, Y, Z as well as X_j for $j = 0, \dots, r$ denote Banach spaces.

Lemma 6.1. a) [Leibniz rule] *Let* $T \in C^m(\mathbb{R}^n, \mathcal{L}(X, Y))$, $S \in C^m(\mathbb{R}^n, \mathcal{L}(Y, Z))$. *Then* $ST \in C^m(\mathbb{R}^n, \mathcal{L}(X, Z))$ *and for* $\alpha \in \mathbb{N}_0^n$ *such that* $|\alpha| \leq m$ *it holds that*

$$D^\alpha(ST) = \sum_{\beta \leq \alpha} \binom{\alpha}{\beta} (D^{\alpha-\beta} S)(D^\beta T).$$

b) *Let* $S_j \in C^1(\mathbb{R}^n, \mathcal{L}(X_{j-1}, X_j))$ *for* $j = 1, \dots, r$.
Then $\prod_{j=1}^r S_j \in C^1(\mathbb{R}^n, \mathcal{L}(X_0, X_r))$ *and*

$$D^{e_i}\left(\prod_{j=1}^r S_j\right) = \sum_{l=1}^r \left(\prod_{j=1}^{l-1} S_j\right)(D^{e_i} S_l)\left(\prod_{j=l+1}^r S_j\right).$$

c) *Let* $S \in C^m(\mathbb{R}^n, \mathcal{L}(X, Y))$ *such that* $S^{-1}(x) := (S(x))^{-1}$ *exists for all* $x \in \mathbb{R}^n$.
Then $S^{-1} \in C^m(\mathbb{R}^n, \mathcal{L}(Y, X))$ *and for* $\alpha \in \mathbb{N}_0^n$ *such that* $|\alpha| \leq m$ *it holds that*

$$D^\alpha(S^{-1}) = \sum_{\mathcal{W} \in \mathcal{Z}_\alpha} (-1)^{r_{\mathcal{W}}} S^{-1} \prod_{j=1}^{r_{\mathcal{W}}} \left((D^{\omega^j} S)S^{-1}\right).$$

Proof. The formulas follow by induction from the basic formulas for derivatives of vector-valued functions as presented e.g. in [Ama03, Section 2.4].   □

In the sequel let E denote a Banach space and

$$A: E \supset D(A) \to E$$

a closed operator. We turn our attention to multipliers of resolvent type.

Definition 6.2. Let P and Q define $\mathbb{C}$-valued polynomials. We call a continuous Fourier multiplier a *continuous multiplier of resolvent type* if it is built by sums and compositions of operator-valued functions

$$\xi \mapsto P(\xi) + Q(\xi)A: D(A) \to E$$

and their pointwise inverses

$$\xi \mapsto \left(P(\xi) + Q(\xi)A\right)^{-1}: E \to D(A).$$

In the situation described above $D := \left(D(A), \|\cdot\|_A\right)$ is a Banach space, where $\|\cdot\|_A$ denotes the graph norm of A. If we set $A_D: D \to E;\ x \mapsto Ax$, then $A_D \in \mathcal{L}(D, E)$. Right after the following lemma we will no longer distinguish between A and A_D

Lemma 6.3. *Let P and Q define $\mathbb{C}$-valued polynomials. Set*

$$S: \mathbb{R}^n \to \mathcal{L}(D, E);\ \xi \mapsto P(\xi) + Q(\xi)A_D.$$

Then $S \in C^\infty(\mathbb{R}^n, \mathcal{L}(D, E))$ and

$$D^\alpha S(\xi) = D^\alpha P(\xi) + D^\alpha Q(\xi)A_D \qquad (\alpha \in \mathbb{N}_0^n).$$

To establish a suitable class of resolvent type multipliers, we will consider elliptic polynomials P and Q. Recall the *principal part* $P^{\#}(\xi) := \sum_{|\alpha|=m} a_\alpha \xi^\alpha$ of an arbitrary polynomial $P(\xi) := \sum_{|\alpha| \leq m} a_\alpha \xi^\alpha$ with $a_\alpha \in \mathbb{C}$, $|\alpha| \leq m$ as well as the degree of P given by $\deg P := m$.

Definition 6.4. Let $P: \mathbb{R}^n \to \mathbb{C};\ \xi \mapsto P(\xi)$ define a polynomial and let $P^{\#}$ denote its principal part.
a) P is called *elliptic* if $P^{\#}(\xi) \neq 0$ for $\xi \in \mathbb{R}^n \setminus \{0\}$.
b) Let $\phi \in (0, \pi)$. Then P is called *parameter-elliptic* in $\overline{\Sigma}_{\pi-\phi}$ if $\lambda + P^{\#}(\xi) \neq 0$ for $(\lambda, \xi) \in \overline{\Sigma}_{\pi-\phi} \times \mathbb{R}^n \setminus \{(0, 0)\}$. In this case,

$$\varphi_P := \inf\{\phi \in (0, \pi);\ P \text{ is parameter-elliptic in } \overline{\Sigma}_{\pi-\phi}\}$$

is called the angle of parameter-ellipticity of P.

Lemma 6.5. a) P *is parameter-elliptic in* $\overline{\Sigma}_{\pi-\phi}$ *if and only if for all polynomials* N *with* $\deg N \leq \deg P$ *there exist* $C > 0$ *and a bounded subset* $G \subset \mathbb{R}^n$ *such that the estimate* $|\xi|^m |N(\xi)| \leq C|\lambda + P(\xi)|$ *holds for all* $\lambda \in \overline{\Sigma}_{\pi-\phi}$, *all* $\xi \in \mathbb{R}^n \setminus G$, *and all* $0 \leq m \leq \deg P - \deg N$.
b) P *is elliptic if and only if the assertion in* a) *is valid for* $\lambda = 0$.
c) *If* P *is parameter-elliptic in* $\overline{\Sigma}_{\pi-\phi}$, *for each polynomial* N *with* $\deg N \leq \deg P$ *and each* $\varepsilon > 0$ *there exists* $C > 0$ *such that the estimate* $|\xi|^m |N(\xi)| \leq C|\lambda + P^{\#}(\xi)|$ *holds for all* $\lambda \in \overline{\Sigma}_{\pi-\phi}$, *all* $\xi \in \mathbb{R}^n \setminus B_\varepsilon(0)$, *and all* $0 \leq m \leq \deg P - \deg N$.
d) *If* P *is elliptic, the assertion in* c) *is valid for* $\lambda = 0$.

Proof. a) First assume P to be parameter-elliptic in $\overline{\Sigma}_{\pi-\phi}$. By the triangle inequality we can assume that N is given as a monomial. The function

$$\kappa \colon \left(\overline{\Sigma}_{\pi-\phi} \times \mathbb{R}^n\right) \setminus (0,0) \to \mathbb{C}; \ (\lambda, \xi) \mapsto \frac{|\xi|^m N(\xi)}{\lambda + P^{\#}(\xi)}$$

is continuous and quasi-$(\deg P, 1)$-homogeneous of degree $m + \deg N - \deg P$. If $m + \deg N$ equals $\deg P$, it is therefore quasi-$(\deg P, 1)$-homogeneous of degree zero, i.e.

$$\kappa(s^\rho \lambda, s\xi) = \kappa(\lambda, \xi) \quad (s > 0),$$

where $\rho := \deg P$. Hence, it is bounded. To see this, we set

$$K := \{(\lambda, \xi) \in \overline{\Sigma}_{\pi-\phi} \times \mathbb{R}^n; \ |\lambda| + |\xi|^\rho = 1\}.$$

By the parameter-ellipticity condition we obtain

$$\lambda + P^{\#}(\xi) \neq 0 \quad ((\lambda, \xi) \in K).$$

Consequently, κ is a continuous function on the compact set K and we obtain

$$|\kappa(\lambda, \xi)| \leq M \quad ((\lambda, \xi) \in K).$$

By the quasi-homogeneity of κ this implies

$$|\kappa(s^\rho \lambda, s\xi)| \leq M \quad ((\lambda, \xi) \in K, \ s > 0).$$

Due to

$$|s^\rho \lambda| + |s\xi|^\rho = s^\rho(|\lambda| + |\xi|^\rho) \quad ((\lambda, \xi) \in K, \ s > 0)$$

and the particular choice $s = (|\lambda| + |\xi|^\rho)^{-1/\rho}$ we thus deduce

$$(s^\rho \lambda, s\xi) \in K \quad ((\lambda, \xi) \in \overline{\Sigma}_{\pi-\phi} \times \mathbb{R}^n).$$

Therefore

$$|\kappa(\lambda, \xi)| = |\kappa(s^\rho \lambda, s\xi)| \leq M \quad ((\lambda, \xi) \in \overline{\Sigma}_{\pi-\phi} \times \mathbb{R}^n).$$

If $m + \deg N < \deg P$, for each $\varepsilon > 0$ boundedness remains true for κ restricted to $\overline{\Sigma}_{\pi-\phi} \times (\mathbb{R}^n \setminus B_\varepsilon(0))$. Moreover, the upper bound tends to zero as ε is getting large. In particular, for $L := P - P^\#$ and ε large enough

$$|L(\xi)| \leq \tfrac{1}{2}|\lambda + P^\#(\xi)| \quad ((\lambda, \xi) \in \overline{\Sigma}_{\pi-\phi} \times (\mathbb{R}^n \setminus B_\varepsilon(0)))$$

follows. Therefore

$$|\lambda + P(\xi)| = |\lambda + P^\#(\xi) + L(\xi)| \geq \big||\lambda + P^\#(\xi)| - |L(\xi)|\big| \geq \tfrac{1}{2}|\lambda + P^\#(\xi)|$$

and for each fixed $\varepsilon > 0$ there exists $C > 0$ such that

$$|\xi|^m |N(\xi)| \leq C|\lambda + P^\#(\xi)| \leq 2C|\lambda + P(\xi)|$$

holds true for all $(\lambda, \xi) \in \overline{\Sigma}_{\pi-\phi} \times (\mathbb{R}^n \setminus B_\varepsilon(0))$.

Conversely, let this estimate hold true for some $\varepsilon > 0$. Assume $\lambda_0 + P^\#(\xi_0) = 0$ for a fixed $(\lambda_0, \xi_0) \in \big(\overline{\Sigma}_{\pi-\phi} \times \mathbb{R}^n\big) \setminus (0, 0)$. Then with $\rho := \deg P$

$$t^\rho(\lambda_0 + P^\#(\xi_0)) = t^\rho \lambda_0 + P^\#(t\xi_0) = 0$$

follows. The particular choice $m = \rho$ and $N \equiv 1$ now implies

$$t^\rho|\xi_0|^m = |t\xi_0|^m |N(t\xi_0)| \leq C|t^\rho \lambda_0 + P(t\xi_0)| = |L(t\xi_0)|$$

which gives a contradiction for large $t > 0$.

Assertion c) has just been proved in part one of the proof of a). With dependence of κ on λ being neglected, assertions b) and d) follow along the lines. $\qquad\square$

Remark 6.6. For $|\alpha| \leq \deg P$ the polynomial $D^\alpha P$ defines a polynomial of degree not greater than $\deg P - |\alpha|$. Hence, the assertions of Lemma 6.5 particularly apply to $m := |\alpha|$ and $N := D^\alpha P$.

Remark 6.7. Recall that $\deg P$ has to be even in case P is elliptic and $n > 1$.

In the following proposition we investigate multipliers of resolvent type involving elliptic polynomials. Note that we do not require $\deg P = \deg Q$ when saying 'P and Q elliptic' and that existence of $(\lambda + \mu A)^{-1}$ for $\lambda, \mu \in \mathbb{C}$ is meant to imply both $(\lambda + \mu A)^{-1}(E) = D(A)$ and $(\lambda + \mu A)^{-1} \in \mathcal{L}(E)$.

Proposition 6.8. *Let A be a closed operator in a Banach space E of class $\mathcal{HT}$ and let $P, Q: \mathbb{R}^n \to \mathbb{C}$ denote elliptic polynomials such that $\big(P(\xi) + Q(\xi)A\big)^{-1}$ is well-defined for all $\xi \in \mathbb{R}^n \setminus \{0\}$. Let N define an arbitrary polynomial subject to $\deg N \leq \deg P$ and assume that the families*

$$\Big\{P(\xi)\big(P(\xi) + Q(\xi)A\big)^{-1};\ \xi \in \mathbb{R}^n \setminus \{0\}\Big\}, \quad \Big\{\big(P(\xi) + Q(\xi)A\big)^{-1};\ \xi \in \mathbb{R}^n \setminus \{0\}\Big\},$$

and

$$\Big\{A\big(P(\xi) + Q(\xi)A\big)^{-1};\ \xi \in \mathbb{R}^n \setminus \{0\}\Big\}$$

are $\mathcal{R}$-bounded. Then for $1 < p < \infty$

$$m: \mathbb{R}^n \setminus \{0\} \to \mathcal{L}(E);\ \xi \mapsto N(\xi)\big(P(\xi) + Q(\xi)A\big)^{-1}$$

defines an L^p-multiplier.

Proof. Since the subset of invertible operators in $\mathcal{L}(E)$ is known to be open, existence of $\big(P(\xi) + Q(\xi)A\big)^{-1}$ for all $\xi \in \mathbb{R}^n \setminus \{0\}$ implies smoothness of m.

Lemma 6.1 and Lemma 6.3 therefore yield

$$|\xi|^{|\gamma|}D^\gamma m(\xi) = \sum_{\beta \leq \gamma} \binom{\gamma}{\beta} \sum_{w \in \mathcal{Z}_\beta} (-1)^{r_w} |\xi|^{|\gamma-\beta|} (D^{\gamma-\beta}N)(\xi)\big(P(\xi) + Q(\xi)A\big)^{-1}$$

$$\cdot \prod_{j=1}^{r_w} |\xi|^{|\omega_j|}\big(D^{\omega_j}P(\xi) + D^{\omega_j}Q(\xi)A\big)\big(P(\xi) + Q(\xi)A\big)^{-1}.$$

Recall $\deg D^{\gamma-\beta}N \leq \deg N - |\gamma-\beta|$ from Remark 6.6. Hence, ellipticity of P and Lemma 6.5 imply $|\xi|^{|\gamma-\beta|}|D^{\gamma-\beta}N(\xi)| \leq C|P(\xi)|$ for $\xi \in \mathbb{R}^n \setminus G_N$ with a bounded set $G_N \subset \mathbb{R}^n$. By Kahane's contraction principle we obtain the $\mathcal{R}$-boundedness of

$$\Big\{|\xi|^{|\gamma-\beta|}D^{\gamma-\beta}N(\xi)\big(P(\xi) + Q(\xi)A\big)^{-1}; \ \xi \in \mathbb{R}^n \setminus G_N\Big\}.$$

Along the same lines $\mathcal{R}$-boundedness of

$$\Big\{|\xi|^{|\omega_j|}D^{\omega_j}P(\xi)\big(P(\xi) + Q(\xi)A\big)^{-1}; \ \xi \in \mathbb{R}^n \setminus G_P\Big\}$$

follows. Since

$$Q(\xi)A\big(P(\xi) + Q(\xi)A\big)^{-1} = \mathrm{id}_E - P(\xi)\big(P(\xi) + Q(\xi)A\big)^{-1},$$

$\mathcal{R}$-boundedness of

$$\Big\{|\xi|^{|\omega_j|}D^{\omega_j}Q(\xi)A\big(P(\xi) + Q(\xi)A\big)^{-1}; \ \xi \in \mathbb{R}^n \setminus G_Q\Big\}$$

finally follows from the ellipticity of Q.

Setting $G := G_N \cup G_P \cup G_Q$, it is left to prove $\mathcal{R}$-boundedness of the above families for $\xi \in G \setminus \{0\}$ for then the assertion follows from Lemma 3.2 and Theorem 3.17. To this end, we make use of the additional $\mathcal{R}$-boundedness conditions for the families

$$\Big\{\big(P(\xi) + Q(\xi)A\big)^{-1}; \ \xi \in \mathbb{R}^n \setminus \{0\}\Big\}$$

and

$$\Big\{A\big(P(\xi) + Q(\xi)A\big)^{-1}; \ \xi \in \mathbb{R}^n \setminus \{0\}\Big\}.$$

By means of the contraction principle of Kahane, for a fixed $M > 0$, they show the families

$$\Big\{\big(P(\xi) + \delta\big)\big(P(\xi) + Q(\xi)A\big)^{-1}; \ \xi \in \mathbb{R}^n \setminus \{0\}, \ |\delta| \leq M\Big\}$$

and

$$\Big\{\big(Q(\xi) + \delta\big)A\big(P(\xi) + Q(\xi)A\big)^{-1}; \ \xi \in \mathbb{R}^n \setminus \{0\}, \ |\delta| \leq M\Big\}$$

to be $\mathcal{R}$-bounded as well. Now let $|\delta|$ be large enough, such that $P(\xi) + \delta \neq 0$ and $Q(\xi) + \delta \neq 0$ for all $\xi \in \overline{G}$. Then there exists $C > 0$ such that

$$\frac{|\xi|^{|\gamma - \beta|}|D^{\gamma - \beta}N(\xi)|}{|P(\xi) + \delta|} \leq C, \qquad \frac{|\xi|^{|\omega_j|}|D^{\omega_j}P(\xi)|}{|P(\xi) + \delta|} \leq C,$$

and

$$\frac{|\xi|^{|\omega_j|}|D^{\omega_j}Q(\xi)|}{|Q(\xi) + \delta|} \leq C$$

for all $\xi \in \overline{G}$ due to continuity. Applying the contraction principle of Kahane, the proof is complete. $\qquad\square$

Corollary 6.9. *Let A be a closed operator in a Banach space E of class $\mathcal{HT}$, let $P, Q \colon \mathbb{R}^n \to \mathbb{C}$ denote homogeneous elliptic polynomials, and let $\big(P(\xi) + Q(\xi)A\big)^{-1}$ be well-defined for all $\xi \in \mathbb{R}^n \setminus \{0\}$. Let N define a homogeneous polynomial subject to $\deg N = \deg P$ and assume that the family*

$$\Big\{P(\xi)\big(P(\xi) + Q(\xi)A\big)^{-1}; \ \xi \in \mathbb{R}^n \setminus \{0\}\Big\}$$

is $\mathcal{R}$-bounded. Then for $1 < p < \infty$

$$m \colon \mathbb{R}^n \setminus \{0\} \to \mathcal{L}(E); \ \xi \mapsto N(\xi)\big(P(\xi) + Q(\xi)A\big)^{-1}$$

defines an L^p-multiplier.

Proof. Because of $P = P^{\#}$, $N = N^{\#}$ subject to $\deg N = \deg P$, and $Q = Q^{\#}$ we can set $G = G_N = G_P = G_Q = \{0\}$ (cf. Lemma 6.5). Therefore we can neglect $\mathcal{R}$-boundedness of

$$\Big\{\big(P(\xi) + Q(\xi)A\big)^{-1}; \ \xi \in \mathbb{R}^n \setminus \{0\}\Big\}$$

and

$$\Big\{A\big(P(\xi) + Q(\xi)A\big)^{-1}; \ \xi \in \mathbb{R}^n \setminus \{0\}\Big\}.$$

$\qquad\square$

Example 6.10. Let $A \in \Psi\mathcal{R}S(E)$, i.e. $(0, \infty) \subset \rho(-A)$ and

$$\mathcal{R}\big(\{t(t + A)^{-1}; \ t \in (0, \infty)\}\big) < \infty.$$

Let $\alpha \in \mathbb{N}_0^n$ such that $|\alpha| = 2$. Then, due to Corollary 6.9

$$m \colon \mathbb{R}^n \setminus \{0\} \to \mathcal{L}(E); \ \xi \mapsto \xi^\alpha (|\xi|^2 + A)^{-1}$$

defines a Fourier multiplier.

In view of Lemma 3.7, the previous example indicates that parameter-ellipticity of P and pseudo-$\mathcal{R}$-sectoriality of A are likely to define an appropriate setting to apply the theory of resolvent type multipliers successfully.

6.2 $\mathcal{R}$-sectoriality and $\mathcal{RH}^\infty$-calculus

Let E be a Banach space of class $\mathcal{HT}$ and let A be a closed operator in E. In what follows we consider the parameter-dependent and A-dependent problem given by

$$(6.2) \qquad \lambda u + \mathcal{A}(D)u = f \quad \text{in } \mathbb{R}^n.$$

Here

$$\mathcal{A}(D) := P(D) + Q(D)A := \sum_{|\alpha| \le m_1} p_\alpha D^\alpha + \sum_{|\alpha| \le m_2} q_\alpha D^\alpha A$$

with $m_1, m_2 \in \mathbb{N}_0$ and constant coefficients $p_\alpha, q_\alpha \in \mathbb{C}$.

The following results are restricted to homogeneous polynomials. To some extent, they can be improved in order to cover non-homogeneous polynomials as well. However, as indicated by Proposition 6.8, the assumptions needed are less convenient to present. Moreover, in view of the applications we have in mind we can relinquish full generality.

Lemma 6.11. *Let $\varphi_P, \varphi_Q, \phi, \vartheta \in [0, \pi)$ be given. If*

$$\frac{\lambda + P(\xi)}{Q(\xi)} \in \Sigma_{\pi - \phi} \quad (\lambda \in \Sigma_{\pi - \vartheta},\ \xi \in \mathbb{R}^n \setminus \{0\})$$

holds for all homogeneous polynomials P and Q which are parameter-elliptic with angles of parameter-ellipticity φ_P and φ_Q, then $\varphi_P + \varphi_Q + \phi < \pi$.
In that case, $\vartheta \in (\max\{\varphi_P, \varphi_Q + \phi\}, \pi - \min\{\varphi_P, \varphi_Q + \phi\})$ implies

$$\frac{\lambda + P(\xi)}{Q(\xi)} \in \Sigma_{\pi - \phi} \quad (\lambda \in \Sigma_{\pi - \vartheta},\ \xi \in \mathbb{R}^n \setminus \{0\}).$$

Proof. Let $\varphi > \varphi_P$. By definition, parameter-ellipticity of P yields $P(\xi) + \lambda \ne 0$ for $(\lambda, \xi) \in \overline{\Sigma}_{\pi - \varphi} \times \mathbb{R}^n \setminus \{(0,0)\}$, that is, $P(\xi) \ne \lambda$ for $(\lambda, \xi) \in (\mathbb{C} \setminus \Sigma_\varphi) \times \mathbb{R}^n \setminus \{(0,0)\}$ respectively $P(\xi) \in \Sigma_\varphi$. Since the range of P is closed, this ensures $P(\xi) \in \overline{\Sigma}_{\varphi_P}$. Accordingly we deduce $Q(\xi) \in \overline{\Sigma}_{\varphi_Q}$ and moreover

$$\frac{1}{Q(\xi)} = \frac{\overline{Q(\xi)}}{|Q(\xi)|^2} \in \overline{\Sigma}_{\varphi_Q} \quad (\xi \in \mathbb{R}^n \setminus \{0\}).$$

Recall that e.g. $\xi \mapsto \rho e^{\pm i\theta} |\xi|^2$ with $\rho > 0$ and $\theta \in [0, \pi)$ define parameter-elliptic polynomials with angles of parameter-ellipticity equal to θ. For fixed $\vartheta \in [0, \pi)$ the assertion

$$\frac{\lambda + P(\xi)}{Q(\xi)} \in \Sigma_{\pi - \phi} \quad (\lambda \in \Sigma_{\pi - \vartheta},\ \xi \in \mathbb{R}^n \setminus \{0\})$$

for all parameter-elliptic polynomials P, Q with angles of parameter-ellipticity φ_P and φ_Q is therefore equivalent to

$$\arg(\lambda + e^{\pm i\varphi_P}) + \varphi_Q < \pi - \phi \quad (\lambda \in \Sigma_{\pi - \vartheta}).$$

This implies $\pi - \vartheta + \varphi_P + \varphi_Q \leq \pi - \phi$ with $\pi - \vartheta > 0$. Hence, $\varphi_P + \varphi_Q + \phi < \pi$.

In that case $\vartheta \in (\max\{\varphi_P, \varphi_Q + \phi\}, \pi - \min\{\varphi_P, \varphi_Q + \phi\})$ exists.

First assume that $\varphi_P \geq \varphi_Q + \phi$. This implies $\varphi_P < \vartheta < \pi - (\varphi_Q + \phi)$ respectively $\varphi_Q + \phi < \pi - \vartheta < \pi - \varphi_P$.

If $\varphi_P \geq \frac{\pi}{2}$ then $\pi - \vartheta < \pi - \varphi_P < \varphi_P$ yields $\arg(\lambda + P(\xi)) < \varphi_P$ for $\lambda \in \Sigma_{\pi - \vartheta}$ and $\xi \in \mathbb{R}^n \setminus \{0\}$.

Now let $\varphi_P < \frac{\pi}{2}$. If $\pi - \vartheta \leq \varphi_P$ then $\arg(\lambda + P(\xi)) < \varphi_P$ holds for $\lambda \in \Sigma_{\pi - \vartheta}$ and $\xi \in \mathbb{R}^n \setminus \{0\}$ follows. In case $\varphi_P < \pi - \vartheta$ taking into account $\pi - \vartheta < \pi - \varphi_P$ yields $\arg(\lambda + P(\xi)) < \pi - \vartheta$ for $\lambda \in \Sigma_{\pi - \vartheta}$ and $\xi \in \mathbb{R}^n \setminus \{0\}$.

Since $\varphi_P, \pi - \vartheta < \pi - (\varphi_Q + \phi)$ in both cases

$$\frac{\lambda + P(\xi)}{Q(\xi)} \in \Sigma_{\pi - \phi} \quad (\lambda \in \Sigma_{\pi - \vartheta}, \ \xi \in \mathbb{R}^n \setminus \{0\})$$

follows.

Conversely, assume $\varphi_P < \varphi_Q + \phi$, that is, $\varphi_Q + \phi < \vartheta < \pi - \varphi_P$. Consequently we have $\varphi_P < \pi - \vartheta < \pi - (\varphi_Q + \phi)$ and due to $\varphi_P + \varphi_Q + \phi < \pi$ necessarily $\varphi_P < \frac{\pi}{2}$. Thus, $\varphi_P < \pi - \vartheta < \pi - (\varphi_Q + \phi) < \pi - \varphi_P$ implies $\arg(\lambda + P(\xi)) < \pi - \vartheta$ for $\lambda \in \Sigma_{\pi - \vartheta}$ and $\xi \in \mathbb{R}^n \setminus \{0\}$ as well as $\pi - \vartheta + \varphi_Q < \pi - \phi$, that is,

$$\frac{\lambda + P(\xi)}{Q(\xi)} \in \Sigma_{\pi - \phi} \quad (\lambda \in \Sigma_{\pi - \vartheta}, \ \xi \in \mathbb{R}^n \setminus \{0\}).$$

$\square$

Remark 6.12. The assertion of Lemma 6.11 remains true if $\lambda \in \Sigma_{\pi - \vartheta} \cup \{0\}$.

The $L^p(\mathbb{R}^n, E)$-realization of $\mathcal{A}$ as given in problem (6.2) is defined as

$$D(\mathbb{A}) := \{u \in W^{m_1, p}(\mathbb{R}^n, E); \ Q(D)\mathcal{A}u \in L^p(\mathbb{R}^n, E)\},$$
$$\mathbb{A}u := \mathcal{A}(D)u \quad (u \in D(\mathbb{A})).$$

Proposition 6.13. *Let $1 < p < \infty$, let E be a Banach space of class $\mathcal{HT}$ enjoying property (α), and let $A \in \Psi\mathcal{RS}(E)$. For homogeneous polynomials P and Q assume that*

(i) P is parameter-elliptic with angle $\varphi_P \in [0, \pi)$,

(ii) Q is parameter-elliptic with angle $\varphi_Q \in [0, \pi)$,

(iii) $\varphi_P + \varphi_Q + \phi_A^{\mathcal{R}} < \pi$.

Set $\varphi_0 := \max\{\varphi_P, \varphi_Q + \phi_A^{\mathcal{R}}\}$. Then $\mathbb{A} \in \mathcal{RS}(L^p(\mathbb{R}^n, E))$ and $\phi_{\mathbb{A}}^{\mathcal{R}} \leq \varphi_0$. Moreover, for each $\phi > \varphi_0$ it holds that

$$(6.3) \quad \mathcal{R}\left(\left\{\lambda^{1 - \frac{|\alpha|}{m_1}} D^\alpha (\lambda + \mathbb{A})^{-1}; \ \lambda \in \Sigma_{\pi - \phi}, \ \alpha \in \mathbb{N}_0^n, \ 0 \leq |\alpha| \leq m_1\right\}\right) < \infty.$$

In case $Q \equiv c$, $c \neq 0$ subject to condition (ii), $0 \in \rho(A)$ implies $0 \in \rho(\mathbb{A})$.

Proof. We consider the formal representation $(\lambda + \mathbb{A})^{-1} = T_{m_\lambda}$ of the resolvent of $\mathbb{A}$, where T_{m_λ} denotes the operators associated with

$$m_\lambda(\xi) := \left(\lambda + P(\xi) + Q(\xi)A\right)^{-1}.$$

More generally, with $\alpha \in \mathbb{N}_0^n$ we consider $\lambda^{1 - \frac{|\alpha|}{m_1}} D^\alpha (\lambda + \mathbb{A})^{-1} = T_{m_\lambda^\alpha}$, where

$$m_\lambda^\alpha(\xi) := \lambda^{1 - \frac{|\alpha|}{m_1}} \xi^\alpha \left(\lambda + P(\xi) + Q(\xi)A\right)^{-1}.$$

To justify the representation formulas, we make use of Theorem 3.17 to establish m_λ^α as a Fourier multiplier. By uniqueness of the Fourier transform in $\mathcal{S}'(\mathbb{R}^n, E)$ it follows that $\lambda + \mathbb{A}$ is bijective and that T_{m_λ} defines its inverse operator.

Let $\phi \in (\max\{\varphi_P, \varphi_Q + \phi_A^\mathcal{R}\}, \pi - \min\{\varphi_P, \varphi_Q + \phi_A^\mathcal{R}\})$. We apply Lemma 6.1 in order to calculate $\xi^\gamma D^\gamma m_\lambda^\alpha(\xi)$. As in the proof of Proposition 6.8 it suffices to show that

$$(6.4) \qquad \left\{ \lambda^{1 - \frac{|\alpha|}{m_1}} \xi^\omega D^\omega N(\xi) \left(\lambda + P(\xi) + Q(\xi)A\right)^{-1}; \ \lambda \in \Sigma_{\pi - \phi}, \ \xi \in \mathbb{R}^n \setminus \{0\} \right\}$$

for $N(\xi) := \xi^\alpha$ and arbitrary $\omega \leq \gamma$,

$$(6.5) \qquad \left\{ \xi^\omega D^\omega P(\xi) \left(\lambda + P(\xi) + Q(\xi)A\right)^{-1}; \ \lambda \in \Sigma_{\pi - \phi}, \ \xi \in \mathbb{R}^n \setminus \{0\} \right\}$$

for $0 < \omega \leq \gamma$, and

$$(6.6) \qquad \left\{ \xi^\omega D^\omega Q(\xi)A \left(\lambda + P(\xi) + Q(\xi)A\right)^{-1}; \ \lambda \in \Sigma_{\pi - \phi}, \ \xi \in \mathbb{R}^n \setminus \{0\} \right\}$$

for $0 < \omega \leq \gamma$ are $\mathcal{R}$-bounded.

Due to our assumptions on P and Q, the pseudo-$\mathcal{R}$-sectoriality of A, and Lemma 6.11

$$\left\{ \left(\frac{\lambda + P(\xi)}{Q(\xi)} \right) \left(\frac{\lambda + P(\xi)}{Q(\xi)} + A \right)^{-1}; \ \lambda \in \Sigma_{\pi - \phi}, \ \xi \in \mathbb{R}^n \setminus \{0\} \right\}$$

is $\mathcal{R}$-bounded. This readily yields $\mathcal{R}$-boundedness of

$$(6.7) \qquad \left\{ (\lambda + P(\xi)) \left(\lambda + P(\xi) + Q(\xi)A\right)^{-1}; \ \lambda \in \Sigma_{\pi - \phi}, \ \xi \in \mathbb{R}^n \setminus \{0\} \right\}$$

and further $\mathcal{R}$-boundedness of

$$(6.8) \qquad \left\{ Q(\xi)A \left(\lambda + P(\xi) + Q(\xi)A\right)^{-1}; \ \lambda \in \Sigma_{\pi - \phi}, \ \xi \in \mathbb{R}^n \setminus \{0\} \right\}.$$

Recall that $\lambda + P(\xi) \neq 0$ for $\lambda \in \overline{\Sigma}_{\pi - \phi}$ by our choice of $\phi > \varphi_P$. Therefore, quasi-homogeneity of $(\lambda, \xi) \mapsto \lambda + P(\xi)$ allows to apply the contraction principle of Kahane to prove (6.4) and (6.5) (cf. Lemma 6.5). Similarly, ellipticity and homogeneity of Q proves (6.6). Finally, $Q(D)A(\lambda + \mathbb{A})^{-1} f \in L^p(\mathbb{R}^n, E)$ for arbitrary $f \in L^p(\mathbb{R}^n, E)$ follows from (6.8).

We prove injectivity of $\mathbb{A}$. To this end, consider $u \in D(\mathbb{A})$ such that $\mathbb{A}u = 0$, that is, $(\lambda + \mathbb{A})u = \lambda u$. Hence, for each $\lambda \in \mathbb{R}_+$ the unique solution of (6.2) with right-hand side $f := \lambda u$ is given by u. Due to (6.3) for $\alpha \in \mathbb{N}_0^n$ subject to $|\alpha| = m_1$ we have $\|D^\alpha u\|_p \leq \|\lambda u\|_p$. Since $\lambda \in \mathbb{R}_+$ is arbitrary, $\|D^\alpha u\|_p = 0$ follows. Because of $u \in W^{m,p}(\mathbb{R}^n, E)$, this yields $u = 0$.

In case $Q \equiv c$, $c \neq 0$ subject to condition (ii), and $0 \in \rho(A)$, what has been proved so far remains valid if A is replaced by $A - \delta$ with $\delta > 0$ sufficiently small. This proves $0 \in \rho(\mathbb{A})$. $\qquad\square$

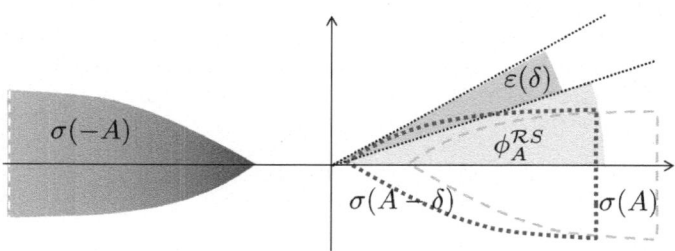

Figure 6.1: Possible shift in case $0 \in \rho(A)$.

Remark 6.14. By ellipticity of Q it holds that

$$D(\mathbb{A}) = \{u \in W^{m_1,p}(\mathbb{R}^n, E); \; D^\alpha Au \in L^p(\mathbb{R}^n, E) \; (|\alpha| = m_2)\}.$$

Consider $\mathcal{A}_\delta(D) := \mathcal{A}(D) + \delta A$, $\delta > 0$. Then all steps of the proof above apply unchanged and we end up with $\mathcal{R}$-boundedness of

$$\left\{ (Q(\xi) + \delta) A (\lambda + P(\xi) + (Q(\xi) + \delta)A)^{-1}; \; \lambda \in \Sigma_{\pi - \phi}, \; \xi \in \mathbb{R}^n \setminus \{0\} \right\}$$

instead of (6.8). Let $\mathbb{A}_\delta$ denote the corresponding L^p-realization of $\mathcal{A}_\delta(D)$. Then the assertion of Proposition 6.13 remains true for $\mathbb{A}_\delta$ and we have

$$D(\mathbb{A}_\delta) = \{u \in W^{m_1,p}(\mathbb{R}^n, E); \; Au \in W^{m_2,p}(\mathbb{R}^n, E)\}$$

due to parameter-ellipticity of Q.

In the sequel we consider $\mathcal{A}_\delta(D)$, where $\delta > 0$ in case $m_2 \neq 0$, and assume it to be *even*, that is,

$$\mathcal{A}_\delta(D) = \sum_{|\alpha|=m_1} p_\alpha D^\alpha + \sum_{|\alpha|=m_2} q_\alpha D^\alpha A + \delta A,$$

where m_1 and m_2 are even. In case $m_2 = 0$ the shift δA can be dropped.

Let $\tilde{\mathbb{R}}^n = \prod_{j=1}^n \mathbb{R}_{j,(+)}$ with $\mathbb{R}_{j,(+)} \in \{\mathbb{R}, \mathbb{R}_+\}$. We consider a boundary value problem

(6.9)
$$\begin{aligned}
\lambda u + \mathcal{A}_\delta(D)u &= f && \text{in } \tilde{\mathbb{R}}^n, \\
\mathcal{B}_1(D)u &= 0 && \text{on } \partial\tilde{\mathbb{R}}^n, \\
\mathcal{B}_2(D)Au &= 0 && \text{on } \partial\tilde{\mathbb{R}}^n.
\end{aligned}$$

Here the boundary $\partial\tilde{\mathbb{R}}^n$ is defined to be the union of all sets

$$\mathbb{R}_{1,(+)} \times \mathbb{R}_{j-1,(+)} \times \{0\} \times \mathbb{R}_{j+1,(+)} \times \mathbb{R}_{n,(+)} \quad (j = 1, \ldots, n; \ \mathbb{R}_{j,(+)} = \mathbb{R}_+).$$

The pairwise intersections of these sets are empty, i.e. the Lebesgue null sets of vertices is neglected. In each direction $j \in \{1, \ldots, n\}$ such that $\mathbb{R}_{j,(+)} = \mathbb{R}_+$, the boundary operator $\mathcal{B}(D) := (\mathcal{B}_1(D), \mathcal{B}_2(D))$ endows problem (6.9) with Dirichlet

(i) $D_j^\ell u|_{x_j=0} = 0 \quad (\ell = 0, 2, \ldots, m_1 - 2)$ and
   $D_j^\ell Au|_{x_j=0} = 0 \quad (\ell = 0, 2, \ldots, m_2 - 2)$

or Neumann

(ii) $D_j^\ell u|_{x_j=0} = 0 \quad (\ell = 1, 3, \ldots, m_1 - 1)$ and
   $D_j^\ell Au|_{x_j=0} = 0 \quad (\ell = 1, 3, \ldots, m_2 - 1)$

boundary conditions. Note that the types may be different in different directions.

Remark 6.15. If $u, Au \in W^{m_1,p}(\tilde{\mathbb{R}}^n, E)$, then $\mathcal{B}_1(D)u = 0$ implies

$$D_j^\ell Au|_{x_j=0} = 0 \quad (\ell = 1, 3, \ldots, m_1 - 1)$$

due to closedness of A. In case $m_1 < m_2$, it is therefore enough to assume

$$D_j^\ell Au|_{x_j=0} = 0 \quad (\ell = m_1 - 2, m_1, \ldots, m_2 - 2)$$

in (i), respectively

$$D_j^\ell Au|_{x_j=0} = 0 \quad (\ell = m_1 - 1, m_1 + 1, \ldots, m_2 - 1)$$

in (ii). Accordingly, $m_2 \le m_1$ renders the boundary condition $\mathcal{B}_2(D)Au = 0$ unnecessary.

For the sake of simplicity we assume $\mathbb{R}_{j,(+)} = \mathbb{R}_+$ for all $j = 1, \ldots, n$ which will be indicated by writing $\mathbb{K}^n$ instead of $\tilde{\mathbb{R}}^n$. The $L^p(\mathbb{K}^n, E)$-realization of problem (6.9) is defined as

$$D(\mathbb{A}_{\delta,\mathcal{B}}) := \{u \in W^{m_1,p}(\mathbb{K}^n, E); \ Au \in W^{m_2,p}(\mathbb{K}^n, E), \ \mathcal{B}(D)u = 0\},$$
$$\mathbb{A}_{\delta,\mathcal{B}}u := \mathcal{A}_\delta(D)u \quad (u \in D(\mathbb{A}_{\delta,\mathcal{B}})).$$

Proposition 6.16. *Given the assumptions of Proposition 6.13, let $\mathcal{A}(D)$ be even. Then $\mathbb{A}_{\delta,\mathcal{B}} \in \mathcal{RS}(L^p(\mathbb{R}^n, E))$ and $\phi^{\mathcal{R}}_{\mathbb{A}_{\delta,\mathcal{B}}} \leq \varphi_0$. Moreover, for each $\phi > \varphi_0$ it holds that*

$$(6.10) \quad \mathcal{R}\left(\left\{\lambda^{1-\frac{|\alpha|}{m_1}} D^\alpha (\lambda + \mathbb{A}_{\delta,\mathcal{B}})^{-1}; \ \lambda \in \Sigma_{\pi-\phi}, \ \alpha \in \mathbb{N}_0^n, \ 0 \leq |\alpha| \leq m_1\right\}\right) < \infty.$$

Proof. For simplicity of notation we consider the case $n = 2$ and boundary conditions of type (i) in direction x_1 and of type (ii) in direction x_2. Following e.g. [PS93], we carry out a reflection procedure.

Let $f \in L^p(\mathbb{K}^n, E)$ be arbitrary and $\lambda \in \rho(-\mathbb{A}_\delta)$. First considering the odd extension of f to $\mathbb{R} \times \mathbb{R}_+$ and afterwards its even extension to $\mathbb{R}^2$, we end up with a function F fulfilling $F(x_1, x_2) = -F(-x_1, x_2)$ as well as $F(x_1, x_2) = F(x_1, -x_2)$ a.e. in $\mathbb{R}^2$. Now we can apply Proposition 6.13 and Remark 6.14 which yield existence of a unique solution $U \in \{u \in W^{m_1,p}(\mathbb{R}^n, E); \ Au \in W^{m_2,p}(\mathbb{R}^n, E)\}$ of

$$\lambda U + \mathcal{A}_\delta(D)U = F \quad \text{in } \mathbb{R}^2.$$

Symmetry of $\mathcal{A}(D)$ now shows that

$$V_1(x_1, x_2) := -U(-x_1, x_2) \quad \text{and} \quad V_2(x_1, x_2) := U(x_1, -x_2) \quad (x \in \mathbb{R}^2)$$

define solutions, too. Thanks to uniqueness the equality $V_1 = U = V_2$ follows. Hence, $U_{x_2} := U(\cdot, x_2) \in W^{m_1,p}(\mathbb{R}, E) \subset C^{m_1-1}(\mathbb{R}, E)$ for a.e. $x_2 \in \mathbb{R}$ is odd which yields

$$U_{x_2}^{(\ell)}(0) = 0 \quad (\ell = 0, 2, \ldots, m_1 - 2).$$

Accordingly, for a.e. $x_1 \in \mathbb{R}$ we have that U_{x_1} is even, hence,

$$U_{x_1}^{(\ell)}(0) = 0 \quad (\ell = 1, 3, \ldots, m_1 - 1).$$

The same arguments applied to AU yield

$$(AU)_{x_2}^{(\ell)}(0) = 0 \quad (\ell = 0, 2, \ldots, m_2 - 2)$$

and

$$(AU)_{x_1}^{(\ell)}(0) = 0 \quad (\ell = 1, 3, \ldots, m_2 - 1).$$

Therefore, $u := U|_{\mathbb{K}^2}$ solves $\lambda u + \mathcal{A}_\delta(D)u = f$ with boundary conditions (i) for $j = 1$ and (ii) for $j = 2$.

For arbitrary $n \in \mathbb{N}$ and boundary conditions of Dirichlet or Neumann type the construction of the solution follows the same lines: we choose odd extensions in directions subject to case (i) and even extensions in directions subject to case (ii).

On the other hand, let u be a solution of the boundary value problem (6.9). We extend u and f to U and F defined on $\mathbb{R}^n$ as described above. From symmetry of $\mathcal{A}_\delta(D)$ we infer that $U \in \{u \in W^{m_1,p}(\mathbb{R}^n, E); \ Au \in W^{m_2,p}(\mathbb{R}^n, E)\}$ solves

$$\lambda U + \mathcal{A}_\delta(D)U = F \quad \text{in } \mathbb{R}^n.$$

Thus, uniqueness of U yields uniqueness of u.

Altogether, $(\lambda + \mathbb{A}_{\delta,\mathcal{B}})^{-1} = \mathfrak{R}(\lambda + \mathbb{A}_\delta)^{-1}\mathfrak{E}$, where $\mathfrak{E}$ defines the operator of extension to $\mathbb{R}^n$ as explained above and $\mathfrak{R}$ defines the operator of restriction to $\mathbb{K}^n$. Thus, Proposition 6.13 and Remark 6.14 prove the claim. $\qquad\square$

Besides $\mathcal{R}$-sectoriality, we are interested in $\mathcal{R}$-boundedness of the $\mathcal{H}^\infty$-calculus of the operators introduced so far.

Lemma 6.17. *Let E be a Banach space and let $A \in \Psi\mathcal{RH}^\infty(E)$. Let further $0 \leq \vartheta, \zeta < \pi$ such that $\vartheta + \zeta < \pi - \phi_A^{\mathcal{R}\infty}$. Then the $\mathcal{H}_0^\infty$-calculus of $(z_1 + z_2 A)$ in E is uniformly $\mathcal{R}$-bounded with respect to $z_1 \in \Sigma_\vartheta$ and $z_2 \in \Sigma_\zeta$ with angle $\phi_{z_1+z_2A}^{\mathcal{R}\infty} \leq \max\{\vartheta, \zeta + \phi_A^{\mathcal{R}\infty}\}$. More precisely, for each $\sigma > \max\{\vartheta, \zeta + \phi_A^{\mathcal{R}\infty}\}$ it holds that*

$$\mathcal{R}(\{h(z_1 + z_2A); \ |h|_\infty^\sigma \leq 1, \ (z_1, z_2) \in \Sigma_\vartheta \times \Sigma_\zeta\}) < \infty.$$

Proof. The proof is carried out in two steps separately. First we consider $z_1 \in \Sigma_\vartheta$ and $z_2 = 1$, i.e. $\zeta = 0$. Afterwards we consider $z_1 = 0$ and $z_2 \in \Sigma_\zeta$, i.e. $\vartheta = 0$ (see Figure 6.2). Iteratively this proves the assertion.

Let $\phi \in (\max\{\vartheta, \phi_A^{\mathcal{R}\infty}\}, \pi - \min\{\vartheta, \phi_A^{\mathcal{R}\infty}\})$. Let $\sigma > \phi$ and $h \in \mathcal{H}_0^\infty(\Sigma_\sigma)$ with $|h|_\infty^\sigma \leq 1$ be arbitrary. Pick $\psi \in (\phi, \min\{\sigma, \pi - \min\{\vartheta, \phi_A^{\mathcal{R}\infty}\})$ and set

$$(6.11) \qquad \Gamma := (\infty, 0]e^{i\psi} \cup [0, \infty)e^{-i\psi}.$$

Then $\lambda - z \in \rho(A)$ for all $\lambda \in \Gamma$ and $z \in \Sigma_\vartheta$ by our choice of ψ and ϑ.

For arbitrary $z \in \Sigma_\vartheta$ choose $0 < \delta < \text{dist}(z, \Gamma)$. With $\psi' \in (\phi_A^{\mathcal{R}\infty}, \min\{\sigma, \pi - \vartheta\})$ set

$$\Gamma'_\delta := (\infty, \delta]e^{i\psi'} \cup \delta e^{i(\psi', 2\pi - \psi')} \cup [\delta, \infty)e^{-i\psi'}.$$

Then

$$h(z + A) = \int_\Gamma h(\lambda)(\lambda - z - A)^{-1}d\lambda = \int_{\Gamma'_\delta + z} h(\lambda)(\lambda - z - A)^{-1}d\lambda$$

since both h and $R(\lambda, z + A)$ are holomorphic in the area between Γ and $\Gamma'_\delta + z$. By means of the transformation $\lambda \mapsto \lambda + z$ we end up with

$$h(z + A) = \int_{\Gamma'_\delta + z} h(\lambda)(\lambda - z - A)^{-1}d\lambda = \int_{\Gamma'_\delta} h(\lambda + z)(\lambda - A)^{-1}d\lambda$$

$$= \int_{\Gamma'_\delta} h_z(\lambda)(\lambda - A)^{-1}d\lambda = h_z(A),$$

where $h_z(\mu) := h(\mu + z)$ fulfills $h_z \in \mathcal{E}(\Sigma_\sigma)$ for all $z \in \Sigma_\vartheta$. By Remark 4.13 this proves the claim.

Now let $\phi \in (\zeta + \phi_A^{\mathcal{R}\infty}, \pi)$, $\sigma > \phi$, and $h \in \mathcal{H}_0^\infty(\Sigma_\sigma)$ such that $|h|_\infty^\sigma \leq 1$. Pick $\psi \in (\phi, \sigma)$ and set

$$(6.12) \qquad \Gamma := (\infty, 0]e^{i\psi} \cup [0, \infty)e^{-i\psi}.$$

Then $\lambda \in \rho(zA)$ and $\frac{\lambda}{z} \in \rho(A)$ for all $\lambda \in \Gamma$ and $z \in \Sigma_\zeta$ by our choice of ψ and ζ. Pick $\psi' \in (\phi_A^{\mathcal{R}\infty}, \sigma - \zeta)$ and set

$$\Gamma' := (\infty, 0]e^{i\psi'} \cup [0, \infty)e^{-i\psi'}.$$

For arbitrary $z \in \Sigma_\zeta$ set

$$\Gamma'_z := (\infty, 0]e^{i(\psi' + \arg(z))} \cup [0, \infty)e^{i(-\psi' + \arg(z))}.$$

Then

$$h(zA) = \int_\Gamma h(\lambda)(\lambda - zA)^{-1}d\lambda = \int_{\Gamma'_z} h(\lambda)(\lambda - zA)^{-1}d\lambda$$

since both h and $R(\lambda, zA)$ are holomorphic in the area between Γ and Γ'_z. By means of the transformation $\lambda \mapsto \lambda z$ we end up with

$$h(zA) = \int_{\Gamma'_z} h(\lambda)(\lambda - zA)^{-1}d\lambda = \int_{\Gamma'} h(\lambda z)(\lambda - A)^{-1}d\lambda$$
$$= \int_{\Gamma'} h_z(\lambda)(\lambda - A)^{-1}d\lambda = h_z(A),$$

where $h_z(\mu) := h(\mu z)$ fulfills $h_z \in \mathcal{E}(\Sigma_\sigma)$ for all $z \in \Sigma_\zeta$. In virtue of Remark 4.13 the claim follows. $\qquad\square$

Lemma 6.18. *Let $A \in \Psi\mathcal{R}\mathcal{H}^\infty(E)$, ϑ, ζ, and σ as in Lemma 6.17. Let Γ_1 and Γ_2 be the paths defined in (6.11) and (6.12), respectively, and let $h \in \mathcal{H}_0^\infty(\Sigma_\sigma)$ with $|h|_\infty^\sigma \leq 1$ be arbitrary. Set*

$$H_1(z) := \int_{\Gamma_1} h(\lambda)(\lambda - z - A)^{-1}d\lambda \quad and \quad H_2(z) := \int_{\Gamma_2} h(\lambda)(\lambda - zA)^{-1}d\lambda.$$

Then $H_1 \in \mathcal{H}^\infty(\Sigma_\vartheta, \mathcal{L}(E))$ and $H_2 \in \mathcal{H}^\infty(\Sigma_\zeta, \mathcal{L}(E))$.

Proof. Let Γ_0 denote any closed curve in Σ_ϑ. Then

$$\int_{\Gamma_0} H_1(z)dz = \int_{\Gamma_0} \int_\Gamma h(\lambda)(\lambda - z - A)^{-1}d\lambda dz = \int_\Gamma h(\lambda) \int_{\Gamma_0} (\lambda - z - A)^{-1}dz d\lambda = 0$$

since $z \mapsto R(\lambda, z + A)$ is holomorphic in Σ_ϑ. Here Fubini's theorem can be applied for we have

$$\int_\Gamma \left\| h(\lambda)(\lambda - z - A)^{-1} \right\|_{\mathcal{L}(E)} d\lambda < \infty \quad (z \in \Gamma_0).$$

Due to Morera's theorem H_1 is holomorphic in Σ_ϑ. The assertion on H_2 is proved similarly. $\qquad\square$

As a final ingredient, we need Cauchy's integral formula in the following extended version (cf. [KW04, Remark 9.3]).

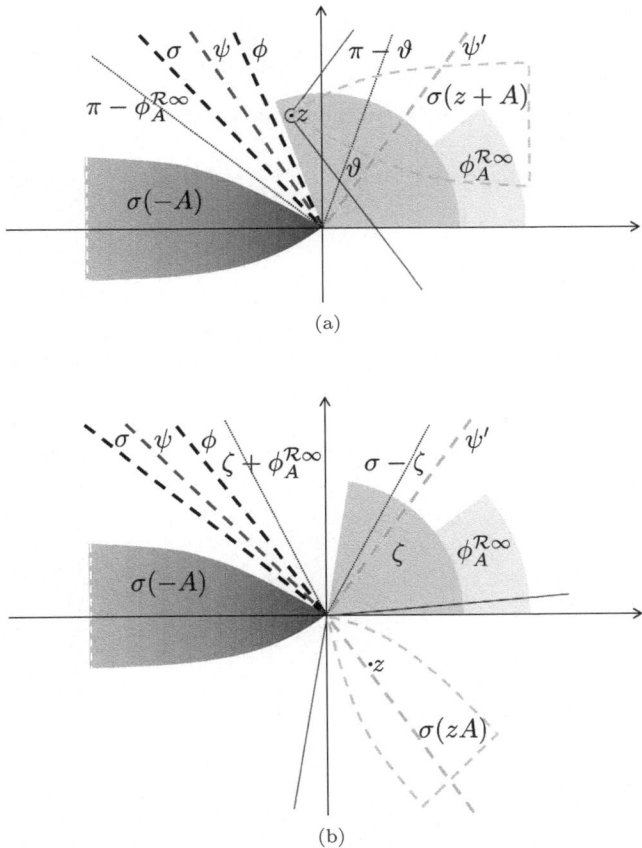

(a)

(b)

Figure 6.2: Different paths of integration. The path defined by ψ is deformed to a path defined by means of ψ' without any change of the integral. In Figure (a) a possible deformation is illustrated for the case that $(z + A)$ is considered. Figure (b) shows a possible deformation for the case that (zA) is considered.

Lemma 6.19. *Let $\vartheta' < \psi' < \vartheta < \pi$, let $H \in \mathcal{H}^{\infty}(\Sigma_{\vartheta}, \mathcal{L}(E))$, and $k \in \mathbb{N}$. Set*

$$\Gamma_{\psi'} := (\infty, 0]e^{i\psi'} \cup [0, \infty)e^{-i\psi'}.$$

Then

$$z^k H^{(k)}(z) = \frac{k!}{2\pi i} \int_{\Gamma_{\psi'}} \frac{z^k}{(\mu - z)^{k+1}} m_h(\mu) d\mu \quad (z \in \Sigma_{\vartheta'}).$$

Proof. Let $z \in \Sigma_{\vartheta'}$ be arbitrary. Pick $0 < r < |z| < R$ and set

$$\Gamma_{\psi',r,R} := [R,r]e^{i\psi'} \cup re^{i(\psi',-\psi')} \cup [r,R]e^{-i\psi'} \cup Re^{i(-\psi',\psi')}$$

(see Figure 6.3). Then by Cauchy's formula for closed rectifiable curves we have

$$H(z) = \frac{1}{2\pi i} \int_{\Gamma_{\psi',r,R}} \frac{1}{\mu - z} H(\mu)d\mu$$

and

$$z^k H^{(k)}(z) = \frac{k!}{2\pi i} \int_{\Gamma_{\psi',r,R}} \frac{z^k}{(\mu - z)^{k+1}} H(\mu)d\mu.$$

Since $g_z(\mu) := \frac{z^k}{(\mu-z)^{k+1}}$ fulfills $g_z \in L^1(\Gamma_{\psi'})$, for each $z \in \Sigma_{\vartheta'}$ the integral on the right-hand side exists. Recall boundedness of H on Σ_ϑ.

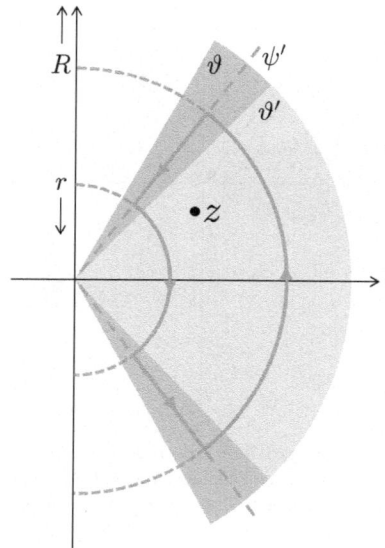

Figure 6.3: The path of integration in the extended Cauchy integral formula.

We estimate the integrals over the two arcs in the representation formula for $z^k H^{(k)}(z)$ and find

$$\left\| \int_{re^{i(\psi',-\psi')}} \frac{z^k}{(\mu-z)^{k+1}} H(\mu)d\mu \right\|_{\mathcal{L}(E)} \leq C \int_{re^{i(\psi',-\psi')}} \frac{|z|^k}{|\mu - z|^{k+1}} d\mu$$

$$\leq C \int_{re^{i(\psi',-\psi')}} \frac{|z|^k}{(r - |z|)^{k+1}} d\mu \leq C \frac{r|z|^k}{(r - |z|)^{k+1}} d\mu \to 0 \quad (r \to 0)$$

as well as

$$\left\| \int_{Re^{i(-\psi',\psi')}} \frac{z^k}{(\mu-z)^{k+1}} H(\mu)d\mu \right\|_{\mathcal{L}(E)} \leq C \int_{Re^{i(-\psi',\psi')}} \frac{|z|^k}{|\mu-z|^{k+1}} d\mu$$

$$\leq C \int_{Re^{i(-\psi',\psi')}} \frac{|z|^k}{(R-|z|)^{k+1}} d\mu \leq C \frac{R|z|^k}{(R-|z|)^{k+1}} d\mu \to 0 \quad (R \to \infty).$$

This yields

$$\lim_{n\to\infty} \int_{\Gamma_{\psi',\frac{1}{n},n}} \frac{z^k}{(\mu-z)^{k+1}} H(\mu)d\mu$$

$$= \lim_{n\to\infty} \int_{\Gamma_{\psi',\frac{1}{n},n}} \frac{z^k}{(\mu-z)^{k+1}} H(\mu)d\mu - \int_{ne^{i(\psi',-\psi')} \cup \frac{1}{n}e^{i(-\psi',\psi')}} \frac{z^k}{(\mu-z)^{k+1}} H(\mu)d\mu$$

$$= \lim_{n\to\infty} \int_{[n,\frac{1}{n}]e^{i\psi'} \cup [n^{-1},n]e^{-i\psi'}} \frac{z^k}{(\mu-z)^{k+1}} H(\mu)d\mu = \int_{\Gamma_{\psi'}} \frac{z^k}{(\mu-z)^{k+1}} H(\mu)d\mu$$

which proves the claim. $\qquad\qquad\square$

With these results at hand, we return to the the investigation of the operator

$$\mathcal{A}(D) := P(D) + Q(D)A := \sum_{|\alpha|\leq m_1} p_\alpha D^\alpha + \sum_{|\alpha|\leq m_2} q_\alpha D^\alpha A.$$

Proposition 6.20. *Let $1 < p < \infty$, let E be a Banach space of class $\mathcal{HT}$ enjoying property (α), and let $A \in \Psi\mathcal{R}\mathcal{H}^\infty(E)$. For homogeneous polynomials P and Q assume that*

(i) P is parameter-elliptic with angle $\varphi_P \in [0,\pi)$,

(ii) Q is parameter-elliptic with angle $\varphi_Q \in [0,\pi)$,

(iii) $\varphi_P + \varphi_Q + \phi_A^{\mathcal{R}\infty} < \pi$.

Then $\mathbb{A} \in \mathcal{R}\mathcal{H}^\infty(L^p(\mathbb{R}^n, E))$ and $\phi_\mathbb{A}^{\mathcal{R}\infty} \leq \max\{\varphi_P, \varphi_Q + \phi_A^{\mathcal{R}\infty}\}$.

Proof. Let $\phi \in (\max\{\varphi_P, \varphi_Q + \phi_A^{\mathcal{R}\infty}\}, \pi - \min\{\varphi_P, \varphi_Q + \phi_A^{\mathcal{R}\infty}\})$ and $\sigma > \phi$. Define $\Gamma := (\infty,0]e^{i\psi} \cup [0,\infty)e^{-i\psi}$ where $\psi \in (\phi, \min\{\sigma, \pi - \min\{\varphi_P, \varphi_Q + \phi_A^{\mathcal{R}\infty}\}\})$ and consider an arbitrary $h \in \mathcal{H}_0^\infty(\Sigma_\sigma)$ such that $|h|_\infty^\sigma \leq 1$.

We formally apply the Fourier transform to the Dunford integral representation of $h(\mathbb{A})$ to the result

$$h(\mathbb{A}) = \int_\Gamma h(\lambda)(\lambda - \mathbb{A})^{-1}d\lambda$$

$$= \mathcal{F}^{-1} \int_\Gamma h(\lambda)(\lambda - P(\cdot) - Q(\cdot)A)^{-1}d\lambda\mathcal{F} = \mathcal{F}^{-1}(m_h \circ (P,Q))\mathcal{F}.$$

Here

$$m_h(z_1, z_2) := \int_\Gamma h(\lambda)(\lambda - z_1 - z_2 A)^{-1} d\lambda \quad ((z_1, z_2) \in \Sigma_\vartheta \times \Sigma_\zeta),$$

where $\vartheta \geq \varphi_P$, $\zeta \geq \varphi_Q$ such that $\vartheta + \zeta + \phi_A^{\mathcal{R}\infty} < \pi$. Thus, Lemma 6.17 applies to the result

$$\mathcal{R}(\{m_h(z_1, z_2); \ |h|_\infty^\sigma \leq 1, \ (z_1, z_2) \in \Sigma_\vartheta \times \Sigma_\zeta\}) < \infty.$$

Now choose $\psi_p \in (\varphi_P, \vartheta)$ and $\psi_q \in (\varphi_Q, \zeta)$ to define

$$\Gamma^p := (\infty, 0]e^{i\psi_p} \cup [0, \infty)e^{-i\psi_p} \quad \text{and} \quad \Gamma^q := (\infty, 0]e^{i\psi_q} \cup [0, \infty)e^{-i\psi_q}.$$

Due to Lemma 6.18 m_h is holomorphic in each variable separately. Hence, we can apply Cauchy's integral formula for polydiscs (see e.g. [Ran86, Theorem 1.3]) which again can be extended along the lines of Lemma 6.19. For $\alpha \in \mathbb{N}_0^2$ this yields

$$z^\alpha \partial^\alpha m_h(z_1, z_2) = \frac{|\alpha|!}{(2\pi i)^2} \int_{\Gamma^p} \int_{\Gamma^q} \frac{z^\alpha}{(\mu_1 - z_1)^{\alpha_1+1}(\mu_2 - z_2)^{\alpha_2+1}} m_h(\mu_1, \mu_2) d\mu_2 d\mu_1$$

$$= \frac{|\alpha|!}{(2\pi i)^2} \int_{\Gamma^p} \frac{z_1^{\alpha_1}}{(\mu_1 - z_1)^{\alpha_1+1}} \int_{\Gamma^q} \frac{z_2^{\alpha_2}}{(\mu_2 - z_2)^{\alpha_2+1}} m_h(\mu_1, \mu_2) d\mu_2 d\mu_1.$$

For $\vartheta' \in (\varphi_P, \psi_p)$ and $\zeta' \in (\varphi_Q, \psi_q)$ we have

$$\sup_{z_1 \in \Sigma_{\vartheta'}} \|g_{z_1}\|_{L^1(\Gamma^p)} \leq C \quad \text{and} \quad \sup_{z_2 \in \Sigma_{\zeta'}} \|g_{z_2}\|_{L^1(\Gamma^q)} \leq C,$$

where we have set

$$g_{z_1}(\mu_1) := \frac{z_1^{\alpha_1}}{(\mu_1 - z_1)^{\alpha_1+1}} \quad \text{and} \quad g_{z_2}(\mu_2) := \frac{z_2^{\alpha_2}}{(\mu_2 - z_2)^{\alpha_2+1}}.$$

Thus, Lemma 3.3 applies in iteration to the result

$$\mathcal{R}(\{z^\alpha \partial^\alpha m_h(z_1, z_2); \ |h|_\infty^\sigma \leq 1, \ (z_1, z_2) \in \Sigma_\vartheta \times \Sigma_\zeta\}) < \infty.$$

In particular,

$$\mathcal{R}(\{P(\xi)^{\alpha_1} Q(\xi)^{\alpha_2}(\partial^\alpha m_h)(P(\xi), Q(\xi)); \ |h|_\infty^\sigma \leq 1, \xi \in \mathbb{R}^n \setminus \{0\}\}) < \infty$$

follows due to parameter-ellipticity of P and Q. Employing Lemma 6.1 we see that the weighted derivatives

$$\xi^\gamma D^\gamma (m_h \circ (P, Q))(\xi)$$

for $0 \leq \gamma \leq 1$ are the sum of terms of the form

$$\xi^\gamma (\partial^\alpha m_h)(P(\xi), Q(\xi)) \cdot (D^{\kappa_1} P)(\xi) \cdots (D^{\kappa_r} P)(\xi) \cdot (D^{\iota_1} Q)(\xi) \cdots (D^{\iota_s} Q)(\xi),$$

where $\alpha_1 + \alpha_2 \leq |\gamma|$, $r = \alpha_1$ and $s = \alpha_2$. For $\beta^{(1)} := \sum_{i=1}^r \kappa_i$ and $\beta^{(2)} := \sum_{i=1}^s \iota_i$, it holds that $\beta^{(1)} + \beta^{(2)} = \gamma$. By parameter-ellipticity of P and Q we thus can estimate

$$|\xi^{\beta^{(1)}}(D^{\kappa_1}P)(\xi)\cdots(D^{\kappa_r}P)(\xi)| \leq C|P(\xi)^r| = C|P(\xi)^{\alpha_1}|$$

and

$$|\xi^{\beta^{(2)}}(D^{\iota_1}Q)(\xi)\cdots(D^{\iota_s}Q)(\xi)| \leq C|Q(\xi)^s| = C|Q(\xi)^{\alpha_2}|.$$

Hence, the contraction principle of Kahane yields

$$\mathcal{R}(\{\xi^\gamma D^\gamma(m_h \circ (P,Q))(\xi); \ |h|_\infty^\sigma \leq 1, \ \xi \in \mathbb{R}^n \setminus \{0\}, \ 0 \leq \gamma \leq 1\}) < \infty$$

which shows $m_h \circ (P,Q)$ to be a Fourier multiplier. Thus, $\mathbb{A} \in \Psi\mathcal{RH}^\infty(L^p(\mathbb{R}^n, E))$ with $\phi_\mathbb{A}^{\mathcal{R}\infty} \leq \max\{\varphi_P, \varphi_Q + \phi_A^{\mathcal{R}\infty}\}$. Because of $\mathbb{A} \in \mathcal{R}S(L^p(\mathbb{R}^n, E))$ by Proposition 6.13 we have $\mathbb{A} \in \mathcal{RH}^\infty(L^p(\mathbb{R}^n, E))$. $\qquad\square$

Proposition 6.21. *Given the assumptions of Proposition 6.20, let $\delta > 0$ and let $\mathcal{A}_\delta(D)$ be even. Then the assertion of Proposition 6.20 carries over to $\mathbb{A}_{\delta,\mathcal{B}}$.*

Proof. Observe that Proposition 6.20 also applies to $\mathbb{A}_\delta$ for $\delta \geq 0$. Hence, the assertion follows immediately from $(\lambda + \mathbb{A}_{\delta,\mathcal{B}})^{-1} = \mathfrak{R}(\lambda + \mathbb{A}_\delta)^{-1}\mathfrak{E}$. Recall that $\mathfrak{E}$ and $\mathfrak{R}$ define bounded operators. Hence, they commute with the integral sign in the Dunford integral representation. $\qquad\square$

Remark 6.22. Note that related results on sectoriality and a bounded $\mathcal{H}^\infty$-calculus of $\mathbb{A}_\delta$ can still be deduced if E does not enjoy property (α).

7 Fourier series approach to operator-dependent problems

This chapter can be seen as the counterpart of the previous chapter. This time operator-dependent partial differential equations in the cube $(0, 2\pi)^n$ are treated. Here generalized periodic and mixed Dirichlet-Neumann boundary conditions are imposed. This chapter further completes the abstract background for the upcoming treatment of cylindrical boundary value problems.

7.1 Preliminaries

In view of Theorems 3.19 and 3.24, respectively in view of our intermediate condition (3.12), the following lemma provides useful formulas for discrete derivatives. It can be seen as an analogue to Lemma 6.1. In the discrete case, however, the proof is more tedious since one has to keep track of each shift that comes from one of the single derivations. To do so, we will find it convenient to define

$$\omega_*^j := \sum_{l=j+1}^{r} \omega^l$$

for $\mathcal{W} \in \mathcal{Z}_\alpha$, $r = r_\mathcal{W}$, and $\mathcal{W} = (\omega^1, \ldots, \omega^r)$. Here $\mathcal{W}$ and $\mathcal{Z}_\alpha$ are defined as in (6.1). Also recall the discrete differential operator Δ^α from (3.5).

Lemma 7.1. a) [Leibniz rule] *Given* $m \in \mathbb{N}$, *let* $(T_k)_{k \in \mathbb{Z}^n}$ *denote a sequence of operators* $X \to Y$ *and* $(S_k)_{k \in \mathbb{Z}^n}$ *a sequence of operators* $Y \to Z$, *such that the composition* $S_{k^{(1)}} T_{k^{(2)}} : X \to Z$ *is well-defined for all* $k^{(1)}, k^{(2)} \in \mathbb{Z}^n$. *For* $\alpha \in \mathbb{N}_0^n$ *such that* $|\alpha| \le m$ *we then have*

$$\Delta^\alpha (ST)_k = \sum_{\beta \le \alpha} \binom{\alpha}{\beta} (\Delta^{\alpha-\beta} S)_{k-\beta} (\Delta^\beta T)_k \quad (k \in \mathbb{Z}^n).$$

b) *Let* $((S_j)_k)_{k \in \mathbb{Z}^n}$ *for* $j = 1, \ldots, r$ *denote sequences of operators* $X_{j-1} \to X_j$ *such that* $\prod_{j=1}^{r} (S_j)_{k^{(j)}} : X_0 \to X_r$ *is well-defined for all* $k^{(1)}, \ldots, k^{(r)} \in \mathbb{Z}^n$. *Then we have*

$$\Delta^{e_i} \Big(\prod_{j=1}^{r} S_j \Big)_k = \sum_{l=1}^{r} \Big(\prod_{j=1}^{l-1} S_j \Big)_{k-e_i} (\Delta^{e_i} S_l)_k \Big(\prod_{j=l+1}^{r} S_j \Big)_k.$$

c) *Given $m \in \mathbb{N}$, let $(S_k)_{k \in \mathbb{Z}^n}$ denote a sequence of operators $X \to Y$ such that the inverse $(S^{-1})_k := (S_k)^{-1}$ exists for all $k \in \mathbb{Z}^n$. Then, for $\alpha \in \mathbb{N}_0^n$ such that $|\alpha| \leq m$ we have*

$$\Delta^\alpha (S^{-1})_k = \sum_{W \in \mathcal{Z}_\alpha} (-1)^{r_W} (S^{-1})_{k-\alpha} \prod_{j=1}^{r_W} \left((\Delta^{\omega^j} S) S^{-1} \right)_{k-\omega_*^j} \quad (k \in \mathbb{Z}^n).$$

Proof. a) We proof the assertion by induction on n. First let $n = 1$, i.e. $\alpha = a \in \mathbb{N}$, the case $a = 0$ being obvious. By definition for $\Delta = \Delta^1$ we have

$$\Delta(ST)_k = (ST)_k - (ST)_{k-1} = S_{k-1}(\Delta T)_k + (\Delta S)_k T_k$$

and by induction on a we obtain

$$\Delta^a(ST)_k = \Delta \sum_{b \leq a-1} \binom{a-1}{b} (\Delta^{a-1-b} S)_{k-b}(\Delta^b T)_k$$

$$= \sum_{b \leq a-1} \binom{a-1}{b} \left((\Delta^{a-1-b} S)_{k-b-1}(\Delta^{b+1} T)_k + (\Delta^{a-b} S)_{k-b}(\Delta^b T)_k \right).$$

We split the sum to the result

$$\sum_{b \leq a-1} \binom{a-1}{b} (\Delta^{a-1-b} S)_{k-b-1}(\Delta^{b+1} T)_k + \sum_{b \leq a-1} \binom{a-1}{b} (\Delta^{a-b} S)_{k-b}(\Delta^b T)_k.$$

The first sum equals

$$\sum_{0 < b \leq a} \binom{a-1}{b-1} (\Delta^{a-b} S)_{k-b}(\Delta^b T)_k$$

$$= \sum_{0 < b \leq a-1} \binom{a-1}{b-1} (\Delta^{a-b} S)_{k-b}(\Delta^b T)_k + (S)_{k-a}(\Delta^a T)_k$$

while the second sum equals

$$\sum_{0 < b \leq a-1} \binom{a-1}{b} (\Delta^{a-b} S)_{k-b}(\Delta^b T)_k + (\Delta^a S)_k(T)_k.$$

Since

$$\binom{a-1}{b} + \binom{a-1}{b-1} = \binom{a}{b},$$

summing up all terms yields

$$\Delta^a(ST)_k = \sum_{b \leq a} \binom{a}{b} (\Delta^{a-b} S)_{k-b}(\Delta^b T)_k.$$

Now let $n \in \mathbb{N}$ be arbitrary and $\alpha = (a, \alpha') \in \mathbb{N}_0 \times \mathbb{N}_0^{n-1}$. Then

$$
\begin{aligned}
\Delta^\alpha (ST)_k &= \Delta^{(a,0)} \Delta^{(0,\alpha')} (ST)_k \\
&= \Delta^{(a,0)} \sum_{\beta' \leq \alpha'} \binom{\alpha'}{\beta'} (\Delta^{(0,\alpha'-\beta')} S)_{k-(0,\beta')} (\Delta^{(0,\beta')} T)_k \\
&= \sum_{\beta' \leq \alpha'} \sum_{b \leq a} \binom{\alpha'}{\beta'} \binom{a}{b} (\Delta^{(a-b,\alpha'-\beta')} S)_{k-(b,\beta')} (\Delta^{(b,\beta')} T)_k \\
&= \sum_{\beta \leq \alpha} \binom{\alpha}{\beta} (\Delta^{\alpha-\beta} S)_{k-\beta} (\Delta^\beta T)_k .
\end{aligned}
$$

b) We prove the assertion by induction on r, the case $r = 1$ being obvious, the case $r = 2$ being included in a). Let $r \in \mathbb{N}$ be arbitrary. Then

$$
\begin{aligned}
\Delta^{e_i} \Big(\prod_{j=1}^r S_j \Big)_k &= (\Delta^{e_i} S_1)_k \Big(\prod_{j=2}^r S_j \Big)_k + (S_1)_{k-e_i} \Delta^{e_i} \Big(\prod_{j=2}^r S_j \Big)_k \\
&= (\Delta^{e_i} S_1)_k \Big(\prod_{j=2}^r S_j \Big)_k + (S_1)_{k-e_i} \sum_{l=2}^r \Big(\prod_{j=2}^{l-1} S_j \Big)_{k-e_i} (\Delta^{e_i} S_l)_k \Big(\prod_{j=l+1}^r S_j \Big)_k \\
&= \sum_{l=1}^r \Big(\prod_{j=1}^{l-1} S_j \Big)_{k-e_i} (\Delta^{e_i} S_l)_k \Big(\prod_{j=l+1}^r S_j \Big)_k .
\end{aligned}
$$

c) We prove the assertion by induction on $|\alpha|$, the case $\alpha = 0$ being obvious due to the definition of $\mathcal{Z}_0$ and $r_\emptyset$. For $\alpha = e_i$ we easily compute

$$
\begin{aligned}
\Delta^{e_i} (S^{-1})_k &= (S^{-1})_k - (S^{-1})_{k-e_i} \\
&= -(S^{-1})_{k-e_i} (S_k - S_{k-e_i})(S^{-1})_k = -(S^{-1})_{k-e_i} (\Delta^{e_i} S)_k (S^{-1})_k .
\end{aligned}
$$

Now let $\alpha \in \mathbb{N}_0^n \setminus \{0\}$ and $i \in \{1, \ldots, n\}$ be arbitrary such that $\alpha_i \neq 0$ and let $\alpha' := \alpha - e_i$. Then

$$
\begin{aligned}
\Delta^\alpha (S^{-1})_k &= \Delta^{e_i} \Delta^{\alpha'} (S^{-1})_k \\
&= \sum_{w \in \mathcal{Z}_{\alpha'}} (-1)^{r_w} \Delta^{e_i} \Big((S^{-1})_{k-\alpha'} \prod_{j=1}^{r_w} ((\Delta^{\omega^j} S) S^{-1})_{k-\omega_*^j} \Big) .
\end{aligned}
$$

By means of b) we get

$$\sum_{w \in \mathcal{Z}_{\alpha'}} (-1)^{r_W} \Delta^{e_i} \left((S^{-1})_{k-\alpha'} \prod_{j=1}^{r_W} ((\Delta^{\omega^j} S) S^{-1})_{k-\omega_*^j} \right)$$

$$= \sum_{w \in \mathcal{Z}_{\alpha'}} (-1)^{r_W} \left((\Delta^{e_i} S^{-1})_{k-\alpha'} \prod_{j=1}^{r_W} ((\Delta^{\omega^j} S) S^{-1})_{k-\omega_*^j} \quad + \quad (S^{-1})_{k-\alpha'-e_i} \right.$$

$$\left. \cdot \sum_{l=1}^{r_W} (\prod_{j=1}^{l-1} (\Delta^{\omega^j} S) S^{-1})_{k-\omega_*^j-e_i} (\Delta^{e_i} ((\Delta^{\omega^l} S) S^{-1}))_{k-\omega_*^l} (\prod_{j=l+1}^{r_W} (\Delta^{\omega^j} S) S^{-1})_{k-\omega_*^j} \right).$$

Since

$$\Delta^{e_i} (S^{-1})_{k-\alpha'} = -(S^{-1})_{k-\alpha} (\Delta^{e_i} S)_{k-\alpha'} (S^{-1})_{k-\alpha'}$$

as well as

$$\left(\Delta^{e_i} ((\Delta^{\omega^l} S) S^{-1}) \right)_{k-\omega_*^l}$$
$$= (\Delta^{\omega^l} S)_{k-\omega_*^l-e_i} (\Delta^{e_i} (S^{-1}))_{k-\omega_*^l} + (\Delta^{\omega^l+e_i} S)_{k-\omega_*^l} (S^{-1})_{k-\omega_*^l}$$
$$= -(\Delta^{\omega^l} S)_{k-\omega_*^l-e_i} (S^{-1})_{k-\omega_*^l-e_i} (\Delta^{e_i} S)_{k-\omega_*^l} (S^{-1})_{k-\omega_*^l}$$
$$+ (\Delta^{\omega^l+e_i} S)_{k-\omega_*^l} (S^{-1})_{k-\omega_*^l},$$

we infer that $\Delta^{\alpha}(S^{-1})_k$ is given as the sum of the three terms

$$\sum_{w \in \mathcal{Z}_{\alpha'}} (-1)^{r_W} (-1)(S^{-1})_{k-\alpha} ((\Delta^{e_i} S) S^{-1})_{k-\alpha'} \prod_{j=1}^{r_W} ((\Delta^{\omega^j} S) S^{-1})_{k-\omega_*^j},$$

$$\sum_{w \in \mathcal{Z}_{\alpha'}} (-1)^{r_W} (S^{-1})_{k-\alpha'-e_i} \sum_{l=1}^{r_W} (\prod_{j=1}^{l-1} (\Delta^{\omega^j} S) S^{-1})_{k-\omega_*^j-e_i}$$

$$\cdot (-1) ((\Delta^{\omega^l} S) S^{-1})_{k-\omega_*^l-e_i} ((\Delta^{e_i} S) S^{-1})_{k-\omega_*^l} (\prod_{j=l+1}^{r_W} (\Delta^{\omega^j} S) S^{-1})_{k-\omega_*^j},$$

and

$$\sum_{w \in \mathcal{Z}_{\alpha'}} (-1)^{r_W} (S^{-1})_{k-\alpha'-e_i} \sum_{l=1}^{r_W} (\prod_{j=1}^{l-1} (\Delta^{\omega^j} S) S^{-1})_{k-\omega_*^j-e_i}$$

$$\cdot ((\Delta^{\omega^l+e_i} S) S^{-1})_{k-\omega_*^l} (\prod_{j=l+1}^{r_W} (\Delta^{\omega^j} S) S^{-1})_{k-\omega_*^j}.$$

Summing up yields the desired representation formula. $\square$

Of course, a) and b) can be presented simultaneously in terms of a Leibniz rule for finitely many factors in n dimensions for derivatives of arbitrary order. This is done in the following remark. However, for the applications we have in mind the formulas given in a) and c) are sufficient.

Remark 7.2. For $\alpha \in \mathbb{N}_0^n$ and $r \in \mathbb{N}_0$ let

$$\mathcal{X}_\alpha^r := \left\{ \mathcal{W} = (\omega^1, \ldots, \omega^r); \ 0 \le \omega^j \le \alpha, \ \sum_{j=1}^r \omega^j = \alpha \right\}$$

denote the set of all additive decompositions of α into r multi-indices. Note that contrary to the definition of $\mathcal{Z}_\alpha$, $\mathcal{W} \in \mathcal{X}_\alpha^r$ has a fixed number of components which may as well be given as zeros. Then, more generally than stated in Lemma 7.1, it holds that

$$\Delta^\alpha \Big(\prod_{j=1}^r S_j\Big)_k = \sum_{\mathcal{W} \in \mathcal{X}_\alpha^r} \prod_{j=1}^r \binom{w_*^{j-1}}{w^j} (\Delta^{w^j} S_j)_{k-w_*^j}.$$

Note that

$$\prod_{j=1}^r \binom{w_*^{j-1}}{w^j} = \alpha! \Big(\prod_{j=1}^r (w^j!)\Big)^{-1}$$

and therefore

$$\Delta^\alpha \Big(\prod_{j=1}^r S_j\Big)_k = \alpha! \sum_{\mathcal{W} \in \mathcal{X}_\alpha^r} \prod_{j=1}^r (w^j!)^{-1} (\Delta^{w^j} S_j)_{k-w_*^j}.$$

Proof. Again we prove the assertion by induction on r, the case $r = 1$ being obvious, the case $r = 2$ being included in Lemma 7.1a). Let $r \in \mathbb{N}$ be arbitrary. Then

$$\Delta^\alpha \Big(\prod_{j=1}^r S_j\Big)_k = \sum_{\beta \le \alpha} \binom{\alpha}{\beta} (\Delta^{\alpha-\beta} S_1)_{k-\beta} (\Delta^\beta (S_2 \cdots S_r))_k$$

$$= \sum_{\beta \le \alpha} \binom{\alpha}{\beta} (\Delta^{\alpha-\beta} S_1)_{k-\beta} \sum_{\mathcal{W} \in \mathcal{X}_\beta^{r-1}} \prod_{j=2}^r \binom{w_*^{j-2}}{w^{j-1}} (\Delta^{w^{j-1}} S_j)_{k-w_*^{j-1}}$$

$$= \sum_{\beta \le \alpha} \sum_{\mathcal{W} \in \mathcal{X}_\beta^{r-1}} \binom{\alpha}{\beta} (\Delta^{\alpha-\beta} S_1)_{k-\beta} \prod_{j=2}^r \binom{w_*^{j-2}}{w^{j-1}} (\Delta^{w^{j-1}} S_j)_{k-w_*^{j-1}}$$

$$= \sum_{\mathcal{W} \in \mathcal{X}_\alpha^r} \prod_{j=1}^r \binom{w_*^{j-1}}{w^j} (\Delta^{w^j} S_j)_{k-w_*^j}.$$

$\square$

Definition 7.3. We call a discrete Fourier multiplier a *discrete multiplier of resolvent type* if it arises from a continuous multiplier of resolvent type by discretization.

The notion of (parameter-) ellipticity is transferred to discrete polynomials by means of discretization, too.

Definition 7.4. We call a discrete polynomial $P\colon \mathbb{Z}^n \to \mathbb{C};\; k \mapsto P(k)$ *elliptic* (respectively *parameter-elliptic*) if it is given as restriction $P = \tilde{P}|_{\mathbb{Z}^n}$ of a (parameter-)elliptic polynomial $\tilde{P}\colon \mathbb{R}^n \to \mathbb{C}$.

As an immediate consequence of Lemma 6.5 we have the following implications.

Lemma 7.5.   a) *For all polynomials N with $\deg N \leq \deg P$ there exist $C > 0$ and a finite subset $G \subset \mathbb{Z}^n$ such that the estimate $|k|^m |N(k)| \leq C|\lambda + P(k)|$ holds for all $\lambda \in \overline{\Sigma}_{\pi-\phi}$, all $0 \leq m \leq \deg P - \deg N$ and all $k \in \mathbb{Z}^n \setminus G$ if P is parameter-elliptic in $\overline{\Sigma}_{\pi-\phi}$.*
b) *The assertion in a) is valid for $\lambda = 0$ if P is elliptic.*

Remark 7.6. By induction one can see that for $|\alpha| \leq \deg P$ the discrete polynomial $\Delta^\alpha P(k)$ defines a polynomial of degree not greater than $\deg P - |\alpha|$. If P is elliptic, this implies $|k|^{|\alpha|}|\Delta^\alpha P(k)| \leq C|P(k)|$ for all $k \in \mathbb{Z}^n \setminus G$ with a finite set $G \subset \mathbb{Z}^n$.

Remark 7.7. Note that the condition $P^\#(k) \neq 0$ for $k \in \mathbb{Z}^n \setminus \{0\}$ is no appropriate definition of ellipticity of a discrete polynomial P. The reason lies in the fact that the important assertions of Lemma 7.5 are no longer valid if this definition is taken as a basis. To see this, consider $P(k) := k_1 - \zeta k_2$, where

$$\zeta = 10 - 9\sum_{j=1}^{\infty} 10^{-\left((j-1)+\frac{j(j+1)}{2}\right)} = 9,0990999099990999990\ldots.$$

Here $P^\#(k) \neq 0$ for $k \in \mathbb{Z}^n \setminus \{0\}$ obviously holds, whereas no $C > 0$ can exist such that $1 \leq C|P(k)|$ is valid for $k \in \mathbb{Z}^n \setminus G$ with a finite set $G \subset \mathbb{Z}^n$. To see this, let $j_0 \in \mathbb{N}$ be arbitrary and set $\varepsilon := 10^{-j_0}$. By definition of ζ for each $N \in \mathbb{N}$ there exists $i_0 \in \mathbb{N}$ such that $k_2 := 10^{i_0} > N$ and

$$\zeta k_2 = M + 9\sum_{j=1}^{j_0} 10^{-j} + \delta,$$

where $M = \lfloor \zeta k_2 \rfloor$ and $0 < \delta < \varepsilon$. Hence, if we pick $k_1 := M + 1$, we obtain the inequality

$$\zeta k_2 < k_1 < \varepsilon + \zeta k_2 \iff 0 < k_1 - \zeta k_2 < \varepsilon.$$

In what follows the assumption that $(\lambda + \mu A)^{-1}$ exists for $\lambda, \mu \in \mathbb{C}$ is meant to imply both $(\lambda + \mu A)^{-1} \in \mathcal{L}(E)$ and $(\lambda + \mu A)^{-1}(E) = D(A)$.

Proposition 7.8. *Let A be a closed operator in a Banach space E of class $\mathcal{HT}$ and let $P, Q\colon \mathbb{Z}^n \to \mathbb{C}$ denote elliptic polynomials. Let N define an arbitrary polynomial* subject to $\deg N \leq \deg P$ and let $\big(P(k) + Q(k)A\big)^{-1} \in \mathcal{L}(E)$ exist for all $k \in \mathbb{Z}^n \setminus \{0\}$. If the set

$$\Big\{ P(k)\big(P(k) + Q(k)A\big)^{-1};\ k \in \mathbb{Z}^n \Big\}$$

is $\mathcal{R}$-bounded, then for each $1 < p < \infty$

$$M \colon \mathbb{Z}^n \to \mathcal{L}(E),\ k \mapsto N(k)\big(P(k) + Q(k)A\big)^{-1}$$

defines an L^p-multiplier.

Proof. Lemma 7.1 yields

$$|k|^{|\gamma|} \Delta^\gamma M(k) =$$

$$\sum_{\beta \leq \gamma} \binom{\gamma}{\beta} \sum_{w \in \mathcal{Z}_\beta} (-1)^{r_w} |k|^{|\gamma - \beta|} (\Delta^{\gamma - \beta} N)(k - \beta)\big(P(k - \beta) + Q(k - \beta)A\big)^{-1}$$

$$\cdot \prod_{j=1}^{r_w} |k|^{|\omega_j|} \big(\Delta^{\omega_j} P(k - \omega_j^*) + \Delta^{\omega_j} Q(k - \omega_j^*)A\big)\big(P(k - \omega_j^*) + Q(k - \omega_j^*)A\big)^{-1}.$$

By Remark 7.6 we know that $\deg(\Delta^{\gamma - \beta} N) \leq \deg N - |\gamma - \beta|$. This and the ellipticity of P imply $|k|^{|\gamma - \beta|} |\Delta^{\gamma - \beta} N(k)| \leq C |P(k)|$ for $k \in \mathbb{Z}^n \setminus G$ with a finite set $G \subset \mathbb{Z}^n$. By Kahane's contraction principle we obtain the $\mathcal{R}$-boundedness of

$$\Big\{ |k|^{|\gamma - \beta|} \Delta^{\gamma - \beta} N(k - \beta)\big(P(k - \beta) + Q(k - \beta)A\big)^{-1};\ k \in \mathbb{Z}^n \setminus G \Big\}.$$

Since

$$Q(k)A\big(P(k) + Q(k)A\big)^{-1} = \mathrm{id}_E - P(k)\big(P(k) + Q(k)A\big)^{-1},$$

in the same way the $\mathcal{R}$-boundedness of

$$\Big\{ |k|^{|\omega_j|} \Delta^{\omega_j} Q(k - \omega_j^*)A\big(P(k - \omega_j^*) + Q(k - \omega_j^*)A\big)^{-1};\ k \in \mathbb{Z}^n \setminus G \Big\}$$

follows from the ellipticity of Q. Now the assertion follows from the fact that G is finite, from Lemma 3.2, and from Theorem 3.19. $\qquad\square$

Proposition 7.8 is closely related to the concept of 1-regularity of complex-valued sequences considered in [KL04] for the one dimensional case $n = 1$. There the notion of 1-regularity as introduced by Prüss in [Prü93] for functions of a complex variable (cf. Definition 5.7) is transferred to sequences $(a_k)_{k \in \mathbb{N}} \subset \mathbb{C} \setminus \{0\}$ by assuming

$$\left(\frac{k(a_{k+1} - a_k)}{a_k} \right)_{k \in \mathbb{Z}}$$

to be bounded. It turns out that 1-regularity is the right concept to investigate multipliers of type $M(k) = a_k (b_k - A)^{-1}$ with $(a_k)_{k \in \mathbb{Z}}, (b_k)_{k \in \mathbb{Z}} \subset \mathbb{C}$. Indeed, by

1-regularity, $\mathcal{R}$-boundedness of discrete derivatives of an operator family can be deduced from $\mathcal{R}$-boundedness of the family itself. This allows for an easy application of the multiplier theorem due to Arendt and Bu. In fact, it is shown in [KLP09, Proposition 5.3] that $\mathcal{R}$-boundedness of $\{a_k(b_k - A)^{-1};\ k \in \mathbb{Z}\}$ implies that M defined as above defines a multiplier for $1 < p < \infty$, where both $(a_k)_{k \in \mathbb{Z}}$ and $(b_k)_{k \in \mathbb{Z}}$ are 1-regular sequences such that $(b_k/a_k)_{k \in \mathbb{Z}}$ is bounded and $(b_k)_{k \in \mathbb{Z}} \subset \rho(A)$. If $Q(k) \neq 0$ for all $k \in \mathbb{Z}^n$, we may write

$$M(k) = \frac{N(k)}{Q(k)} \left(\frac{P(k)}{Q(k)} + A \right)^{-1} \quad (k \in \mathbb{Z}^n).$$

Hence, for $n = 1$ we enter the framework of 1-regularity, i.e. $M(k) = a_k(b_k - A)^{-1}$ with $(a_k)_{k \in \mathbb{Z}}, (b_k)_{k \in \mathbb{Z}} \subset \mathbb{C}$ defined by

$$a_k := \frac{N(k)}{Q(k)} \quad \text{and} \quad b_k := \frac{P(k)}{Q(k)}.$$

In the subsequent lines, we will give a generalization of this concept to arbitrary n and briefly indicate the connection to the results above.

Definition 7.9. We call a pair of sequences $(a_k, b_k)_{k \in \mathbb{Z}^n} \subset \mathbb{C}^2$ *1-regular* if for all $0 \leq \gamma \leq 1$ there exist a finite set $K \subset \mathbb{Z}^n$ and a constant $C > 0$ such that

$$(7.1) \qquad |k^\gamma| \max\{|(\Delta^\gamma a)_k|, |(\Delta^\gamma b)_k|\} \leq C|b_k| \quad (k \in \mathbb{Z}^n \setminus K).$$

We say the pair $(a_k, b_k)_{k \in \mathbb{Z}^n}$ is *strictly 1-regular* if $|k^\gamma|$ can be replaced by $|k|^{|\gamma|}$ in (7.1). A sequence $(a_k)_{k \in \mathbb{Z}^n}$ is called (strictly) 1-regular if $(a_k, a_k)_{k \in \mathbb{Z}^n}$ has this property.

Remark 7.10. a) In the case $n = 1$ a sequence $(a_k)_{k \in \mathbb{Z}} \subset \mathbb{C} \setminus \{0\}$ is 1-regular in $\mathbb{Z}$ in the sense of Definition 7.9 if and only if the sequence $\left(\frac{k(a_{k+1} - a_k)}{a_k} \right)_{k \in \mathbb{Z}}$ is bounded. Hence, our definition extends the one from [KL04] for a sequence $(a_k)_{k \in \mathbb{Z}}$.
b) With $\gamma = 0$ the definition especially requests $|a_k| \leq C|b_k|$ for $k \in \mathbb{Z}^n \setminus K$.
c) Strict 1-regularity implies 1-regularity. If $n = 1$ both concepts are equivalent.

(Strict) 1-regularity by definition allows to apply the contraction principle of Kahane to deduce the sufficient conditions of the multiplier theorem for resolvent type multipliers from $\mathcal{R}$-boundedness of the range of the multiplier itself. It is therefore not surprising that the ellipticity condition used in Proposition 7.8 leads to strictly 1-regular sequences.

Lemma 7.11. *Subject to the assumptions of Proposition 7.8, let $Q(k) \neq 0$ for $k \in \mathbb{Z}^n$. Then the pair $\left(\frac{N(k)}{Q(k)}, \frac{P(k)}{Q(k)} \right)_{k \in \mathbb{Z}^n}$ is strictly 1-regular.*

Proof. We have to show the existence of $C > 0$ and a finite set $K \subset \mathbb{Z}^n$ such that

$$|k|^{|\gamma|} \max \left\{ \left| \Delta^\gamma \frac{N}{Q}(k) \right|, \left| \Delta^\gamma \frac{P}{Q}(k) \right| \right\} \leq C \left| \frac{P}{Q}(k) \right| \quad (k \in \mathbb{Z}^n \setminus K).$$

Due to Lemma 7.1 for any polynomial M we have

$$|k|^{|\gamma|}\Delta^{\gamma}\frac{M}{Q}(k) =$$

$$\sum_{\beta\leq\gamma}\binom{\gamma}{\beta}\sum_{w\in\mathcal{Z}_{\beta}}(-1)^{w}|k|^{|\gamma-\beta|}\frac{\Delta^{\gamma-\beta}M}{Q}(k-\beta)\prod_{j=1}^{r_w}|k|^{|\omega_j|}\frac{\Delta^{\omega_j}Q}{Q}(k-\omega_j^*).$$

Ellipticity of P now implies that there exists a finite set $G \subset \mathbb{Z}^n$ such that $P(k) \neq 0$ for all $k \in \mathbb{Z}^n \setminus G$. Therefore it suffices to show the existence of $C > 0$ and a finite set $K \supset G$ such that

$$|k|^{|\gamma-\beta|}\left|\frac{\Delta^{\gamma-\beta}M}{Q}(k-\beta)\prod_{j=1}^{r_w}|k|^{|\omega_j|}\frac{\Delta^{\omega_j}Q}{Q}(k-\omega_j^*)\frac{Q}{P}(k)\right| \leq C.$$

If $\deg M \leq \deg P$ the assertion follows from ellipticity of

$$Q(\cdot - \beta)\prod_{j=1}^{r_w}Q(\cdot - \omega_j^*)P,$$

which is given since P and Q are assumed to be elliptic.   □

With the aid of Lemma 7.1 we deduce the following more abstract version of Proposition 7.8. Here $(a_k)_{k\in\mathbb{Z}^n}$ and $(b_k)_{k\in\mathbb{Z}^n}$ no longer have to be given as quotients of discrete polynomials.

Proposition 7.12. *Let $(a_k, b_k)_{k\in\mathbb{Z}^n}$ be strictly 1-regular, let $b_k \in \rho(A)$ for all $k \in \mathbb{Z}^n$, and let*

$$\mathcal{R}(\{b_k(b_k - A)^{-1};\ k \in \mathbb{Z}^n \setminus G\}) < \infty$$

for some finite subset $G \subset \mathbb{Z}^n$. Then $M(k) := a_k(b_k - A)^{-1}$ defines a Fourier multiplier.

Proof. By Lemma 7.1 we have

$$|k|^{|\gamma|}\Delta^{\gamma}M(k) = \sum_{\beta\leq\gamma}\binom{\gamma}{\beta}\sum_{w\in\mathcal{Z}_{\beta}}(-1)^{r_w}|k|^{|\gamma-\beta|}(\Delta^{\gamma-\beta}a)_{k-\beta}(b_{k-\beta} - A)^{-1}$$

$$\cdot\prod_{j=1}^{r_w}|k|^{|\omega_j|}(\Delta^{\omega_j}b)_{k-\omega_j^*}(b_{k-\omega_j^*} - A)^{-1}.$$

Now the claim follows at once from the contraction principle of Kahane.   □

7.2 ν-periodic problems in cubical domains

Let E be a Banach space of class $\mathcal{HT}$ and A be a closed operator in E. With $n \in \mathbb{N}$ and $\nu \in \mathbb{C}^n$ we consider the boundary value problem in $\mathcal{Q}_n$ given by (7.2)

$$
\begin{aligned}
\mathcal{A}(D)u &= f & (x \in \mathcal{Q}_n), \\
(D^\beta u)|_{x_j=2\pi} - e^{2\pi\nu_j}(D^\beta u)|_{x_j=0} &= 0 & (j=1,\ldots,n;\ |\beta| < m_1), \\
(D^\beta A u)|_{x_j=2\pi} - e^{2\pi\nu_j}(D^\beta A u)|_{x_j=0} &= 0 & (j=1,\ldots,n;\ |\beta| < m_2).
\end{aligned}
$$

In view of the boundary conditions, we refer to the boundary value problem (7.2) as ν-periodic. Note that $\nu_j = 0$ corresponds to periodic boundary conditions and $\nu_j = \frac{i}{2}$ to antiperiodic boundary conditions with respect to the j-th coordinate direction. Here

$$
\mathcal{A}(D) := P(D) + Q(D)A := \sum_{|\alpha| \le m_1} p_\alpha D^\alpha + \sum_{|\alpha| \le m_2} q_\alpha D^\alpha A
$$

with $m_1, m_2 \in \mathbb{N}$ and $p_\alpha, q_\alpha \in \mathbb{C}$. Note that $(D^\beta u)|_{x_j=2\pi} = e^{2\pi\nu_j}(D^\beta u)|_{x_j=0}$ implies $(D^\beta A u)|_{x_j=2\pi} = e^{2\pi\nu_j}(D^\beta A u)|_{x_j=0}$ by closedness of A. Thus $m_2 \le m_1$ renders the second boundary condition unnecessary. With $m := \max\{m_1, m_2\}$ in what follows we frequently write $\mathcal{A}(D) = \sum_{|\alpha| \le m}(p_\alpha D^\alpha + q_\alpha D^\alpha A)$, where additional coefficients are understood to be equal to zero. Besides that we define the complex polynomials $P(z) := \sum_{|\alpha| \le m_1} p_\alpha z^\alpha$ and $Q(z) := \sum_{|\alpha| \le m_2} q_\alpha z^\alpha$ for $z \in \mathbb{C}^n$.

Definition 7.13. A solution of the ν-periodic boundary value problem (7.2) is a function $u \in W^{m_1,p}_{\nu,per}(\mathcal{Q}_n, E)$ such that $A u \in W^{m_2,p}_{\nu,per}(\mathcal{Q}_n, E)$ and $\mathcal{A}(D)u(x) = f(x)$ for almost every $x \in \mathcal{Q}_n$.

Remark 7.14. Since the trace operator with respect to one direction and tangential derivation commute, the ν-periodic boundary conditions as imposed in (7.2) are equivalent to

$$
\begin{aligned}
(D^\ell_j u)|_{x_j=2\pi} - e^{2\pi\nu_j}(D^\ell_j u)|_{x_j=0} &= 0 & (j=1,\ldots,n,\ 0 \le \ell < m_1), \\
(D^\ell_j A u)|_{x_j=2\pi} - e^{2\pi\nu_j}(D^\ell_j A u)|_{x_j=0} &= 0 & (j=1,\ldots,n,\ 0 \le \ell < m_2).
\end{aligned}
$$

Again the assumption that $(\lambda + \mu A)^{-1}$ exists for $\lambda, \mu \in \mathbb{C}$ is meant to imply both $(\lambda + \mu A)^{-1} \in \mathcal{L}(E)$ and $(\lambda + \mu A)^{-1}(E) = D(A)$.

Theorem 7.15. *Let $1 < p < \infty$ and assume P and Q to be elliptic. Then the following assertions are equivalent:*

(i) For each $f \in L^p(\mathcal{Q}_n, E)$ there exists a unique solution of (7.2).

(ii) $\left(P(k - i\nu) + Q(k - i\nu)A\right)^{-1}$ exists for $k \in \mathbb{Z}^n$ and

$$
M_\alpha(k) := k^\alpha \left(P(k - i\nu) + Q(k - i\nu)A\right)^{-1}
$$

defines a Fourier multiplier for every $|\alpha| = m_1$.

(iii) $\left(P(k - i\nu) + Q(k - i\nu)A\right)^{-1}$ exists for $k \in \mathbb{Z}^n$ and for all $|\alpha| = m_1$ there exists a finite subset $G \subset \mathbb{Z}^n$ such that the sets $\{M_\alpha(k); \ k \in \mathbb{Z}^n \setminus G\}$ are $\mathcal{R}$-bounded.

(iv) $\left(P(k - i\nu) + Q(k - i\nu)A\right)^{-1}$ exists for $k \in \mathbb{Z}^n$ and there exists a finite subset $G \subset \mathbb{Z}^n$ such that the set

$$\left\{ P(k - i\nu)\left(P(k - i\nu) + Q(k - i\nu)A\right)^{-1}; \ k \in \mathbb{Z}^n \setminus G \right\}$$

is $\mathcal{R}$-bounded.

Proof. (i) $\Rightarrow$ (ii): Let $f \in L^p(\mathcal{Q}_n, E)$ be arbitrary and let u be a solution of (7.2) with right-hand side $e^{\nu \cdot} f$. Then $e^{-\nu \cdot} \mathcal{A}(D)u = f$. In order to compute the Fourier coefficients, we first remark that

$$\left(e^{-\nu \cdot} P(D)u\right)^{\hat{}}(k) = P(k - i\nu)(e^{-\nu \cdot} u)^{\hat{}}(k)$$

and

$$\left(e^{-\nu \cdot} Q(D)Au\right)^{\hat{}}(k) = Q(k - i\nu)(e^{-\nu \cdot} Au)^{\hat{}}(k) = Q(k - i\nu)A(e^{-\nu \cdot} u)^{\hat{}}(k)$$

by Lemma 2.16. Writing $k_\nu := k - i\nu$ for short, we obtain

(7.3) $$\left(P(k_\nu) + Q(k_\nu)A\right)\left(e^{-\nu \cdot} u\right)^{\hat{}}(k) = \hat{f}(k).$$

For arbitrary $y \in E$ and $k \in \mathbb{Z}^n$ the choice $f := e^{ik \cdot} y$ shows $\left(P(k_\nu) + Q(k_\nu)A\right)$ to be surjective. Let $z \in D(A)$ such that $\left(P(k_\nu) + Q(k_\nu)A\right)z = 0$. For fixed $k \in \mathbb{Z}^n$ set $v := e^{ik \cdot} z$ and $u := e^{\nu \cdot} v$. Then

$$P(k_\nu)\left(e^{-\nu \cdot} u\right)^{\hat{}}(k) + Q(k_\nu)A\left(e^{-\nu \cdot} u\right)^{\hat{}}(k) = 0.$$

As $(e^{-\nu \cdot} u)^{\hat{}}(\ell) = 0$ for all $\ell \neq k$, this gives $\mathcal{A}(D)u = 0$, hence $v = u = 0$ and $z = 0$. Altogether we have shown bijectivity of $P(k_\nu) + Q(k_\nu)A$ for $k \in \mathbb{Z}^n$. The closedness of A yields $\left(P(k_\nu) + Q(k_\nu)A\right)^{-1} \in \mathcal{L}(E)$.

For $f \in L^p(\mathcal{Q}_n, E)$ let u be a solution of (7.2) with right hand side $e^{\nu \cdot} f$ and $v := e^{-\nu \cdot} u$. Then $v \in W_{per}^{m_1, p}(\mathcal{Q}_n, E)$, $Av \in W_{per}^{m_2, p}(\mathcal{Q}_n, E)$, and (7.3) implies

$$\hat{v}(k) = \left(P(k_\nu) + Q(k_\nu)A\right)^{-1} \hat{f}(k).$$

This shows

$$M_0 \colon \mathbb{Z}^n \to \mathcal{L}(L^p(\mathcal{Q}_n, E)); \ k \mapsto \left(P(k_\nu) + Q(k_\nu)A\right)^{-1}$$

to be a Fourier multiplier such that T_{M_0} maps $L^p(\mathcal{Q}_n, E)$ into $W_{per}^{m_1, p}(\mathcal{Q}_n, E)$. Due to Lemma 3.11 we have that M_α is a Fourier multiplier for all $|\alpha| = m_1$.

(ii) $\Rightarrow$ (iii): This follows from Proposition 3.9.

(iii) $\Leftrightarrow$ (iv): Let e_j, $j = 1, \ldots, n$ denote the j-th unit vector. For $k \neq 0$ on the one hand it holds that

$$P(k_\nu)\big(P(k_\nu) + Q(k_\nu)A\big)^{-1} = \frac{P(k_\nu)}{\sum\limits_{j=1}^{n} k^{m_1 e_j}} \left(\sum_{j=1}^{n} k^{m_1 e_j} \big(P(k_\nu) + Q(k_\nu)A\big)^{-1} \right).$$

On the other hand for $|\alpha| = m_1$ and $k \in \mathbb{Z}^n$ such that $P(k_\nu) \neq 0$ we have

$$k^\alpha \big(P(k_\nu) + Q(k_\nu)A\big)^{-1} = \frac{k^\alpha}{P(k_\nu)} \left(P(k_\nu)\big(P(k_\nu) + Q(k_\nu)A\big)^{-1} \right).$$

By ellipticity of $k \mapsto \sum_{j=1}^{n} k^{m_1 e_j}$ there exist $c, C > 0$ such that the estimate

$$c|k^\alpha| \leq |P(k_\nu)| \leq C| \sum_{j=1}^{n} k^{m_1 e_j} |$$

is valid for $k \in \mathbb{Z}^n \setminus G$ with suitably chosen finite $G \subset \mathbb{Z}^n$. Recall that m_1 has to be even in case $n > 1$ due to ellipticity of P. Lemma 3.2 now shows the claim.

(iii) $\Rightarrow$ (i): Since (iii) $\Leftrightarrow$ (iv), by Proposition 7.8 it follows that M_α for $|\alpha| = m_1$ as well as $P(\cdot - i\nu)M_0$ are Fourier multipliers. For arbitrary $f \in L^p(\mathcal{Q}_n, E)$ we therefore get $v := T_{M_0}(e^{-\nu} f) \in W_{per}^{m_1, p}(\mathcal{Q}_n, E)$. As

$$(7.4) \qquad Q(k_\nu)A\big(P(k_\nu) + Q(k_\nu)A\big)^{-1} = \mathrm{id}_E - P(k_\nu)\big(P(k_\nu) + Q(k_\nu)A\big)^{-1},$$

$Q(\cdot - i\nu)AM_0$ is a Fourier multiplier, too. By ellipticity of Q and Lemma 3.2 again the same holds for $k^\alpha A\big(P(k_\nu) + Q(k_\nu)A\big)^{-1}$, $|\alpha| \leq m_2$.

By construction $u := e^\nu v = e^{\nu} T_{M_0} e^{-\nu} f$ solves (7.2) and Lemma 3.11 yields $u \in W_{\nu, per}^{m_1, p}(\mathcal{Q}_n, E)$ and $Au \in W_{\nu, per}^{m_2, p}(\mathcal{Q}_n, E)$. Finally, uniqueness of u follows immediately from the uniqueness of the representation as a Fourier series. $\square$

Theorem 7.15 enables us to treat Dirichlet-Neumann type boundary conditions on $\tilde{\mathcal{Q}}_n := (0, \pi)^n$ for even operators. More precisely, for all $|\alpha| \leq m$ we assume $p_\alpha = q_\alpha = 0$ or $\alpha \in 2\mathbb{N}_0^n$ in the representation $\mathcal{A}(D) = \sum_{|\alpha| \leq m}(p_\alpha D^\alpha + q_\alpha D^\alpha A)$. In particular, m_1 and m_2 are even.

Given an even operator $\mathcal{A}(D)$ we consider the boundary value problem

$$(7.5) \qquad \begin{aligned} \mathcal{A}(D)u &= f && \text{in } \tilde{\mathcal{Q}}_n, \\ \mathcal{B}_1(D)u &= 0 && \text{on } \partial\tilde{\mathcal{Q}}_n, \\ \mathcal{B}_2(D)Au &= 0 && \text{on } \partial\tilde{\mathcal{Q}}_n. \end{aligned}$$

Here the boundary $\partial\tilde{\mathcal{Q}}_n$ is defined to be the union of all sets

$$\tilde{\mathcal{Q}}_{j-1} \times \{0, 2\pi\} \times \tilde{\mathcal{Q}}_{n-j} \quad (j = 1, \ldots, n).$$

The pairwise intersections of these sets are empty, i.e. the Lebesgue null sets of vertices are neglected. In each direction $j \in \{1, \ldots, n\}$, the boundary operator $\mathcal{B}(D) := (\mathcal{B}_1(D), \mathcal{B}_2(D))$ represents one of the following boundary conditions:

(i) $D_j^\ell u|_{x_j=0} = D_j^\ell u|_{x_j=\pi} = 0$ $(\ell = 0, 2, \ldots, m_1 - 2)$ and
$D_j^\ell Au|_{x_j=0} = D_j^\ell Au|_{x_j=\pi} = 0$ $(\ell = 0, 2, \ldots, m_2 - 2)$,

(ii) $D_j^\ell u|_{x_j=0} = D_j^\ell u|_{x_j=\pi} = 0$ $(\ell = 1, 3, \ldots, m_1 - 1)$ and
$D_j^\ell Au|_{x_j=0} = D_j^\ell Au|_{x_j=\pi} = 0$ $(\ell = 1, 3, \ldots, m_2 - 1)$,

(iii) $D_j^\ell u|_{x_j=0} = D_j^{\ell+1} u|_{x_j=\pi} = 0$ $(\ell = 0, 2, \ldots, m_1 - 2)$ and
$D_j^\ell Au|_{x_j=0} = D_j^{\ell+1} Au|_{x_j=\pi} = 0$ $(\ell = 0, 2, \ldots, m_2 - 2)$,

(iv) $D_j^{\ell+1} u|_{x_j=0} = D_j^\ell u|_{x_j=\pi} = 0$ $(\ell = 0, 2, \ldots, m_1 - 2)$ and
$D_j^{\ell+1} Au|_{x_j=0} = D_j^\ell Au|_{x_j=\pi} = 0$ $(\ell = 0, 2, \ldots, m_2 - 2)$.

Note that for a second-order operator, (i) is of Dirichlet type, (ii) is of Neumann type, and (iii) and (iv) are of mixed type. For instance, in case (iii) we have $u|_{x_j=0} = 0$ and $D_j u|_{x_j=\pi} = 0$. In general, the types may be different in different directions. Therefore, we refer to these boundary conditions as conditions of Dirichlet-Neumann type. Also recall that the boundary operator might $\mathcal{B}_1$ render some or even all boundary conditions represented by $\mathcal{B}_2$ unnecessary. This depends on m_1 and m_2 (see Remark 6.15).

Proposition 7.16. *Let $\mathcal{A}(D)$ be even, let P and Q be elliptic, and let the boundary conditions in each coordinate direction be of Dirichlet-Neumann type as explained above. Define $\nu \in \mathbb{C}^n$ by setting $\nu_j := 0$ in cases (i) and (ii) and $\nu_j := i/2$ in cases (iii) and (iv). If for this ν one of the equivalent conditions of Theorem 7.15 is fulfilled, then for each $f \in L^p(\tilde{\mathcal{Q}}_n, E)$ there exists a unique solution $u \in W^{m_1,p}(\tilde{\mathcal{Q}}_n, E)$ with $Au \in W^{m_2,p}(\tilde{\mathcal{Q}}_n, E)$ of problem (7.5).*

Proof. Following an idea from [AB02], the solution is constructed by a suitable even or odd extension of the right-hand side from $(0, \pi)^n$ to $(0, 2\pi)^n$. For simplicity of notation we consider the case $n = 2$ and boundary conditions of type (ii) in direction x_1 and of type (iii) in direction x_2. By definition this leads to $\nu_1 = 0$ and $\nu_2 = \frac{i}{2}$.

Let $f \in L^p(\tilde{\mathcal{Q}}_2, E)$ be arbitrary. First considering the even extension of f to the rectangle $(0, 2\pi) \times (0, \pi)$ and afterwards its odd extension to $(0, 2\pi) \times (0, 2\pi)$ we end up with a function F on $\mathcal{Q}_2$ fulfilling $F(x_1, x_2) = F(2\pi - x_1, x_2)$ and $F(x_1, x_2) = -F(x_1, 2\pi - x_2)$ a.e. in $\mathcal{Q}_2$. Now we can apply Theorem 7.15 with $\nu = (\nu_1, \nu_2)^T$ as above. This yields a unique solution U of

$$(7.6) \quad \begin{aligned} \mathcal{A}(D)U &= F & &\text{in } \mathcal{Q}_2, \\ D_1^\ell U|_{x_1=0} &= D_1^\ell U|_{x_1=2\pi} & &(\ell = 0, \ldots, m_1 - 1), \\ -D_2^\ell U|_{x_2=0} &= D_2^\ell U|_{x_2=2\pi} & &(\ell = 0, \ldots, m_1 - 1), \\ D_1^\ell AU|_{x_1=0} &= D_1^\ell AU|_{x_1=2\pi} & &(\ell = 0, \ldots, m_2 - 1), \\ -D_2^\ell AU|_{x_2=0} &= D_2^\ell AU|_{x_2=2\pi} & &(\ell = 0, \ldots, m_2 - 1). \end{aligned}$$

Symmetry of $\mathcal{A}(D)$ now shows that

$$V_1(x_1, x_2) := U(2\pi - x_1, x_2) \quad \text{and} \quad V_2(x_1, x_2) := -U(x_1, 2\pi - x_2) \quad (x \in \mathcal{Q}_2)$$

define solutions of the boundary value problem (7.6) as well. Due to uniqueness $V_1 = U = V_2$ and $U_{x_2} := U(\cdot, x_2) \in W^{m_1, p}((0, 2\pi), E) \subset C^{m_1 - 1}((0, 2\pi), E)$ is even for a.e. $x_2 \in (0, 2\pi)$. Together with periodicity of U_{x_2} due to (7.6) this yields

$$U_{x_2}^{(\ell)}(0) = U_{x_2}^{(\ell)}(\pi) = 0 \quad (\ell = 1, 3, \ldots, m_1 - 1).$$

Accordingly, for a.e. $x_1 \in (0, 2\pi)$ we have that U_{x_1} is odd, and antiperiodicity due to (7.6) gives

$$U_{x_1}^{(\ell)}(0) = U_{x_1}^{(\ell+1)}(\pi) = 0 \quad (\ell = 0, 2, \ldots, m_1 - 2).$$

The same arguments applied to AU yield

$$(AU)_{x_2}^{(\ell)}(0) = (AU)_{x_2}^{(\ell)}(\pi) = 0 \quad (\ell = 1, 3, \ldots, m_2 - 1)$$

and

$$(AU)_{x_1}^{(\ell)}(0) = (AU)_{x_1}^{(\ell+1)}(\pi) = 0 \quad (\ell = 0, 2, \ldots, m_2 - 2).$$

Therefore, $u := U|_{(0,\pi)^2}$ solves $\mathcal{A}(D)u = f$ with boundary conditions (ii) for $j = 1$ and (iii) for $j = 2$.

For arbitrary $n \in \mathbb{N}$ and arbitrary boundary conditions of Dirichlet-Neumann type the construction of the solution follows the same lines. Here we choose even extensions in the cases (ii) and (iv) and odd extensions in the cases (i) and (iii).

On the other hand, let u be a solution of $\mathcal{A}(D)u = f$ satisfying boundary conditions of Dirichlet-Neumann type. We extend u and f to U and F on $(0, 2\pi)^n$ as described above. Then $U \in W^{m_1, p}_{\nu, per}((0, 2\pi)^n, E)$, $AU \in W^{m_2, p}_{\nu, per}((0, 2\pi)^n, E)$ and due to symmetry of $\mathcal{A}(D)$ we see that U solves (7.2) with right-hand side F and ν defined as above. Thus, uniqueness of U yields uniqueness of u and the proof is complete. $\qquad \square$

We close this section with a discussion of the previous results for the case of *homogeneous* polynomials P and Q.

Remark 7.17. a) Consider $\nu = 0$ in Theorem 7.15 and *homogeneous* polynomials P and Q. First assume both polynomials to be non-constant. Then $M_0(0)$ is no longer well-defined. However, we can overcome this problem quickly by changing the underlying space from $L^p(\mathcal{Q}_n, E)$ to

$$L^p_{(0)}(\mathcal{Q}_n, E) := \left\{ f \in L^p(\mathcal{Q}_n, E); \int_{\mathcal{Q}_n} f(x)dx = 0 \right\}.$$

Now assume only P to be non-constant. Without loss of generality let $Q \equiv 1$ which yields $M_0(0) = A^{-1}$. Hence, $0 \in \rho(A)$ is necessary for unique solvability in $L^p(\mathbb{R}^n, E)$. If this assumption is violated, existence and uniqueness of a solution can still be deduced with the aid of Theorem 7.15, provided that $N(A) = \{0\}$ and that the right-hand side f belongs to $L^p_{(0)}(\mathcal{Q}_n, E)$.

b) The observations in part a) affect Proposition 7.16 as soon as boundary conditions of type (i) or (ii) are imposed in every coordinate direction since then results from Theorem 7.15 on pure periodic boundary conditions are employed. Here it is important to distinguish.

First consider pure Neumann boundary conditions on the whole of $\partial \mathcal{Q}_n$. In that case, the extension of the right-hand side f is carried out by means of even reflections throughout all coordinate directions. Given non-constant polynomials P and Q, the operator $M(0)$ is not well-defined and the underlying space has to be changed to $L^p_{(0)}(\tilde{\mathcal{Q}}_n, E)$. If $Q \equiv 1$ a change of the underlying space is unnecessary provided $0 \in \rho(A)$.

However, if Dirichlet conditions are imposed with respect to at least one coordinate direction, the extension of the right-hand side in that direction is carried out by means of an odd reflection. As a consequence, the resulting right-hand side F fulfills $F \in L^p_{(0)}(\mathcal{Q}_n, E)$ automatically. By virtue of part a) the operator $M(0)$ does no longer affect the outcome of Theorem 7.15. Consequently, if existence and uniqueness of a solution U of the pure periodic problem subject to $U \in L^p_{(0)}(\mathcal{Q}_n, E)$ can be deduced, this in turn shows existence and uniqueness of the solution u of the Dirichlet-Neumann problem in $L^p(\mathcal{Q}_n, E)$.

Remark 7.18. In case $n = 1$ ellipticity does no longer force m_1 and m_2 to be even. Therefore, the same results as in Proposition 7.16 can be achieved if $\mathcal{A}(D)$ is odd in the obvious sense, e.g. $\mathcal{A}(D_t) := D_t^3 + D_t + D_t A$. Moreover, we can as well treat a combination of ν-periodic and Dirichlet-Neumann-type boundary conditions on $\mathcal{Q}_{n_1} \times \tilde{\mathcal{Q}}_{n_2}$ as long as the structures of P and Q allow for an application of Proposition 7.16 with respect to $\tilde{\mathcal{Q}}_{n_2}$.

7.3 $\mathcal{R}$-sectoriality and $\mathcal{RH}^\infty$-calculus

Let E be a Banach space of class $\mathcal{HT}$ and A be a closed operator in E. We expand the boundary value problems as considered in the previous section by an additional complex parameter λ. In the sequel we allow for a shift δA, where $\delta \geq 0$ (cf. Remark 6.14).

More precisely, we treat the parameter-dependent ν-periodic boundary value problem given by
(7.7)
$$
\begin{aligned}
\lambda u + \mathcal{A}_\delta(D)u &= f & (x \in \mathcal{Q}_n), \\
(D^\beta u)|_{x_j=2\pi} - e^{2\pi \nu_j}(D^\beta u)|_{x_j=0} &= 0 & (j = 1, \ldots, n;\ |\beta| < m_1), \\
(D^\beta A u)|_{x_j=2\pi} - e^{2\pi \nu_j}(D^\beta A u)|_{x_j=0} &= 0 & (j = 1, \ldots, n;\ |\beta| < m_2).
\end{aligned}
$$

Here $\mathcal{A}_\delta(D)$ is defined as in Remark 6.14.

We further define the $L^p(\mathcal{Q}_n, E)$-realization of the boundary value problem (7.7) to be

$$D(\mathbb{A}_\delta) := \left\{ u \in W^{m_1,p}_{\nu,per}(\mathcal{Q}_n, E);\ Au \in W^{m_2,p}_{\nu,per}(\mathcal{Q}_n, E) \right\},$$
$$\mathbb{A}_\delta u := \mathcal{A}_\delta(D)u \quad (u \in D(\mathbb{A}_\delta)).$$

If $\delta = 0$ we frequently write $\mathbb{A}$ instead of $\mathbb{A}_\delta$. In order to emphasize the ν under consideration, we also employ the notation $\mathbb{A}_{\delta,\nu}$ and $\mathbb{A}_\nu$, respectively.

Remark 7.19. If $m_2 \leq m_1$ it holds that

$$D(\mathbb{A}_\delta) = W^{m_1,p}_{\nu,per}(\mathcal{Q}_n, E) \cap W^{m_2,p}_{\nu,per}(\mathcal{Q}_n, D(A)).$$

In the sequel we restrict ourselves to $\nu = 0$ or purely imaginary components of ν. In that case, it is sufficient to consider $\nu \in i(-1,1)^n$. Note that periodic as well as antiperiodic boundary conditions are still captured. The reason for this constraint lies in the fact that $k_\nu \in \mathbb{R}^n$ has real components only. Hence, (parameter-)ellipticity conditions on P and Q need not to be extended to $P = P(z)$ and $Q = Q(z)$ for z out of a whole open sector of the complex plane. Furthermore, we agree to consider homogeneous parameter-elliptic polynomials P and Q, that is, $P = P^\#$ and $Q = Q^\#$, for elliptic estimates involving lower order terms of polynomials are valid up to $k \in G$ in finite sets $G \subset \mathbb{Z}^n$ only. In view of $(\mathcal{R}\text{-})$sectoriality, however, the additional parameter λ leads to uncountably many multiplier functions M_λ which in turn come along with the challenge to keep control of uncountably many sets G_λ.

Proposition 7.20. *Let $1 < p < \infty$, let E be a Banach space of class $\mathcal{HT}$ enjoying property (α), let $\nu \in i(-1,1)^n$, and let $A \in \Psi\mathcal{RS}(E)$. For homogeneous polynomials P and Q assume that*

(i) P is parameter-elliptic with angle $\varphi_P \in [0,\pi)$,

(ii) Q is parameter-elliptic with angle $\varphi_Q \in [0,\pi)$,

(iii) $\varphi_P + \varphi_Q + \phi^{\mathcal{R}}_A < \pi$.

Set $\varphi_0 := \max\{\varphi_P, \varphi_Q + \phi^{\mathcal{R}}_A\}$. Then $\mathbb{A}_\delta \in \Psi\mathcal{RS}(L^p(\mathcal{Q}_n, E))$ for each $\delta > 0$, $\phi^{\mathcal{R}}_{\mathbb{A}_\delta} \leq \varphi_0$ and for each $\phi > \varphi_0$ it holds that

$$(7.8) \quad \mathcal{R}\left(\left\{ \lambda^{1-\frac{|\alpha|}{m_1}} D^\alpha (\lambda + \mathbb{A}_\delta)^{-1};\ \lambda \in \Sigma_{\pi-\phi},\ \alpha \in \mathbb{N}_0^n,\ 0 \leq |\alpha| \leq m_1 \right\} \right) < \infty.$$

In case $\nu \neq 0$ moreover $\mathbb{A}_\delta \in \mathcal{RS}(L^p(\mathcal{Q}_n, E))$ and $0 \in \rho(\mathbb{A}_\delta)$ for $\delta \geq 0$.
In case $A \in \mathcal{RS}(E)$ moreover $\mathbb{A}_\delta \in \mathcal{RS}(L^p(\mathcal{Q}_n, E))$ for $\delta > 0$.
In case $\nu = 0$ and $Q \equiv c$, $c \neq 0$ subject to condition (ii) and $A \in \mathcal{RS}(E)$ moreover $\mathbb{A}_\delta \in \mathcal{RS}(L^p(\mathcal{Q}_n, E))$ for $\delta \geq 0$. In that case, $0 \in \rho(A)$ implies $0 \in \rho(\mathbb{A}_\delta)$ for $\delta \geq 0$.

Proof. We consider the following formal representation

$$(\lambda + \mathbb{A}_\delta)^{-1} = e^{\nu \cdot} T_{M_\lambda} e^{-\nu \cdot}$$

of the resolvent of $\mathbb{A}_\delta$, where T_{M_λ} denotes the operator associated with

$$M_\lambda(k) := \big(\lambda + P(k - i\nu) + Q_\delta(k - i\nu)A)\big)^{-1}.$$

More generally, for $\alpha \in \mathbb{N}_0^n$, the Leibniz rule shows

$$D^\alpha(\lambda + \mathbb{A}_\delta)^{-1} f = D^\alpha e^{\nu \cdot} T_{M_\lambda} e^{-\nu \cdot} f = \sum_{\beta \le \alpha} g_\beta(\nu) e^{\nu \cdot} T_{M_\lambda^\beta}\big(e^{-\nu \cdot} f\big),$$

where g_β is a polynomial depending on β. Here $T_{M_\lambda^\beta}$ denotes the operator associated with

$$M_\lambda^\beta(k) := k^\beta\big(\lambda + P(k - i\nu) + Q_\delta(k - i\nu)A\big)^{-1}$$

where $\beta \le \alpha$. In case $\nu = 0$ we simply have

$$D^\alpha(\lambda + \mathbb{A}_\delta)^{-1} = T_{M_\lambda^\alpha}.$$

First consider $\nu \ne 0$ and $\delta \ge 0$. Given $\alpha \in \mathbb{N}_0^n$ with $0 \le |\alpha| \le m_1$ let $0 \le \beta \le \alpha$, $0 \le \gamma \le 1$, and $\phi > \varphi_0$. To prove (7.8), we apply Lemma 7.1 in order to calculate $k^\gamma \Delta^\gamma M_\lambda^\beta(k)$. In what follows we write $k_\nu := k - i\nu$ and $Q_\delta(k_\nu) := Q(k_\nu) + \delta$ for short. Note that $k_\nu \ne 0$ for all $k \in \mathbb{Z}^n$.

As in the proof of Theorem 7.15 it suffices to show that

$$(7.9) \qquad \Big\{\lambda^{1 - \frac{|\alpha|}{m_1}} k^\omega \Delta^\omega N(k)\big(\lambda + P(k_\nu) + Q_\delta(k_\nu)A\big)^{-1};\ \lambda \in \Sigma_{\pi - \phi},\ k \in \mathbb{Z}^n\Big\}$$

for $N(k) := k^\beta$ and arbitrary $\omega \le \gamma$,

$$(7.10) \qquad \Big\{k^\omega \Delta^\omega P(k_\nu)\big(\lambda + P(k_\nu) + Q_\delta(k_\nu)A\big)^{-1};\ \lambda \in \Sigma_{\pi - \phi},\ k \in \mathbb{Z}^n\Big\}$$

for $0 < \omega \le \gamma$, and

$$(7.11) \qquad \Big\{k^\omega \Delta^\omega Q_\delta(k_\nu)A\big(\lambda + P(k_\nu) + Q_\delta(k_\nu)A\big)^{-1};\ \lambda \in \Sigma_{\pi - \phi},\ k \in \mathbb{Z}^n\Big\}$$

for $0 < \omega \le \gamma$ are $\mathcal{R}$-bounded.

Due to our assumptions on P and Q, the pseudo-$\mathcal{R}$-sectoriality of A, and Lemma 6.11

$$\left\{\left(\frac{\lambda + P(k_\nu)}{Q_\delta(k_\nu)}\right)\left(\frac{\lambda + P(k_\nu)}{Q_\delta(k_\nu)} + A\right)^{-1};\ \lambda \in \Sigma_{\pi - \phi},\ k \in \mathbb{Z}^n\right\}$$

is $\mathcal{R}$-bounded. Recall that $k_\nu \in \mathbb{R}^n$. Hence, we can employ our previous results on the continuous extensions of the discrete polynomials under consideration. Consequently, we get $\mathcal{R}$-boundedness of

$$(7.12) \qquad \Big\{\big(\lambda + P(k_\nu)\big)\big(\lambda + P(k_\nu) + Q_\delta(k_\nu)A\big)^{-1};\ \lambda \in \Sigma_{\pi - \phi},\ k \in \mathbb{Z}^n\Big\}$$

and by means of the resolvent equation $\mathcal{R}$-boundedness of

$$(7.13) \qquad \left\{ Q_\delta(k_\nu) A \big(\lambda + P(k_\nu) + Q_\delta(k_\nu) A \big)^{-1}; \ \lambda \in \Sigma_{\pi - \phi}, \ k \in \mathbb{Z}^n \right\}.$$

To prove (7.9) and (7.10) we cannot copy the arguments of Proposition 6.13 directly since the continuous extensions of $\Delta^\omega N$ respectively $\Delta^\omega P$ do no longer define homogeneous polynomials. However, since ν is supposed to have at least one non-zero component, there exists $\varepsilon > 0$ such that $|k_\nu| > \varepsilon$ holds true for all $k \in \mathbb{Z}^n$. In particular (7.12) remains valid for $\lambda = 0$. Moreover, the ideas in part c) of Lemma 6.5 yield the existence of $C > 0$ such that

$$\frac{|k|^{|\omega|} |\Delta^\omega N(k)|}{|\lambda + P(k_\nu)|} \leq C, \quad \frac{\lambda^{1 - \frac{|\alpha|}{m_1}} |k|^{|\omega|} |\Delta^\omega N(k)|}{|\lambda + P(k_\nu)|} \leq C, \quad \text{and} \quad \frac{|k|^{|\omega|} |\Delta^\omega P(k_\nu)|}{|\lambda + P(k_\nu)|} \leq C$$

for all $k \in \mathbb{Z}^n$ and all $\lambda \in \overline{\Sigma}_{\pi - \phi}$. Note that Lemma 6.11 applies to λ taken from the closed sector $\overline{\Sigma}_{\pi - \phi}$, in particular $\lambda = 0$, since $|k_\nu| > 0$. Again we apply the contraction principle of Kahane to prove (7.9) and (7.10). Similarly, parameter-ellipticity of Q and part c) of Lemma 6.5 proves (7.11) as well as

$$\left\{ k^\omega A \big(\lambda + P(k_\nu) + Q_\delta(k_\nu) A \big)^{-1}; \ \lambda \in \Sigma_{\pi - \phi}, \ k \in \mathbb{Z}^n \right\}$$

for $|\omega| \leq m_2$. Since $\lambda = 0$ was considered, too, sectoriality of $\mathbb{A}_\delta$ and $0 \in \rho(\mathbb{A}_\delta)$ for $\delta \geq 0$ follows.

Now consider the case $\nu = 0$ and $\delta > 0$. With what has been proved so far, the sufficient intermediate condition (3.12) for the multiplier theorem can be applied successfully. Note that the ideas of the first part of the proof carry over to this situation only if $k \neq 0$. Thus, two $\mathcal{R}$-boundedness statements have to be proved. Firstly, the $\mathcal{R}$-boundedness of

$$\left\{ \lambda^{1 - \frac{|\alpha|}{m_1}} M_\lambda^\alpha; \ \lambda \in \Sigma_{\pi - \phi}, \ k \in \mathbb{Z}^n \right\}.$$

This follows immediately due to homogeneity arguments. Recall the structure of M_λ^α, in particular the fact that we no longer have to consider M_λ^β with $|\beta| < |\alpha|$ and that

$$\lambda^{1 - \frac{|\alpha|}{m_1}} M_\lambda^\alpha(0) = \begin{cases} 0, & \alpha \neq 0, \\ \lambda(\lambda + \delta A)^{-1}, & \alpha = 0. \end{cases}$$

Secondly, we have to prove the $\mathcal{R}$-boundedness of (7.9), (7.10) and (7.11). This time, however, it suffices to consider $k \in \mathbb{Z}^n \setminus \{0\}$ instead of $k \in \mathbb{Z}^n$. Thus, part one of the proof applies verbatim, except for the statement on $\lambda = 0$. In particular, we have shown $\mathbb{A}_\delta \in \Psi\mathcal{R}S(L^p(\mathcal{Q}_n, E))$.

Finally let $A \in \mathcal{R}S(E)$, i.e. we additionally have $N(A) = \{0\}$. Assume $\mathbb{A}_\delta u = 0$ for some $u \in D(\mathbb{A}_\delta)$. Calculating Fourier coefficients gives

$$\big(P(k) + (Q(k) + \delta) A \big) \hat{u}(k) = 0 \quad (k \in \mathbb{Z}^n).$$

Due to injectivity of A and parameter-ellipticity of P and Q

$$\hat{u}(k) = 0 \quad (k \in \mathbb{Z}^n)$$

follows. Hence, $u = 0$ and $\mathbb{A}_\delta \in \mathcal{RS}(L^p(\mathcal{Q}_n, E))$ for $\delta > 0$. In case $Q \equiv c$, $c \neq 0$ subject to condition (ii) we can obviously allow $\delta = 0$ by the same arguments. The assertion on $0 \in \rho(\mathbb{A}_\delta)$ follows as in Proposition 6.13. $\qquad \square$

Remark 7.21. Note that additionally we have proved

$$(7.14) \qquad \mathcal{R}(\{D^\alpha A(\lambda + \mathbb{A}_\delta)^{-1}; \ \lambda \in \Sigma_{\pi-\phi}, \ \alpha \in \mathbb{N}_0^n, \ 0 \le |\alpha| \le m_2\}) < \infty.$$

Remark 7.22. a) The shift $\delta > 0$ cannot be neglected in case $Q \not\equiv c$ and $\nu = 0$. To see this, take a right-hand side $f \in L^p(\mathcal{Q}_n, E)$ which is given as a constant function $\eta \in E \setminus D(A)$. If $\lambda u + \mathcal{A}(D)u = f$, then $\lambda \hat{u}(0) = \hat{f}(0) = \eta$ by parameter-ellipticity of P and Q. Hence, $u \notin D(\mathbb{A})$. On the other hand, $\delta = 0$ and $\nu = 0$ can simultaneously be handled if the underlying space is changed to $L^p_{(0)}(\mathcal{Q}_n, E)$.

b) Likewise we cannot omit the assumption $N(A) = \{0\}$ in case $Q \equiv c$ and $\delta = 0$ in order to deduce $\mathbb{A} \in \mathcal{RS}(L^p(\mathcal{Q}_n, E))$. To see this, let $\eta \in N(A) \setminus \{0\}$. Then u defined by $u(x) := \eta$ for $x \in \mathcal{Q}_n$ fulfills $u \in D(\mathbb{A}) \setminus \{0\}$ and $\mathbb{A}u = 0$.

For what follows assume $\mathcal{A}_\delta(D)$ to be even. Let $\mathbb{A}_{\delta,\mathcal{B}}$ denote the L^p-realization of the parameter-dependent boundary value problem

$$(7.15) \qquad \begin{aligned} \lambda u + \mathcal{A}_\delta(D)u &= f & \text{in } \tilde{\mathcal{Q}}_n, \\ \mathcal{B}_1(D)u &= 0 & \text{on } \partial\tilde{\mathcal{Q}}_n, \\ \mathcal{B}_2(D)Au &= 0 & \text{on } \partial\tilde{\mathcal{Q}}_n, \end{aligned}$$

where the boundary operator $\mathcal{B}(D) := (\mathcal{B}_1(D), \mathcal{B}_2(D))$ represents boundary conditions as in problem (7.5), cf. Remark 6.15. Since $\lambda + \mathcal{A}_\delta(D)$ is even, too, we have the following immediate consequence of Proposition 7.20.

Proposition 7.23. *Let the assumptions of Proposition 7.20 on $\mathcal{A}_\delta(D)$ be fulfilled and let $\mathcal{A}_\delta(D)$ be even. Then $\mathbb{A}_{\delta,\mathcal{B}} \in \Psi\mathcal{RS}(L^p(\tilde{\mathcal{Q}}_n, E))$ for each $\delta > 0$, $\phi^{\mathcal{R}}_{\mathbb{A}_{\delta,\mathcal{B}}} \le \varphi_0$ and for each $\phi > \varphi_0$ it holds that*

$$(7.16) \qquad \mathcal{R}\left(\left\{\lambda^{1-\frac{|\alpha|}{m_1}} D^\alpha(\lambda + \mathbb{A}_{\delta,\mathcal{B}})^{-1}; \ \lambda \in \Sigma_{\pi-\phi}, \ \alpha \in \mathbb{N}_0^n, \ 0 \le |\alpha| \le m_1\right\}\right) < \infty.$$

In case $\mathcal{B}$ includes non-pure Neumann boundary conditions in at least one direction moreover $\mathbb{A}_{\delta,\mathcal{B}} \in \mathcal{RS}(L^p(\tilde{\mathcal{Q}}_n, E))$ and $0 \in \rho(\mathbb{A}_{\delta,\mathcal{B}})$ for $\delta \ge 0$.
In case $A \in \mathcal{RS}(E)$ moreover $\mathbb{A}_{\delta,\mathcal{B}} \in \mathcal{RS}(L^p(\tilde{\mathcal{Q}}_n, E))$ for $\delta > 0$.
In case $\nu = 0$ and $Q \equiv c$, $c \neq 0$ subject to condition (ii) of Proposition 7.20 and $A \in \mathcal{RS}(E)$ moreover $\mathbb{A}_{\delta,\mathcal{B}} \in \mathcal{RS}(L^p(\tilde{\mathcal{Q}}_n, E))$ for $\delta \ge 0$.

Proof. Let E denote the operator of extension and R the operator of restriction as described in the proof of Proposition 7.16. Let further $\mathbb{A}_\delta = \mathbb{A}_{\delta,\nu}$ with ν as described in Proposition 7.16. Then the assertion on pseudo-$\mathcal{R}$-sectoriality follows due to the representation $(\lambda + \mathbb{A}_{\delta,\mathcal{B}})^{-1} = \mathfrak{R}(\lambda + \mathbb{A}_{\delta,\nu})^{-1}\mathfrak{E}$ and Proposition 7.20.

The results on $\mathcal{R}$-sectoriality of $\mathbb{A}_{\delta,\mathcal{B}}$, i.e. $N(\mathbb{A}_{\delta,\mathcal{B}}) = \{0\}$, can also be deduced from Proposition 7.20. To see this, observe that $u \in D(\mathbb{A}_{\delta,\mathcal{B}})$ subject to $\mathbb{A}_{\delta,\mathcal{B}}u = 0$ implies $U := \mathfrak{E}u \in D(\mathbb{A}_{\delta,\nu})$ and $\mathbb{A}_{\delta,\nu}U = 0$ with ν as described in Proposition 7.16. Clearly $U = 0$ implies $u = 0$. See further Remark 7.17. $\qquad\square$

It remains to prove $\mathcal{R}$-boundedness of the $\mathcal{H}^\infty$-calculus for the operators under consideration. To do so, we more or less adopt the proofs of the continuous setting. First recall the holomorphic, operator-valued functions $H_1 \in \mathcal{H}^\infty(\Sigma_\vartheta, \mathcal{L}(E))$ and $H_2 \in \mathcal{H}^\infty(\Sigma_\zeta, \mathcal{L}(E))$ from Lemma 6.18. In the continuous setting a representation of e.g. $z^k H_1^{(k)}(z)$ based on Cauchy's integral formula in an extended version entered. It was employed to deduce $\mathcal{R}$-bounds of derivatives $D^\gamma(H_1 \circ P)(\xi)$, whereas in the present situation $\mathcal{R}$-bounds of discrete derivatives $\Delta^\gamma(H_1 \circ P)(k)$ have to be inferred. However, shifts of H_1 itself can not be carried out since the sector Σ_σ is not invariant under shifting. Consequently, we have to treat $m_h \circ P$ at once. This is done in the subsequent lines.

Lemma 7.24. *Let $\nu \in i(-1,1)^n$, $0 \neq \gamma \leq 1$, and let $0 < \vartheta' < \psi' < \vartheta < \pi$. Let $H \in \mathcal{H}^\infty(\Sigma_\vartheta, \mathcal{L}(E))$ and let $P \colon \mathbb{Z}^n \to \mathbb{C}$ be homogeneous and parameter-elliptic with angle $\varphi_P < \vartheta'$. Set*

$$\Gamma_{\psi'} := (\infty, 0]e^{i\psi'} \cup [0,\infty)e^{-i\psi'}.$$

Then there exist $g_k \in L^1(\Gamma_{\psi'})$ such that

$$\sup_{k \in \mathbb{Z}^n \setminus \{-1,0,1\}^n} \|k^\gamma g_k\|_{L^1(\Gamma_{\psi'})} < \infty$$

and

$$\Delta^\gamma(H \circ P)(k_\nu) = \frac{1}{2\pi i}\int_{\Gamma_{\psi'}} g_k(\mu)H(\mu)d\mu \quad (k \in \mathbb{Z}^n \setminus \{-1,0,1\}^n).$$

Proof. Let $k \in \mathbb{Z}^n \setminus \{-1,0,1\}^n$ be arbitrary and set $z := P(k_\nu) \in \Sigma_{\vartheta'}$. Due to homogeneity and parameter-ellipticity of P and our choice of k and ν there exist $0 < r < R$ such that $0 < r < |P(k_\nu - \omega)| < R$ for all $0 \leq \omega \leq 1$. Set

$$\Gamma_{\psi',r,R} := [R,r]e^{i\psi'} \cup re^{i(\psi',-\psi')} \cup [r,R]e^{-i\psi'} \cup Re^{i(-\psi',\psi')}.$$

Then Cauchy's integral formula for closed rectifiable curves yields

$$H(z) = \frac{1}{2\pi i}\int_{\Gamma_{\psi',r,R}} \frac{1}{\mu - z}H(\mu)d\mu.$$

With $g_k(\mu) := \Delta^\gamma(\mu - P)^{-1}(k_\nu)$ we further have

$$\Delta^\gamma(H \circ P)(k_\nu) = \frac{1}{2\pi i}\int_{\Gamma_{\psi',r,R}} g_k(\mu)H(\mu)d\mu$$

and from Lemma 7.1 we infer that each value $g_k(\mu)$ equals

$$\sum_{\mathcal{W}\in\mathcal{Z}_\gamma}(-1)^{r_\mathcal{W}}\left(\mu - P(k_\nu - \gamma)\right)^{-1}\prod_{j=1}^{r_\mathcal{W}}(\Delta^{\omega^j}P)(k_\nu - \omega^j_*)\left(\mu - P(k_\nu - \omega^j_*)\right)^{-1}.$$

Recall that $0 < r < |P(k_\nu - \omega)| < R$ for all $0 \leq \omega \leq 1$ thanks to $k \notin \{-1,0,1\}^n$. Since $\gamma \neq 0$, we have $r_\mathcal{W} \geq 1$ in this representation formula of $g_k(\mu)$. Therefore, $g_k \in L^1(\Gamma_{\psi'})$ and $\int_{\Gamma_{\psi'}} g_k(\mu)H(\mu)d\mu$ exists due to boundedness of H on Σ_ϑ. As in Lemma 6.19 we estimate the integrals over the two arcs in the representation formula for $\Delta^\gamma(H \circ P)(k_\nu)$ to find

$$\Delta^\gamma(H \circ P)(k_\nu) = \frac{1}{2\pi i}\int_{\Gamma_{\psi'}} g_k(\mu)H(\mu)d\mu.$$

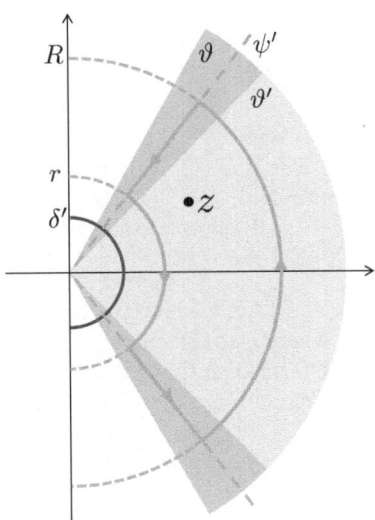

Figure 7.1: The path of integration, discrete case. All values $P(k_\nu - \omega)$, where $0 \leq \omega \leq 1$ and $k \in \mathbb{Z}^n \setminus \{-1,0,1\}^n$ stay in the exterior of the red half circle.

In order to establish the uniform L^1-bound for $k^\gamma g_k$, consider $\delta > 0$ such that $|k_\nu - \omega| > \delta$ for all $0 \leq \omega \leq 1$ and all $k \in \mathbb{Z}^n \setminus \{-1,0,1\}^n$. By homogeneity of P

we observe existence of a $\delta' > 0$ such that $|P(k_\nu - \omega)| > \delta'$ for all $0 \leq \omega \leq 1$ and all $k \in \mathbb{Z}^n \setminus \{-1, 0, 1\}^n$.

Moreover, there exists $C > 0$ such that we can estimate

$$|\xi^\omega (\Delta^\omega P)(\xi)| \leq C|P(\xi)| \quad (\xi \in \mathbb{R}^n, \, |\xi| \geq \delta, \, 0 \leq \omega \leq 1)$$

for the continuous extensions of P and $\Delta^\omega P$. For $g_{z_1, z_2}^{(\gamma)}(\mu) := \dfrac{z_1^{|\gamma|}}{(\mu - z_1)^{|\gamma|}(\mu - z_2)}$ it therefore holds that

$$\|k^\gamma g_k\|_{L^1(\Gamma_{\vartheta'})} \leq \sup_{z_1, z_2 \in \Sigma_{\vartheta'}, |z_1|, |z_2| > \delta'} \|g_{z_1, z_2}^{(\gamma)}\|_{L^1(\Gamma_{\vartheta'})} < \infty.$$

□

As already indicated by the previous lemma, our intermediate condition (3.12) plays a crucial role in the upcoming result.

Proposition 7.25. *Let $1 < p < \infty$, let E be a Banach space of class $\mathcal{HT}$ enjoying property (α), let $\nu \in i(-1, 1)^n$, and let $A \in \Psi\mathcal{RH}^\infty(E)$. For the homogeneous polynomials P and Q assume that*

(i) P is parameter-elliptic with angle $\varphi_P \in [0, \pi)$,

(ii) Q is parameter-elliptic with angle $\varphi_Q \in [0, \pi)$,

(iii) $\varphi_P + \varphi_Q + \phi_A^{\mathcal{R}\infty} < \pi$.

*Then $\mathbb{A}_\delta \in \Psi\mathcal{RH}^\infty(L^p(\mathcal{Q}_n, E))$ and $\phi_{\mathbb{A}_\delta}^{\mathcal{R}\infty} \leq \max\{\varphi_P, \varphi_Q + \phi_A^{\mathcal{R}\infty}\}$ for each $\delta > 0$.
In case $\nu \neq 0$ moreover $\mathbb{A}_\delta \in \mathcal{RH}^\infty(L^p(\mathcal{Q}_n, E))$ for $\delta \geq 0$.
In case $A \in \mathcal{RH}^\infty(E)$ moreover $\mathbb{A}_\delta \in \mathcal{RH}^\infty(L^p(\mathcal{Q}_n, E))$ for $\delta > 0$.
In case $\nu = 0$ and $Q \equiv c$, $c \neq 0$ subject to condition (ii) and $A \in \mathcal{RH}^\infty(E)$ moreover $\mathbb{A}_\delta \in \mathcal{RH}^\infty(L^p(\mathcal{Q}_n, E))$ for $\delta \geq 0$.*

Proof. Let $\phi \in (\max\{\varphi_P, \varphi_Q + \phi_A^{\mathcal{R}\infty}\}, \pi - \min\{\varphi_P, \varphi_Q + \phi_A^{\mathcal{R}\infty}\})$ and $\sigma > \phi$. Define $\Gamma := (\infty, 0]e^{i\psi} \cup [0, \infty)e^{-i\psi}$ where $\psi \in (\phi, \min\{\sigma, \pi - \min\{\varphi_P, \varphi_Q + \phi_A^{\mathcal{R}\infty}\}\})$ and consider an arbitrary $h \in \mathcal{H}_0^\infty(\Sigma_\sigma)$ such that $|h|_\infty^\sigma \leq 1$.

We formally have $h(\mathbb{A}_\delta) = T_{m_h \circ (P, Q_\delta)}$. Here

$$m_h(z_1, z_2) := \int_\Gamma h(\lambda)(\lambda - z_1 - z_2 A)^{-1} d\lambda \quad ((z_1, z_2) \in \Sigma_\vartheta \times \Sigma_\zeta),$$

where $\vartheta \geq \varphi_P$, $\zeta \geq \varphi_Q$ such that $\vartheta + \zeta + \phi_A^{\mathcal{R}\infty} < \pi$. From Lemma 6.17 we deduce

$$\mathcal{R}(\{m_h(z_1, z_2); \, |h|_\infty^\sigma \leq 1, \, (z_1, z_2) \in \Sigma_\vartheta \times \Sigma_\zeta\}) < \infty.$$

In particular,

$$\mathcal{R}(\{(m_h \circ (P, Q_\delta))(k_\nu); \, |h|_\infty^\sigma \leq 1, \, k \in \mathbb{Z}^n\}) < \infty.$$

Now choose $\psi_p \in (\varphi_P, \vartheta)$ and $\psi_q \in (\varphi_Q, \zeta)$ to define

$$\Gamma^p := (\infty, 0]e^{i\psi_p} \cup [0, \infty)e^{-i\psi_p} \quad \text{and} \quad \Gamma^q := (\infty, 0]e^{i\psi_q} \cup [0, \infty)e^{-i\psi_q}.$$

Due to Lemma 6.18 m_h is holomorphic in each variable separately. By Cauchy's integral formula for polydiscs (see e.g. [Ran86, Theorem 1.3]) extended along the lines of Lemma 6.19, we infer

$$m_h(z_1, z_2) = \frac{1}{(2\pi i)^2} \int_{\Gamma^p} \int_{\Gamma^q} \frac{1}{(\mu_1 - z_1)(\mu_2 - z_2)} m_h(\mu_1, \mu_2) d\mu_2 d\mu_1$$

$$= \frac{1}{(2\pi i)^2} \int_{\Gamma^p} \frac{1}{(\mu_1 - z_1)} \int_{\Gamma^q} \frac{1}{(\mu_2 - z_2)} m_h(\mu_1, \mu_2) d\mu_2 d\mu_1.$$

Let $0 \neq \gamma \leq 1$ and set $g_k(\mu_1, \mu_2) := \Delta^\gamma \big((\mu_1 - P)^{-1}(\mu_2 - Q_\delta)^{-1}\big)(k_\nu)$. Employing Lemma 7.1 yields

$$k^\gamma g_k(\mu_1, \mu_2) = \sum_{\theta \leq \gamma} \binom{\gamma}{\theta} k^{\gamma-\theta}(\Delta^{\gamma-\theta}(\mu_1 - P)^{-1})(k_\nu - \theta) k^\theta (\Delta^\theta (\mu_2 - Q_\delta)^{-1})(k_\nu)$$

for $k \in \mathbb{Z}^n \setminus \{-1, 0, 1\}^n$. Applying Δ^γ we get

$$\Delta^\gamma (m_h \circ (P, Q_\delta))(k_\nu)$$

$$= \frac{1}{(2\pi i)^2} \sum_{\theta \leq \gamma} \binom{\gamma}{\theta} \int_{\Gamma^p} h_k^{(1),\theta}(\mu_1) \int_{\Gamma^q} h_k^{(2),\theta}(\mu_2) m_h(\mu_1, \mu_2) d\mu_2 d\mu_1.$$

Here

$$h_k^{(1),\theta}(\mu_1) := (\Delta^{\gamma-\theta}(\mu_1 - P)^{-1})(k_\nu - \theta) \quad \text{and} \quad h_k^{(2),\theta}(\mu_2) := (\Delta^\theta (\mu_2 - Q_\delta)^{-1})(k_\nu)$$

fulfill

$$\sup_{k \in \mathbb{Z}^n \setminus \{-1,0,1\}^n} \left\| k^{\gamma-\theta} h_k^{(1),\theta} \right\|_{L^1(\Gamma^p)} \leq C \quad \text{and} \quad \sup_{k \in \mathbb{Z}^n \setminus \{-1,0,1\}^n} \left\| k^\theta h_k^{(2),\theta} \right\|_{L^1(\Gamma^q)} \leq C.$$

Thus, Lemma 3.3 applies in iteration to the result

$$\mathcal{R}(\{k^\gamma \Delta^\gamma (m_h \circ (P, Q_\delta))(k); \ |h|_\infty^\sigma \leq 1, \ k \in \mathbb{Z}^n \setminus \{-1, 0, 1\}^n, \ 0 \leq \gamma \leq 1\}) < \infty$$

which shows $m_h \circ (P, Q_\delta)$ to be a Fourier multiplier. □

Proposition 7.26. *Given the assumptions of Proposition 7.25, let* $\mathcal{A}_\delta(D)$ *be even. Then* $\mathbb{A}_{\delta,\mathcal{B}} \in \Psi\mathcal{R}\mathcal{H}^\infty(L^p(\tilde{\mathcal{Q}}_n, E))$ *and* $\phi_{\mathbb{A}_{\delta,\mathcal{B}}}^{\mathcal{R}\infty} \leq \max\{\varphi_P, \varphi_Q + \phi_A^{\mathcal{R}\infty}\}$ *for each* $\delta > 0$.

In case $\mathcal{B}$ *includes non-pure Neumann boundary conditions in at least one direction moreover* $\mathbb{A}_{\delta,\mathcal{B}} \in \mathcal{R}\mathcal{H}^\infty(L^p(\tilde{\mathcal{Q}}_n, E))$ *for* $\delta \geq 0$.

In case $A \in \mathcal{R}\mathcal{H}^\infty(E)$ *moreover* $\mathbb{A}_{\delta,\mathcal{B}} \in \mathcal{R}\mathcal{H}^\infty(L^p(\tilde{\mathcal{Q}}_n, E))$ *for* $\delta > 0$.

In case $\nu = 0$ *and* $Q \equiv c$, $c \neq 0$ *subject to condition (ii) of Proposition 7.25 and* $A \in \mathcal{R}\mathcal{H}^\infty(E)$ *moreover* $\mathbb{A}_{\delta,\mathcal{B}} \in \mathcal{R}\mathcal{H}^\infty(L^p(\tilde{\mathcal{Q}}_n, E))$ *for* $\delta \geq 0$.

Remark 7.27. Note that related results on sectoriality and a bounded $\mathcal{H}^\infty$-calculus of $\mathbb{A}_{\delta,\nu}$ respectively $\mathbb{A}_{\delta,\mathcal{B}}$ can still be deduced if E does not enjoy property (α).

8 Application to cylindrical boundary value problems

In this chapter the Fourier transform approach and the Fourier series approach from the previous chapters are employed to investigate cylindrical boundary value problems. With their aid, a model problem for cylindrical boundary value problems containing partially non-constant coefficients can be treated. Thanks to $\mathcal{R}$-bounds derived for the solution operator of the model problem a localization procedure as known for problems in the whole space can be carried out to deal with fully non-constant coefficients. The crucial assumption is that the cylindrical boundary value problem is parameter-elliptic. As a main result, we prove pseudo-$\mathcal{R}$-sectoriality of the according L^p-realizations. Due to the results from Chapter 5 this implies results on the associated parabolic problems. At the end of the chapter we focus on the Laplacian subject to mixed periodic and Dirichlet-Neumann boundary conditions in cylindrical domains.

8.1 Known results for bounded and exterior domains

We start this section with known results on parameter-elliptic boundary value problems in standard domains.

Definition 8.1. Let $m, n \in \mathbb{N}$. The domain $V \subset \mathbb{R}^n$ is called a *standard domain* in $\mathbb{R}^n$ if it is given as the whole space $\mathbb{R}^n$, the half space $\mathbb{R}^n_+$ or as a domain in $\mathbb{R}^n$ with compact boundary, that is, a bounded or an exterior domain. If a standard domain V is of class C^m, it is also called a C^m *standard domain.*

Let F be a Banach space, $m, n \in \mathbb{N}$, and let $V \subset \mathbb{R}^n$ be a C^{2m} standard domain. Set

$$A(x, D)u := \sum_{|\alpha| \leq 2m} a_\alpha(x) D^\alpha u,$$

where $\alpha \in \mathbb{N}_0^n$ and $a_\alpha \colon V \to \mathcal{L}(F)$. Furthermore, let

$$B_j(x, D)u := \sum_{|\beta| \leq m_j} b_{j,\beta}(x) (D^\beta u)|_{\partial V},$$

where $m_j < 2m$, $\beta \in \mathbb{N}_0^n$, and $b_{j,\beta} \colon \partial V \to \mathcal{L}(F)$ for $j = 1, \ldots, m$. We consider the boundary value problem (A, B) given through

$$(8.1) \qquad \begin{aligned} \lambda u + A(x, D)u &= f \text{ in } V, \\ B_j(x, D)u &= 0 \text{ on } \partial V \quad (j = 1, \ldots, m). \end{aligned}$$

The $L^p(V)$-realization of (A, B) will be denoted by

$$D(A) := \left\{ u \in W^{2m,p}(V);\ B_j(\cdot, D)u = 0\ (j = 1, \ldots, m) \right\},$$
$$Au := A(\cdot, D)u \quad (u \in D(A)).$$

The differential operator $A(x, D)$ is assumed to be parameter-elliptic in V with angle of parameter-ellipticity $\varphi \in [0, \pi)$, i.e. φ is the infimum over all $\phi \in (0, \pi)$ such that the principal part of its symbol

$$A^{\#}(x, \xi) := \sum_{|\alpha|=2m} a_\alpha(x)\xi^\alpha$$

for each $x \in \overline{V}$ is parameter-elliptic with angle of parameter-ellipticity $\varphi_x < \phi$. Here, an $\mathcal{L}(F)$-valued homogeneous polynomial

$$a(\xi) := \sum_{|\alpha|=2m} a_\alpha \xi^\alpha \quad (\xi \in \mathbb{R}^n)$$

is called parameter-elliptic if there exists an angle $\phi \in (0, \pi)$ such that the spectrum $\sigma(a(\xi))$ of $a(\xi)$ in $\mathcal{L}(F)$ satisfies

$$(8.2) \qquad\qquad \sigma(a(\xi)) \subset \Sigma_\phi \quad (\xi \in \mathbb{R}^n,\ |\xi| = 1).$$

Then

$$\varphi := \inf\{\phi;\ (8.2)\text{ holds}\}$$

is called angle of parameter-ellipticity of a (see [DHP03]).

Definition 8.2. The boundary value problem (A, B) given through (8.1) is called *parameter-elliptic* in V with *angle of parameter-ellipticity* $\varphi \in [0, \pi)$ if $A(\cdot, D)$ is parameter-elliptic in V with angle of parameter-ellipticity $\varphi \in [0, \pi)$ and if for each $\phi > \varphi$ the Lopatinskii-Shapiro condition holds. In order to indicate that φ is the angle of parameter-ellipticity of the boundary value problem (A, B), we use the subscript notation $\varphi_{(A,B)}$.

We refer to [Wlo87] for an introduction to the Lopatinskii-Shapiro condition for scalar-valued boundary value problems and to [DHP03] for an extensive treatment of the F-valued case.

In the sequel let the boundary value problem (8.1) be parameter-elliptic. Furthermore, assume the following assumptions on the coefficients to hold true:

$$(8.3) \quad \begin{cases} a_\alpha \in C(\overline{V}, \mathcal{L}(F)) \text{ and } a_\alpha(\infty) := \lim_{|x|\to\infty} a_\alpha(x) \text{ exists} \quad (|\alpha| = 2m), \\[2mm] a_\alpha \in [L^\infty + L^{r_k}](V, \mathcal{L}(F)),\ r_k \geq p,\ \dfrac{2m-k}{n} > \dfrac{1}{r_k} \quad (|\alpha| = k < 2m), \\[2mm] b_{j,\beta} \in C^{2m-m_j}(\partial V, \mathcal{L}(F)) \quad (j = 1, \ldots, m;\ |\beta| \leq m_j). \end{cases}$$

The condition on $a_\alpha(\infty)$ can be neglected if V is bounded. By employing finite open coverings of $\overline{V}$ in [DHP03] the following result is proved.

Proposition 8.3. *Let $V \subset \mathbb{R}^n$ be a C^{2m} standard domain. Let the boundary value problem (A, B) be parameter-elliptic and let the assumptions (8.3) on the coefficients be given. Then for each $\phi > \varphi_{(A,B)}$ there exists a $\delta = \delta(\phi) \geq 0$ such that $A + \delta \in \mathcal{RS}(L^p(V, F))$ and $\phi_{A+\delta}^{\mathcal{RS}} \leq \phi$. Moreover, we have*

$$(8.4) \qquad \mathcal{R}\left(\left\{\lambda^{1-\frac{|\gamma|}{2m}} D^\gamma (\lambda + A + \delta)^{-1}; \ \lambda \in \Sigma_{\pi-\phi}, \ 0 \leq |\gamma| \leq 2m\right\}\right) < \infty.$$

As a stronger condition on the coefficients assume

$$(8.5) \quad \begin{cases} a_\alpha \in BUC^\gamma(\overline{V}, \mathcal{L}(F)) \text{ for some } \gamma \in (0, 1), \\[4pt] \qquad a_\alpha(\infty) := \lim_{|x| \to \infty} a_\alpha(x) \text{ exists, and} \\[4pt] \|a_\alpha(x) - a_\alpha(\infty)\| \leq C|x|^{-\gamma} \text{ for } x \in V, \ |x| \geq 1 \quad (|\alpha| = 2m), \\[4pt] a_\alpha \in [L^\infty + L^{r_k}](V, \mathcal{L}(F)), \ r_k \geq p, \ \dfrac{2m-k}{n} > \dfrac{1}{r_k} \quad (|\alpha| = k < 2m), \\[4pt] b_{j,\beta} \in C^{2m-m_j}(\partial V, \mathcal{L}(F)) \quad (j = 1, \ldots, m; \ |\beta| \leq m_j). \end{cases}$$

Again by finite open coverings of $\overline{V}$, in [DDH$^+$04] the following result is proved.

Proposition 8.4. *Let $V \subset \mathbb{R}^n$ be a C^{2m} standard domain. Let the boundary value problem (A, B) be parameter-elliptic and let the assumptions (8.5) on the coefficients be given. Then for each $\phi > \varphi_{(A,B)}$ there exists a $\delta = \delta(\phi) \geq 0$ such that $A + \delta \in \mathcal{RH}^\infty(L^p(V, F))$ and $\phi_{A+\delta}^{\mathcal{R}\infty} \leq \phi$.*

Remark 8.5. In [DHP03] and [DDH$^+$04] formally the case $n \geq 2$ is treated, whereas the case $n = 1$ can be deduced by simplified methods.

For the Laplacian according results for domains rougher than the standard domains as defined in Definition 8.1 are known. More precisely, smoothness of the boundary of G can be reduced to be of graph Lipschitz type.

Definition 8.6. A domain $G \subset \mathbb{R}^n$ is called a *Lipschitz domain* if there exists an $M > 0$ so that every point $x = (x_1, \ldots, x_n) \in \partial G$ has a neighborhood U such that, eventually after an affine change of coordinates, $\partial G \cap U$ is described by the equation $x_n = \varphi(x_1, \ldots, x_{n-1})$, where φ is a Lipschitz continuous function with Lipschitz constant bounded by M and where $G \cap U = \{x \in U; \ x_n > \varphi(x_1, \ldots, n_{n-1})\}$.

This definition is also used in [JK95], for instance. Domains G of the above type are usually termed graph Lipschitz domains or strongly Lipschitz or domains with Lipschitz boundary. Recall that on such domains standard function space theory, trace results, definition of an outer unit normal, etc., are still available (see [Gri85] and [JK95]). For more general classes of Lipschitz domains see also [Gri85] and [HDR09].

In what follows we distinguish between the weak and the strong Dirichlet Laplacian $\Delta_{p,w}^D$ and $\Delta_{p,s}^D$ on $L^p(V)$ defined by

$$D(\Delta_{p,w}^D) := \left\{u \in W_0^{1,p}(V); \ \Delta u \in L^p(V)\right\},$$

$$\Delta_{p,w}^D u := \Delta u \quad (u \in D(\Delta_{p,w}^D))$$

and

$$D(\Delta_{p,s}^D) := W^{2,p}(V) \cap W_0^{1,p}(V),$$
$$\Delta_{p,s}^D u := \Delta u \quad (u \in D(\Delta_{p,s}^D)).$$

The weak and the strong Neumann Laplacian $\Delta_{p,w}^N$ and $\Delta_{p,s}^N$ are defined by

$$D(\Delta_{p,w}^N) := \left\{ u \in W^{1,p}(V); \ \exists v \in L^p(V) \ \forall \varphi \in W^{1,p'}(V) : -\int_V \nabla u \nabla \varphi = \int_V v\varphi \right\},$$
$$\Delta_{p,w}^N u := v \quad (u \in D(\Delta_{p,w}^N))$$

and

$$D(\Delta_{p,s}^N) := \left\{ u \in W^{2,p}(V); \ \partial_{\mathbf{n}} u = 0 \text{ on } \partial V \right\},$$
$$\Delta_{p,s}^N u := \Delta u \quad (u \in D(\Delta_{p,s}^N)).$$

The following proposition gathers results on these operators suitable for our purposes. Recall that $A \in (\Psi)\mathcal{RH}^\infty(L^p(\Omega))$ implies $A \in (\Psi)\mathcal{RS}(L^p(\Omega))$ with $\phi_A^{\mathcal{R}} \le \phi_A^{\mathcal{R}\infty}$.

Proposition 8.7. a) *Let $V \subset \mathbb{R}^n$ be a C^2 standard domain and let $1 < p < \infty$. Then*

(i) $-\Delta_{p,s}^D \in \mathcal{RH}^\infty(L^p(\Omega))$ *and* $\phi_{-\Delta_{p,s}^D}^{\mathcal{R}\infty} < \frac{\pi}{2}$,

(ii) $-\Delta_{p,s}^N \in \Psi\mathcal{RH}^\infty(L^p(\Omega))$ *and* $\phi_{-\Delta_{p,s}^N}^{\mathcal{R}\infty} < \frac{\pi}{2}$,

(iii) $-\Delta_{p,s}^N + \delta \in \mathcal{RH}^\infty(L^p(\Omega))$ *and* $\phi_{-\Delta_{p,s}^N+\delta}^{\mathcal{R}\infty} < \frac{\pi}{2}$ *for each $\delta > 0$.*

b) *Let $V \subset \mathbb{R}^n$, $n \ge 3$, be a bounded Lipschitz domain. Then there exists $\varepsilon > 0$ depending only on the Lipschitz character of V such that for all $(3+\varepsilon)' < p < 3+\varepsilon$*

(i) $-\Delta_{p,w}^D \in \mathcal{RH}^\infty(L^p(\Omega))$ *and* $\phi_{-\Delta_{p,w}^D}^{\mathcal{R}\infty} < \frac{\pi}{2}$,

(ii) $-\Delta_{p,w}^N \in \Psi\mathcal{RH}^\infty(L^p(\Omega))$ *and* $\phi_{-\Delta_{p,w}^N}^{\mathcal{R}\infty} < \frac{\pi}{2}$,

(iii) $-\Delta_{p,w}^N + \delta \in \mathcal{RH}^\infty(L^p(\Omega))$ *and* $\phi_{-\Delta_{p,w}^N+\delta}^{\mathcal{R}\infty} < \frac{\pi}{2}$ *for each $\delta > 0$.*

c) *Let $V \subset \mathbb{R}^n$, $n = 2$, be a bounded Lipschitz domain. Then there exists $\varepsilon > 0$ depending only on the Lipschitz character of V such that the assertion (i) of part b) remains valid for all $(4+\varepsilon)' < p < 4+\varepsilon$.*

Proof. All assertion follow from Proposition 4.16 and Corollary 5.5, see also Corollary 5.4. The assumptions imposed there are proved e.g. in [Duo90, Section 3] for the part a) and [Woo07, Section 4 and 5] for the parts b) and c), respectively. Note that Proposition 4.15 can be applied for $L^p(V)$ enjoys property (α). $\qquad\square$

8.2 Cylindrical boundary value problems

Let $n_1, n_2, n_3 \in \mathbb{N}$, $n := n_1 + n_2 + n_3$, let $V \subset \mathbb{R}^{n_3}$ be a standard domain, and set $\Omega := \mathbb{R}^{n_1} \times (0, 2\pi)^{n_2} \times V$. In this section we investigate a special class of boundary value problems

$$(8.6) \qquad \begin{aligned} \lambda u + \mathcal{A}(\boldsymbol{x}, D)u &= f \quad \text{in } \Omega, \\ \mathcal{B}(\boldsymbol{x}, D)u &= 0 \quad \text{on } \partial\Omega \end{aligned}$$

defined on the cylindrical domain Ω. To some extent, the upcoming results are contained in [NS11b] and [DN11].

In the sequel for $\boldsymbol{x} \in \Omega$ we agree to write $\boldsymbol{x} = (x^1, x^2, x^3) \in \mathbb{R}^{n_1} \times (0, 2\pi)^{n_2} \times V$ and $\alpha = (\alpha^1, \alpha^2, \alpha^3) \in \mathbb{N}_0^{n_1} \times \mathbb{N}_0^{n_2} \times \mathbb{N}_0^{n_3}$ for a multi-index $\alpha \in \mathbb{N}_0^n$, accordingly. Given $m_1, m_2, m_3 \in \mathbb{N}$ and $\nu \in i(-1, 1)^{n_2}$, problem (8.6) is supposed to define a partially ν-periodic boundary value problem $(\mathcal{A}, \mathcal{B}_\nu)$ given through (8.7)

$$\begin{aligned} \lambda u + \mathcal{A}(\boldsymbol{x}, D)u &= f \quad \text{in } \Omega, \\ \mathcal{B}_3(\boldsymbol{x}, D)u &= 0 \quad \text{on } \mathbb{R}^{n_1} \times \mathcal{Q}_{n_2} \times \partial V, \\ (D^\beta u)|_{x_j = 2\pi} - e^{2\pi\nu_j}(D^\beta u)|_{x_j = 0} &= 0 \quad (j = n_1 + 1, \ldots, n_1 + n_2, \; |\beta| < m_2). \end{aligned}$$

Hence, $\mathcal{B} := \mathcal{B}_\nu := (\mathcal{B}_2, \mathcal{B}_3)$, where

$$\mathcal{B}_2 = \{(D^\beta u)|_{x_j = 2\pi} - e^{2\pi\nu_j}(D^\beta u)|_{x_j = 0}, \; j = n_1 + 1, \ldots, n_1 + n_2, \; |\beta| < 2m_2\}.$$

For convenience we set $\partial\mathcal{V} := \mathbb{R}^{n_1} \times \mathcal{Q}_{n_2} \times \partial V$. In particular, we will consider the following class of operators.

Definition 8.8. The boundary value problem (8.7) is called *cylindrical* if the operator $\mathcal{A}(\boldsymbol{x}, D)$ is represented as

$$\mathcal{A}(\boldsymbol{x}, D) = A_1(x^1, D) + A_2(x^2, D) + A_3(x^3, D)$$

with

$$A_i(x^i, D)u := \sum_{|\alpha^i| \leq 2m_i} a_{\alpha^i}^i(x^i) D_{x^i}^{\alpha^1} u$$

and if the boundary operator is given as

$$\mathcal{B}_3(\boldsymbol{x}, D) = \{B_{3,j}(x^3, D); \; j = 1, \ldots, m_3\},$$

where

$$B_{3,j}(x^3, D)u := \sum_{|\beta^3| \leq m_{3,j}} b_{j,\beta^3}^3(x^3)(D_{x^3}^{\beta^3} u)|_{\partial\mathcal{V}} \quad (m_{3,j} < 2m_3, \; j = 1, \ldots, m_3).$$

Thus, the differential operator $\mathcal{A}(x, D)$ resolves completely into parts of which each one acts just on $\mathbb{R}^{n_1}$, $\mathcal{Q}_{n_2}$ or just on V and the boundary operator $\mathcal{B}_3(x, D)$ defined on ∂V in fact acts on ∂V only. Due to this requirement each cylindrical boundary value problem induces a boundary value problem

$$(A_3, B_3) := (A_3(\cdot, D), B_{3,1}(\cdot, D), \ldots, B_{3,m_3}(\cdot, D))$$

on the cross-section V of Ω given through

(8.8)
$$\begin{aligned} \lambda u + A_3(x^3, D)u &= f \text{ in } V, \\ B_{3,j}(x^3, D)u &= 0 \text{ on } \partial V \quad (j = 1, \ldots, m_2). \end{aligned}$$

In view of the results from the previous section, we henceforth assume V to be a C^{2m_3} standard domain. The $L^p(V)$-realization of (A_3, B_3) will be denoted by

$$\begin{aligned} D(A_3) &:= \left\{ u \in W^{2m_3, p}(V); \ B_{3,j}(\cdot, D)u = 0 \ (j = 1, \ldots, m_3) \right\}, \\ A_3 u &:= A_3(\cdot, D)u \quad (u \in D(A_3)). \end{aligned}$$

For the sake of simplicity with $m := \max\{m_1, m_2, m_3\}$ and $\alpha \in \mathbb{N}_0^n$ we henceforth write

$$A_i(x^i, D) = \sum_{|\alpha| \leq 2m} a_\alpha^i(x) D^\alpha \quad (i = 1, 2, 3),$$

where we have set

(8.9)
$$a_\alpha^i(x) = \begin{cases} 0, & \alpha_j \neq 0 \text{ for at least one } j \neq i \text{ or } |\alpha_i| > 2m_i, \\ a_{\alpha^i}^i(x^i), & \text{else.} \end{cases}$$

The notion of parameter-ellipticity for the entire cylindrical boundary value problem from Definition 8.2 (see also [DHP03]) is no longer appropriate for our intentions since any part $A_i(x^i, D)$ of higher order rules out others of lower order. Instead, the differential operators $A_1(x^1, D)$ and $A_2(x^2, D)$ on the one hand and the induced boundary value problem on the other hand can be used to define parameter-ellipticity of a cylindrical boundary value problem efficiently.

Definition 8.9. A cylindrical boundary value problem is called *parameter-elliptic* in Ω if

(i) $A_1(x^1, D)$ is parameter-elliptic in $\mathbb{R}^{n_1}$ with angle $\varphi_1 := \varphi_{A_1} \in [0, \pi)$,

(ii) $A_2(x^2, D)$ is parameter-elliptic in $\mathcal{Q}_{n_2}$ with angle $\varphi_2 := \varphi_{A_2} \in [0, \pi)$,

(iii) the induced boundary value problem (A_3, B_3) on the cross-section V is parameter-elliptic with angle $\varphi_3 := \varphi_{(A_3, B_3)} \in [0, \pi)$, and

(iv) it holds that $\varphi_i + \varphi_j < \pi$ for $i, j = 1, 2, 3$ and $i \neq j$.

We call $\varphi_{(\mathcal{A},\mathcal{B}_\nu)} := \max\{\varphi_i,\ i = 1,2,3\}$ the *angle of parameter-ellipticity* of the cylindrical boundary value problem.

In case $m_1 = m_2 = m_3$ at first sight we could have defined parameter-ellipticity of a cylindrical boundary value problem in Ω simply by subjecting the boundary value problem (8.7) to the common definition of parameter-ellipticity as presented in Definition 8.2. As long as the corresponding angle is less than $\frac{\pi}{2}$ this works out equivalently well since parameter-ellipticity in the sense of Definition 8.9 is implied. Things change considerably in case this constraint is violated, as then we lose track of the Dore-Venni-type condition in (iv) of Definition 8.9. As we have already seen in the previous chapters, however, this condition is crucial in order to carry out a multiplier approach successfully.

As $L^p(\Omega)$-realization of (8.7) we define

$$D(\mathbb{A}) := L^p(\mathbb{R}^{n_1}, L^p(\mathcal{Q}_{n_2}, D(A_3)))\ \cap$$
$$\bigcap_{\frac{\ell_1}{2m_1} + \frac{\ell_2}{2m_2} + \frac{\ell_3}{2m_3} \leq 1} W^{\ell_1,p}(\mathbb{R}^{n_1}, W^{\ell_2,p}_{\nu,per}(\mathcal{Q}_{n_2}, W^{\ell_3,p}(V))),$$
$$\mathbb{A}u := \mathcal{A}(\cdot, D)u \quad (u \in D(\mathbb{A})).$$

Given equality of orders, $m := m_1 = m_2 = m_3$, the domain of definition $D(\mathbb{A})$ equals

$$L^p(\mathbb{R}^{n_1}, L^p(\mathcal{Q}_{n_2}, D(A_3))) \cap \bigcap_{\ell=1}^{2m} W^{\ell_1,p}(\mathbb{R}^{n_1}, \bigcap_{\kappa=1}^{2m-\ell} W^{\kappa,p}_{\nu,per}(\mathcal{Q}_{n_2}, W^{2m-\ell-\kappa,p}(V))),$$

in particular, $D(\mathbb{A}) \subset W^{2m,p}(\Omega)$.

Finally, we require the following smoothness assumptions:

$$(8.10) \quad \begin{cases} a^1_{\alpha^1} \in C(\mathbb{R}^{n_1}) \text{ and } a^1_{\alpha^1}(\infty) := \lim_{|x^1| \to \infty} a^1_{\alpha^1}(x^1) \text{ exists} \quad (|\alpha^1| = 2m_1), \\[2mm] a^2_{\alpha^2} \in C_{per}(\mathcal{Q}_{n_2}) \quad (|\alpha^2| = 2m_2), \\[2mm] a^3_{\alpha^3} \in C(\overline{V}) \text{ and } a^3_{\alpha^3}(\infty) := \lim_{|x^3| \to \infty} a^3_{\alpha^3}(x^3) \text{exists} \quad (|\alpha^3| = 2m_3), \\[2mm] a^1_{\alpha^1} \in [L^\infty + L^{r_\mu}](\mathbb{R}^{n_1})\ r_\mu \geq p,\ \dfrac{2m_1 - \mu}{n_1} > \dfrac{1}{r_\mu} \quad (|\alpha^1| = \mu < 2m_1), \\[2mm] a^2_{\alpha^2} \in L^{r_\mu}(\mathcal{Q}_{n_2})\ r_\mu \geq p,\ \dfrac{2m_2 - \mu}{n_2} > \dfrac{1}{r_\mu} \quad (|\alpha^2| = \mu < 2m_2), \\[2mm] a^3_{\alpha^3} \in [L^\infty + L^{r_\mu}](V)\ r_\mu \geq p,\ \dfrac{2m_3 - \mu}{n_3} > \dfrac{1}{r_\mu} \quad (|\alpha^3| = \mu < 2m_3), \\[2mm] b^3_{j,\beta^3} \in C^{2m_3 - m_{3,j}}(\partial V) \quad (j = 1,\ldots,m_3;\ |\beta^3| \leq m_{3,j}). \end{cases}$$

Here $C_{per}(\mathcal{Q}_{n_2}) := \{f \in C([0,2\pi]^{n_2});\ f|_{x_j=0} = f|_{x_j=2\pi}\ (j = 1,\ldots,n_2)\}$. As a matter of course, the limit condition on $a^3_{\alpha^3}$ for $|\alpha^3| = 2m_3$ drops out in case V

is non-exterior. Note that this limit behavior of the top order coefficients ensures parameter-ellipticity of the limit operators $A_1(\infty, D)$ and $A_2(\infty, D)$ in case V is an exterior domain to be well-defined. Thus, we can extend the notion of parameter-ellipticity of a cylindrical boundary value problem to

$$\overline{\Omega} := \left(\mathbb{R}^n \cup \{\infty\} \right) \times \overline{\mathcal{Q}}_{n_2} \times \overline{V},$$

where we agree on $\{\infty\} \subset \overline{V}$ in case that V is unbounded.

Theorem 8.10. *Let* $1 < p < \infty$ *and* $\nu \in i(-1, 1)^{n_2}$. *Set*

$$\Omega := \mathbb{R}^{n_1} \times \mathcal{Q}_{n_2} \times V \subset \mathbb{R}^n,$$

where V *is a standard domain of class* C^{2m_3} *in* $\mathbb{R}^{n_3}$. *For the boundary value problem* $(\mathcal{A}, \mathcal{B}_\nu)$ *given through* (8.7) *on* Ω *we assume that*

(i) it is cylindrical,

(ii) the coefficients of $A(\cdot, D)$ *and* $B_{3,j}(\cdot, D)$, $j = 1, \ldots, m_3$, *satisfy* (8.10),

(iii) it is (cylindrically) parameter-elliptic in $\overline{\Omega}$ *of angle* $\varphi_{(\mathcal{A}, \mathcal{B}_\nu)} \in [0, \pi)$.

Then for each $\phi > \varphi_{(\mathcal{A}, \mathcal{B}_\nu)}$ *there exists* $\delta = \delta(\phi) \geq 0$ *such that* $\mathbb{A} + \delta \in \mathcal{RS}(L^p(\Omega))$ *and* $\phi^{\mathcal{RS}}_{\mathbb{A}+\delta} \leq \phi$. *Moreover, we have*
(8.11)
$$\mathcal{R}\Big(\Big\{ \lambda^\rho D^\alpha (\lambda + \mathbb{A} + \delta)^{-1};$$

$$\lambda \in \Sigma_{\pi-\phi}, \ \rho \in [0, 1], \ \alpha \in \mathbb{N}_0^n, 0 \leq \rho + \tfrac{|\alpha^1|}{2m_1} + \tfrac{|\alpha^2|}{2m_2} + \tfrac{|\alpha^3|}{2m_3} \leq 1 \Big\} \Big) < \infty.$$

Theorem 8.10 is proved in three steps. First we assume $n_2 = 0$, that is, we consider merely $\Omega_1 := \mathbb{R}^{n_1} \times V$. In a second step, we assume $n_1 = 0$, that is, we consider merely $\Omega_2 := \mathcal{Q}^{n_2} \times V$. Once we have proved Theorem 8.10 for these special cases, in a third step the assertion of Theorem 8.10 without any restriction follows for we can write $\Omega = \mathbb{R}^{n_1} \times \Omega_2$. Note that all relevant properties are transferred from A_3 to $\mathbb{A}$ in each step.

8.2.1 Step 1:

Let $1 < p < \infty$. Since in this step we consider the Cartesian product of two domains only, for the sake of convenience, we shift the index 3 appearing in the boundary value problem (8.7) to 2. That is, from now on we consider

(8.12)
$$\begin{aligned} \lambda u + A_1(x, D)u + A_2(y, D)u &= f \text{ in } \Omega, \\ B_{2,j}(y, D)u &= 0 \text{ on } \partial\Omega. \end{aligned}$$

Here $(x, y) \in \Omega := \mathbb{R}^{n_1} \times V$. Accordingly, the induced boundary value problem

$$(A_2, B_2) := (A_2(\cdot, D), B_{2,1}(\cdot, D), \ldots, B_{2,m_2}(\cdot, D)).$$

reads as

$$
(8.13) \qquad
\begin{aligned}
\lambda u + A_2(y, D)u &= f \text{ in } V, \\
B_{2,j}(y, D)u &= 0 \text{ on } \partial V \quad (j = 1, \ldots, m_2).
\end{aligned}
$$

Clearly, the cross-section V is now assumed to be a C^{2m_2} standard domain. The $L^p(V)$-realization of (A_2, B_2) will be denoted by

$$
\begin{aligned}
D(A_2) &:= \left\{ u \in W^{2m_2, p}(V); \ B_{2,j}(\cdot, D)u = 0 \ (j = 1, \ldots, m_2) \right\}, \\
A_2 u &:= A_2(\cdot, D)u \quad (u \in D(A_2)).
\end{aligned}
$$

Moreover, since ν is no longer involved in (8.7), we write $(\mathcal{A}, \mathcal{B})$ instead of $(\mathcal{A}, \mathcal{B}_\nu)$. As the $L^p(\Omega)$-realization of the cylindrical boundary value problem $(\mathcal{A}, \mathcal{B})$ we accordingly have

$$
\begin{aligned}
D(\mathbb{A}) &:= \Big\{ u \in L^p(\Omega); \ D^\alpha u \in L^p(\Omega) \\
&\qquad \text{for } \tfrac{|\alpha^1|}{2m_1} + \tfrac{|\alpha^2|}{2m_2} \le 1 \text{ and } B_j(\cdot, D)u = 0 \quad (j = 1, \ldots, m_2) \Big\}, \\
\mathbb{A}u &:= \mathcal{A}(\cdot, D)u \quad (u \in D(\mathbb{A})).
\end{aligned}
$$

Note that

$$
D(\mathbb{A}) = L^p(\mathbb{R}^{n_1}, D(A_2)) \cap \bigcap_{\frac{\ell_1}{2m_1} + \frac{\ell_2}{2m_2} \le 1} W^{\ell_1, p}(\mathbb{R}^{n_1}, W^{\ell_2, p}(V)).
$$

First we consider the model problem for the cylindrical boundary value problem (8.12), i.e. we assume $A_1(x, D)$ on $\mathbb{R}^{n_1}$ to be given as homogeneous differential operator

$$
A_1(D) := \sum_{|\alpha| = 2m} a_\alpha^1 D^\alpha
$$

with constant coefficients $a_\alpha^1 \in \mathbb{C}$. We set

$$
\mathcal{A}_0(\cdot, D) := \mathcal{A}_0(y, D) := A_1(D) + A_2(y, D)
$$

and

$$
\mathbb{A}_0 u := \mathcal{A}_0(\cdot, D)u \quad (u \in D(\mathbb{A}_0) := D(\mathbb{A})).
$$

Observe that $A_2(y, D)$ remains unchanged in this model problem and that $\delta_2 \ge 0$ exists such that the assertions of Proposition 8.3 are valid for $A_2 + \delta_2$.

Let $\phi > \varphi_{(A_0, B)}$, $\lambda \in \Sigma_{\pi - \phi}$, and $u \in \mathcal{S}(\mathbb{R}^{n_1}, D(A_2)) \subset D(A_0)$. For the sake of convenience in the sequel we write $E := L^p(V)$ and $X := L^p(\mathbb{R}^{n_1}, E) \cong L^p(\Omega)$. Applying E-valued Fourier transform $\mathcal{F}$ to $f := (\lambda + A_1(D) + A_2 + \delta_2)u$ gives us

$$
(\lambda + a_1(\cdot) + A_2 + \delta_2)\mathcal{F}u = \mathcal{F}f.
$$

Hence, we formally have

$$u = \mathcal{F}^{-1} m_\lambda^0 \mathcal{F} f,$$

where m_λ^0 is given by the operator-valued symbol

$$m_\lambda^0(\xi) := (\lambda + a_1(\xi) + A_2 + \delta_2)^{-1} \quad (\xi \in \mathbb{R}^{n_1}).$$

Note that $m_\lambda^0 \in C^\infty(\mathbb{R}^{n_1}, \mathcal{L}(E))$ is well-defined if

$$-(\lambda + a_1(\xi)) \in \rho(A_2 + \delta_2) \quad (\xi \in \mathbb{R}^{n_1}).$$

In view of our choice of $\lambda \in \Sigma_{\pi-\phi}$, Lemma 6.11, and Proposition 8.3 this is satisfied. The latter and Proposition 6.13 therefore yield

$$\mathcal{R}\left(\left\{\lambda^{1-\frac{|\alpha^1|}{2m_1}} D^{(\alpha^1, 0)}(\lambda + \tilde{A}_0 + \delta_2)^{-1}; \ \lambda \in \Sigma_{\pi-\phi}, \ \alpha^1 \in \mathbb{N}_0^{n_1}, \ 0 \le |\alpha^1| \le 2m_1\right\}\right) < \infty$$

and $N(\tilde{A}_0) = \{0\}$, where $\tilde{A}_0 \supset A_0$ is defined by

$$D(\tilde{A}_0) := \left\{u \in W^{2m_1, p}(\mathbb{R}^{n_1}, E); \ A_2 u \in L^p(\mathbb{R}^{n_1}, E)\right\},$$
$$\tilde{A}_0 u := \mathcal{A}_0(\cdot, D) u \quad (u \in D(\tilde{A}_0)).$$

Towards the stronger estimate (8.11) for A_0, we consider the more involved symbols

$$m_\lambda(\xi) := \lambda^{1-\left(\frac{|\alpha^1|}{2m_1} + \frac{|\alpha^2|}{2m_2}\right)} \xi^{\alpha^1} D^{\alpha^2} m_\lambda^0(\xi)$$
$$= \lambda^{1-\left(\frac{|\alpha^1|}{2m_1} + \frac{|\alpha^2|}{2m_2}\right)} \xi^{\alpha^1} D^{\alpha^2}(\lambda + a_1(\xi) + A_2 + \delta_2)^{-1}$$

for $\lambda \in \Sigma_{\pi-\phi}$, $\xi \in \mathbb{R}^{n_1}$, and $\frac{|\alpha^1|}{2m_1} + \frac{|\alpha^2|}{2m_2} \le 1$. Hence, we restrict ourselves at this point to the special case $\rho = 1 - \left(\frac{|\alpha^1|}{2m_1} + \frac{|\alpha^2|}{2m_2}\right)$ in (8.11) which avoids an additional shift besides δ_2. In turn, (8.11) with arbitrary ρ follows by means of a shift $\delta > \delta_2$.

Proposition 8.11. *For each $\phi > \varphi_{(A_0, B)}$ and $\delta_2 = \delta_2(\phi)$ as in Proposition 8.3 we have $A_0 + \delta_2 \in \mathcal{RS}(X)$ and $\phi_{A_0+\delta_2}^{\mathcal{RS}} \le \phi$. Moreover, it holds that*

(8.14)
$$\mathcal{R}\left(\left\{\lambda^{1-\left(\frac{|\alpha^1|}{2m_1} + \frac{|\alpha^2|}{2m_2}\right)} D^\alpha(\lambda + A_0 + \delta_2)^{-1}; \right.\right.$$
$$\left.\left. \lambda \in \Sigma_{\pi-\phi}, \ \alpha \in \mathbb{N}_0^n, \ 0 \le \frac{|\alpha^1|}{2m_1} + \frac{|\alpha^2|}{2m_2} \le 1\right\}\right) < \infty.$$

Proof. Let $\phi > \varphi_{(A_0, B)}$. For $0 \le \gamma \le 1$ we apply Lemma 6.1 to the result

$$\xi^\gamma D^\gamma m_\lambda(\xi) = \lambda^{1-\left(\frac{|\alpha_1|}{2m_1} + \frac{|\alpha_2|}{2m_2}\right)} \xi^{\alpha^1} D^{\alpha^2}(\lambda + a_1(\xi) + A_2 + \delta_2)^{-1}$$
$$\cdot \sum_{\gamma' \le \gamma} \sum_{\mathcal{W} \in \mathcal{Z}_{\gamma-\gamma'}} C_{\gamma', \alpha^1, \mathcal{W}} \prod_{j=1}^{\mathcal{W}_r} \left(\xi^{\omega^j}(D^{\omega^j} a_1)(\xi)(\lambda + a_1(\xi) + A_2 + \delta_2)^{-1}\right),$$

for all $\lambda \in \Sigma_{\pi-\phi}$, $\xi \in \mathbb{R}^{n_1}$ and all $\alpha \in \mathbb{N}_0^n$ such that $0 \leq \frac{|\alpha^1|}{2m_1} + \frac{|\alpha^2|}{2m_2} \leq 1$. Here $C_{\gamma',\alpha^1,\mathcal{W}} \in \mathbb{Z}$ denotes a constant depending on γ', α^1 and $\mathcal{W}$.

We want to employ Lemma 3.2 to establish m_λ as a Fourier multiplier uniformly in $\lambda \in \Sigma_{\pi-\phi}$. Thanks to (8.4) this is possible if we can show that both

$$\kappa_1(\lambda, \xi) := \frac{\lambda^{1-(\frac{|\alpha_1|}{2m_1}+\frac{|\alpha_2|}{2m_2})}\xi^{\alpha^1}}{\left(\lambda + a_1(\xi)\right)^{1-\frac{|\alpha^2|}{2m_2}}}$$

and

$$\kappa_2(\lambda, \xi) := \frac{\xi^\omega D^\omega a_1(\xi)}{\lambda + a_1(\xi)}$$

are uniformly bounded for $(\lambda, \xi) \in \Sigma_{\pi-\phi} \times \mathbb{R}^{n_1}$. Obviously κ_2 meets the conditions in the proof of Lemma 6.5 perfectly and uniform boundedness follows. While the structure of κ_1 includes fractional powers, quasi-$(2m_1,1)$-homogeneity of degree zero is preserved. We therefore can apply arguments very similar to those in Lemma 6.5 to show that κ_1 is uniformly bounded as well.

Hence, $R_\lambda := \mathcal{F}^{-1} m_\lambda \mathcal{F} \in \mathcal{L}(X)$ exists, where

$$R(R_\lambda) \subset \bigcap_{\frac{\ell_1}{2m_1}+\frac{\ell_2}{2m_2} \leq 1} W^{\ell_1,p}(\mathbb{R}^{n_1}, W^{\ell_2,p}(V)).$$

Therefore, the equality $R_\lambda = (\lambda + \mathbb{A}_0 + \delta_2)^{-1}$ and the validity of (8.14) follow if

$$B_{2,j} R_\lambda f = 0 \quad (f \in L^p(\Omega); \ j = 1, \ldots, m_2).$$

To see this, we represent the resolvent applied to $f \in \mathcal{S}(\mathbb{R}^{n_1}, E)$ as a Bochner integral via

$$(\lambda + \mathbb{A}_0 + \delta_2)^{-1} f(x) = \frac{1}{(2\pi)^{n_1/2}} \int_{\mathbb{R}^{n_1}} e^{ix\xi}(\lambda + a_1(\xi) + A_2 + \delta_2)^{-1} \mathcal{F}f(\xi) d\xi.$$

Since taking the trace acts as a bounded operator on E, it commutes with the integral sign. This yields

$$B_{2,j}(\lambda + \mathbb{A}_0 + \delta_2)^{-1} f = 0 \quad (f \in \mathcal{S}(\mathbb{R}^{n_1}, E); \ j = 1, \ldots, m_2).$$

Employing density of $\mathcal{S}(\mathbb{R}^{n_1}, E)$ in $L^p(\mathbb{R}^{n_1}, E)$ we conclude that

$$R(R_\lambda) = L^p(\mathbb{R}^{n_1}, D(A_2)) \cap \bigcap_{\frac{\ell_1}{2m_1}+\frac{\ell_2}{2m_2} \leq 1} W^{\ell_1,p}(\mathbb{R}^{n_1}, W^{\ell_2,p}(V)) = D(\mathbb{A}_0).$$

$\square$

Note again that the appearing shift δ_2 does not have to be enlarged while carrying over $\mathcal{R}$-sectoriality from A_2 to $\mathbb{A}_0$.

By a perturbation argument we generalize the $\mathcal{R}$-sectoriality for constant coefficients to the case of slightly varying coefficients of A_1. To this end, we will employ the following perturbation result which is based on a standard Neumann series argument.

Lemma 8.12. *Let R be a linear operator in X such that $D(\mathbb{A}_0) \subset D(R)$ and let δ_2 be given as in Proposition 8.11. Assume that there are $\eta > 0$ and $\delta > \delta_2$ such that*

$$\|Ru\|_X \leq \eta \|(\mathbb{A}_0 + \delta)u\|_X \quad (u \in D(\mathbb{A}_0)).$$

Then $\mathbb{A}_0 + R + \delta \in \mathcal{R}S(X)$, $\phi^{\mathcal{R}S}_{\mathbb{A}_0 + R + \delta} \leq \phi^{\mathcal{R}S}_{\mathbb{A}_0 + \delta_2}$, and for every $\phi > \varphi_{(A_0, B)}$ we have

$$(8.15) \quad \mathcal{R}\Big(\Big\{\lambda^\rho D^\alpha (\lambda + \mathbb{A}_0 + R + \delta)^{-1};$$
$$\lambda \in \Sigma_{\pi - \phi},\ \alpha \in \mathbb{N}_0^n,\ \rho \in [0, 1],\ 0 \leq \rho + \frac{|\alpha^1|}{2m_1} + \frac{|\alpha^2|}{2m_2} \leq 1\Big\}\Big) < \infty,$$

whenever $\eta < \mathcal{R}(\{(\mathbb{A}_0 + \delta)(\lambda + \mathbb{A}_0 + \delta)^{-1}\})^{-1}$.

Proof. As

$$\|R(\lambda + \mathbb{A}_0 + \delta)^{-1}\|_{\mathcal{L}(X)} \leq \eta \|(\mathbb{A}_0 + \delta)(\lambda + \mathbb{A}_0 + \delta)^{-1}\|_{\mathcal{L}(X)}$$
$$\leq \eta \mathcal{R}(\{(\mathbb{A}_0 + \delta)(\lambda + \mathbb{A}_0 + \delta)^{-1}\}) < 1$$

by assumption, we see that

$$\lambda + \mathbb{A}_0 + R + \delta = \big(1 + R(\lambda + \mathbb{A}_0 + \delta)^{-1}\big)(\lambda + \mathbb{A}_0 + \delta)$$

is invertible. This implies

$$\lambda^\rho D^\alpha (\lambda + \mathbb{A}_0 + R + \delta)^{-1} = \lambda^\rho D^\alpha (\lambda + \mathbb{A}_0 + \delta)^{-1} \sum_{j=0}^{\infty} (-R(\lambda + \mathbb{A}_0 + \delta)^{-1})^j.$$

By assumption we have $\delta_0 := \delta - \delta_2 > 0$. The fact that

$$|\lambda + \delta_0| \geq c_\phi \delta_0 \quad (\lambda \in \Sigma_{\pi - \phi})$$

for some $c_\phi > 0$ yields the existence of a $M_\phi > 0$ such that

$$\frac{|\lambda^\rho|}{\big|(\lambda + \delta_0)^{1 - (\frac{|\alpha^1|}{2m_1} + \frac{|\alpha^2|}{2m_2})}\big|} \leq M_\phi \quad (\lambda \in \Sigma_{\pi - \phi}).$$

Thanks to the contraction principle of Kahane and Proposition 8.11 we deduce

$$\mathcal{R}(\{\lambda^\rho D^\alpha (\lambda + \mathbb{A}_0 + \delta)^{-1}\})$$
$$\leq C\mathcal{R}\Big(\Big\{(\lambda + \delta_0)^{1 - \big(\frac{|\alpha^1|}{2m_1} + \frac{|\alpha^2|}{2m_2}\big)} D^\alpha ((\lambda + \delta_0) + \mathbb{A}_0 + \delta_2)^{-1}\Big\}\Big) \leq C.$$

Lemma 3.2a) then yields

$$\mathcal{R}(\{\lambda^\rho D^\alpha(\lambda + \mathbb{A}_0 + \delta)^{-1}(-R(\lambda + \mathbb{A}_0 + \delta)^{-1})^j\})$$
$$\leq \mathcal{R}(\{\lambda^\rho D^\alpha(\lambda + \mathbb{A}_0 + \delta)^{-1}\})\mathcal{R}(\{(R(\lambda + \mathbb{A}_0 + \delta)^{-1})^j\})$$
$$\leq C\eta^j\mathcal{R}(\{(\mathbb{A}_0 + \delta)(\lambda + \mathbb{A}_0 + \delta)^{-1})\})^j \leq C\nu^j \quad (j \in \mathbb{N}_0)$$

with $\nu := \eta\mathcal{R}(\{(\mathbb{A}_0 + \delta)(\lambda + \mathbb{A}_0 + \delta)^{-1}\}) < 1$. Employing again Lemma 3.2a), in particular the fact that the $\mathcal{R}$-bound is preserved when taking the closure in the strong operator topology, the assertion follows. $\qquad\square$

For the particular choice $\rho := 1 - \left(\frac{|\alpha^1|}{2m_1} + \frac{|\alpha^2|}{2m_2}\right)$, Lemma 8.12 does not require any enlargement of δ_2. However, for the localization procedure in order it is of importance to be aware of the $\mathcal{R}$-boundedness condition (8.15) for each $\alpha \in \mathbb{N}_0^n$ and all $\ell \in \mathbb{N}_0$ subject to $0 \leq \ell + 2m_2|\alpha^1| + 2m_1|\alpha^2| \leq 4m_1m_2$.

Corollary 8.13. *Let* $R(x, D) := \sum_{|\alpha^1|=2m_1} r_{\alpha^1}(x)D^{\alpha^1}$ *be given such that the condition* $\sum_{|\alpha^1|=2m_1} \|r_{\alpha^1}\|_\infty < \eta$ *is satisfied. Set*

$$(8.16) \qquad \mathcal{A}^{va}(\cdot, D) := \mathcal{A}^{va}(x, y, D) := \mathcal{A}_0(y, D) + R(x, D) \quad ((x, y) \in \Omega)$$

and denote its X*-realization by*

$$\mathbb{A}^{va}u := \mathcal{A}^{va}(\cdot, D)u \quad (u \in D(\mathbb{A}^{va}) := D(\mathbb{A}_0)).$$

Then there exists a $\delta > 0$ *such that* $\mathbb{A}^{va} + \delta \in \mathcal{R}S(X)$ *and* $\phi_{\mathbb{A}^{va}+\delta}^{\mathcal{R}S} \leq \phi_{\mathbb{A}_0+\delta_2}^{\mathcal{R}S}$ *provided that* η *is sufficiently small. In this case, for* $\phi > \varphi_{(\mathbb{A}_0,B)}$ *we have*

$$(8.17) \qquad \mathcal{R}\left(\left\{\lambda^\rho D^\alpha(\lambda + \mathbb{A}^{va} + \delta)^{-1};\right.\right.$$
$$\left.\left.\lambda \in \Sigma_{\pi-\phi},\ \alpha \in \mathbb{N}_0^n,\ \rho \in [0, 1],\ 0 \leq \rho + \frac{|\alpha^1|}{2m_1} + \frac{|\alpha^2|}{2m_2} \leq 1\right\}\right) < \infty.$$

Proof. By Proposition 8.11, in particular by relation (8.14), there exists a $C > 0$ such that

$$\|D^{\alpha^1}(\mathbb{A}_0 + \delta)^{-1}\|_{\mathcal{L}(X)} \leq C \quad (\alpha^1 \in \mathbb{N}_0^{n_1},\ |\alpha^1| = 2m_1)$$

for each $\delta > \delta_2$. For a fixed $\delta > \delta_2$ this implies

$$\|Ru\|_p \leq \sum_{|\alpha^1|=2m_1} \|r_{\alpha^1}\|_\infty\|D^{\alpha^1}(\mathbb{A}_0 + \delta)^{-1}(\mathbb{A}_0 + \delta)u\|_p$$
$$\leq C\eta\|(\mathbb{A}_0 + \delta)u\|_p \quad (u \in D(\mathbb{A}_0)).$$

Thus, if we assume that $\eta < 1/C\mathcal{R}(\{(\mathbb{A}_0+\delta)(\lambda+\mathbb{A}_0+\delta)^{-1}\})$, the assertion follows from Lemma 8.12. $\qquad\square$

In the next lemma we establish estimates that will turn out to be crucial for the localization procedure.

Lemma 8.14. *Let* $1 < p < \infty$, $\beta \in \mathbb{N}_0^{n_1}$, $|\beta| = \mu < 2m_1$, *and* $r_\mu \geq p$ *such that* $2m_1 - \mu > \frac{n_1}{r_\mu}$. *Let* $b \in [L^\infty + L^{r_\mu}](\mathbb{R}^{n_1})$, $\mathbb{A}^{va}$ *be the operator as defined in (8.16), and assume that* $\phi > \varphi_{(\mathcal{A},\mathcal{B})}$.
a) *For every* $\varepsilon > 0$ *there exists* $C(\varepsilon) > 0$ *such that*

$$\|bD^\beta u\|_p \leq \varepsilon\|u\|_{p,2m_1} + C(\varepsilon)\|u\|_p \quad (u \in W^{2m_1,p}(\mathbb{R}^{n_1}, E)).$$

b) *For every* $\varepsilon > 0$ *there exists a* $\delta = \delta(\varepsilon) > 0$ *such that*

$$\mathcal{R}(\{bD^\beta(\lambda + \mathbb{A}^{va} + \delta)^{-1}; \ \lambda \in \Sigma_{\pi-\phi}\}) \leq \varepsilon.$$

Proof. a) Let $\varepsilon > 0$ be arbitrary. For $b \in L^\infty(\mathbb{R}^{n_1})$ we obtain by Hölder's inequality and vector-valued complex interpolation (see e.g. [Ama95]) that

$$\|bD^\beta u\|_p \leq \|b\|_\infty\|u\|_{p,\mu} \leq C\|b\|_\infty\|u\|_{p,2m_1}^{\frac{\mu}{2m_1}}\|u\|_p^{1-\frac{\mu}{2m_1}} \quad (u \in W^{2m_1,p}(\mathbb{R}^{n_1}, E)).$$

With the help of Young's inequality we further deduce

$$\|bD^\beta u\|_p \leq \varepsilon\|u\|_{p,2m_1} + C(\varepsilon)\|u\|_p \quad (u \in W^{2m_1,p}(\mathbb{R}^{n_1}, E)).$$

Now let $b \in L^{r_\mu}(\mathbb{R}^{n_1})$, $r := \frac{r_\mu}{p}$, and $\frac{1}{r} + \frac{1}{r'} = 1$. Then Hölder's inequality and the vector-valued version of the Gagliardo-Nirenberg inequality (see [SS05]) imply

$$\|bD^\beta u\|_p \leq C\|b\|_{pr}\|D^\beta u\|_{pr'} \leq C\|b\|_{r_\mu}\|u\|_{p,2m_1}^\tau\|u\|_p^{1-\tau},$$

where $\tau = \frac{n_1}{r_\mu(2m_1-\mu)} \in (0,1)$ by our assumption on r_μ. Again an application of Young's inequality yields

$$\|bD^\beta u\|_p \leq \varepsilon\|u\|_{p,2m_1} + C(\varepsilon)\|u\|_p \quad (u \in W^{2m_1,p}(\mathbb{R}^{n_1}, E)).$$

b) Let $(\varepsilon_j)_{j\in\mathbb{N}}$ be a family of independent symmetric $\{-1,1\}$-valued random variables on a probability space $([0,1], \mathcal{M}, P)$, $\lambda_j \in \Sigma_{\pi-\phi}$, and $f_j \in X$. For any $b \in L^\infty(\mathbb{R}^{n_1})$, $\delta_0 > 0$, and arbitrary $t \in [0,1]$ we have

$$\|\sum_{j=1}^N \varepsilon_j(t)bD^\beta(\lambda_j + \delta_0 + \mathbb{A}^{va} + \delta)^{-1}f_j\|_p$$

$$\leq \|b\|_\infty\|\sum_{j=1}^N \varepsilon_j(t)D^\beta(\lambda_j + \delta_0 + \mathbb{A}^{va} + \delta)^{-1}f_j\|_p.$$

Note that there is a $c_\phi > 0$ such that

$$|\lambda + \delta_0| \geq c_\phi\delta_0 \quad (\lambda \in \Sigma_{\pi-\phi}, \ \delta_0 > 0).$$

Taking L^p-norm with respect to t and applying the contraction principle of Kahane therefore yields

$$\left\| \sum_{j=1}^{N} \varepsilon_j(\cdot) b D^\beta (\lambda_j + \delta_0 + \mathbb{A}^{va} + \delta)^{-1} f_j \right\|_{L^p([0,1],X)}$$

$$\leq C\|b\|_\infty \left\| \sum_{j=1}^{N} \varepsilon_j(\cdot) \left(\frac{\lambda_j + \delta_0}{\delta_0} \right)^{1 - \frac{|\beta|}{2m_1}} D^\beta (\lambda_j + \delta_0 + \mathbb{A}^{va} + \delta)^{-1} f_j \right\|_{L^p([0,1],X)}.$$

Thanks to (8.17) this implies

$$\left\| \sum_{j=1}^{N} \varepsilon_j(\cdot) b D^\beta (\lambda_j + \delta_0 + \mathbb{A}^{va} + \delta)^{-1} f_j \right\|_{L^p([0,1],X)}$$

$$\leq C\|b\|_\infty \delta_0^{-\left(1 - \frac{|\beta|}{2m_1}\right)} \left\| \sum_{j=1}^{N} \varepsilon_j(\cdot) f_j \right\|_{L^p([0,1],X)}.$$

Thus, for $\delta_0 > (C\|b\|_\infty/\varepsilon)^{1/(1-|\beta|/2m_1)}$ the assertion follows.

In case that $b \in L^{r_\mu}(\mathbb{R}^{n_1})$ Hölder's inequality and the Gagliardo-Nirenberg inequality imply for $\tau(2m_1 - \mu) = \frac{n_1}{r_\mu}$ and arbitrary $t \in [0,1]$ that

$$\left\| \sum_{j=1}^{N} \varepsilon_j(t) b D^\beta (\lambda_j + \delta_0 + \mathbb{A}^{va} + \delta)^{-1} f_j \right\|_p$$

$$\leq \|b\|_{pr} \left\| \sum_{j=1}^{N} \varepsilon_j(t) D^\beta (\lambda_j + \delta_0 + \mathbb{A}^{va} + \delta)^{-1} f_j \right\|_{pr'}$$

$$\leq C\|b\|_{r_\mu} \left(\sum_{|\alpha|=2m_1} \left\| \sum_{j=1}^{N} \varepsilon_j(t) D^\alpha (\lambda_j + \delta_0 + \mathbb{A}^{va} + \delta)^{-1} f_j \right\|_p^p \right)^{\tau/p}$$

$$\cdot \left\| \sum_{j=1}^{N} \varepsilon_j(t) (\lambda_j + \delta_0 + \mathbb{A}^{va} + \delta)^{-1} f_j \right\|_p^{1-\tau}.$$

Taking L^p-norm with respect to t and applying once more Hölder's inequality we deduce

$$\left\| \sum_{j=1}^{N} \varepsilon_j(\cdot) b D^\beta (\lambda_j + \delta_0 + \mathbb{A}^{va} + \delta)^{-1} f_j \right\|_{L^p([0,1],X)}$$

$$\leq C\|b\|_{r_\mu} \left(\sum_{|\alpha|=2m_1} \left\| \sum_{j=1}^{N} \varepsilon_j(\cdot) D^\alpha (\lambda_j + \delta_0 + \mathbb{A}^{va} + \delta)^{-1} f_j \right\|_{L^p([0,1],X)}^p \right)^{\tau/p}$$

$$\cdot \left\| \sum_{j=1}^{N} \varepsilon_j(\cdot) (\lambda_j + \delta_0 + \mathbb{A}^{va} + \delta)^{-1} f_j \right\|_{L^p([0,1],X)}^{1-\tau}.$$

The contraction principle of Kahane then gives us

$$\left\| \sum_{j=1}^{N} \varepsilon_j(\cdot) b D^\beta (\lambda_j + \delta_0 + \mathbb{A}^{va} + \delta)^{-1} f_j \right\|_{L^p([0,1],X)}$$

$$\leq C \|b\|_{r_\mu} \left(\sum_{|\alpha|=2m_1} \left\| \sum_{j=1}^{N} \varepsilon_j(\cdot) D^\alpha (\lambda_j + \delta_0 + \mathbb{A}^{va} + \delta)^{-1} f_j \right\|_{L^p([0,1],X)}^p \right)^{\tau/p}$$

$$\cdot \left\| \sum_{j=1}^{N} \varepsilon_j(\cdot) \left(\frac{\lambda_j + \delta_0}{\delta_0} \right) (\lambda_j + \delta_0 + \mathbb{A}^{va} + \delta)^{-1} f_j \right\|_{L^p([0,1],X)}^{1-\tau}.$$

Taking into account (8.17) we arrive at

$$\left\| \sum_{j=1}^{N} \varepsilon_j(\cdot) b D^\beta (\lambda_j + \delta_0 + \mathbb{A}^{va} + \delta)^{-1} f_j \right\|_{L^p([0,1],X)}$$

$$\leq C \|b\|_{r_\mu} \delta_0^{\tau-1} \left\| \sum_{j=1}^{N} \varepsilon_j(\cdot) f_j \right\|_{L^p([0,1],X)}.$$

Choosing $\delta_0 > (C\|b\|_{r_\mu}/\varepsilon)^{1/(1-\tau)}$ proves the claim. $\square$

Now we are in the position to perform a successful localization procedure. We denote by

$$A_1^\#(x, D) := \sum_{|\alpha|=2m} a_\alpha^1(x) D^\alpha$$

the principal part of $A_1(x, D)$. Freezing the coefficients at $x_0 \in \mathbb{R}^{n_1} \cup \{\infty\}$, Proposition 8.11 applies to $A_1(D) := A_1^\#(x_0, D)$.

So, we first choose a large ball $B_{r_0}(0) \subset \mathbb{R}^{n_1}$ with a fixed radius $r_0 > 0$ such that

$$|a_{\alpha^1}^1(x) - a_{\alpha^1}^1(\infty)| \leq \eta/M_\alpha \quad (|x| \geq r_0, \ |\alpha^1| = 2m_1),$$

were $M_\alpha = |\{\alpha^1 \in \mathbb{N}_0^{n_1}; \ |\alpha^1| = 2m_1, \ a_{\alpha^1} \neq 0\}|$ and $\eta = \eta(\infty)$ is the constant given in Corollary 8.13 applied to $A_1^\#(\infty, D) = \sum_{|\alpha|=2m_1} a_\alpha^1(\infty) D^\alpha$. For every fixed $x_0 \in \overline{B}_{r_0}(0)$ let $\eta = \eta(x_0)$ be the constant given in Corollary 8.13 applied to $A_1^\#(x_0, D)$. Due to the continuity assumptions on the coefficients there exists a radius $r = r(x_0)$ such that

$$|a_{\alpha^1}^1(x) - a_{\alpha^1}^1(x_0)| \leq \eta(x_0)/M_\alpha, \quad (|x - x_0| \leq r(x_0), \ |\alpha^1| = 2m_1).$$

Obviously the collection $\{B_{r(x_0)}(x_0); \ x_0 \in \overline{B}_{r_0}(0)\}$ represents an open covering of $\overline{B}_{r_0}(0)$. Thus, by compactness we have

$$\overline{B}_{r_0}(0) \subseteq \bigcup_{j=1}^{N} B_{r(x_j)}(x_j)$$

for a certain finite set $(x_j)_{j=1}^N$.

For simplicity we set $r_j := r(x_j)$, and $U_j := B_{r_j}(x_j)$ for $j = 1, \ldots, N$, as well as $U_0 := \mathbb{R}^{n_1} \setminus \overline{B}_{r_0}(0)$ and $x_0 := \infty$. For each $j = 0, \ldots, N$ we define coefficients of $A_1^{\#}(x, D)$-localizations

$$A_j^{1,loc}(x, D) := \sum_{|\alpha|=2m} a_{j,\alpha}^1(x) D^\alpha$$

by reflection, that is, we set

$$a_{0,\alpha}^1(x) = \begin{cases} a_\alpha^1(x), & x \notin \overline{B}_{r_0}(0), \\ a_\alpha^1(\frac{r_0^2}{|x|^2}x), & x \in \overline{B}_{r_0}(0) \end{cases}$$

and

$$a_{j,\alpha}^1(x) = \begin{cases} a_\alpha^1(x), & x \in \overline{B}_{r_j}(x_j), \\ a_\alpha^1(x_j + \frac{r_j^2}{|x-x_j|^2}(x - x_j)), & x \notin \overline{B}_{r_j}(x_j). \end{cases}$$

Then by definition we have

$$\sum_{|\alpha^1|=2m_1} |a_{j,\alpha^1}^1(x) - a_{\alpha^1}^1(x_j)| \le \eta(x_j)$$

for $x \in \mathbb{R}^{n_1}$ and $j = 0, \ldots, N$, that is,

$$\mathcal{A}_j^{loc}(x, y, D) := A_j^{1,loc}(x, D) + A_2(y, D)$$

is a small variation of

$$\mathcal{A}^{j,\#}(y, D) := A_1^{\#}(x_j, D) + A_2(y, D)$$

in the sense of (8.16). Hence, Corollary 8.13 applies to

$$\mathbb{A}_j^{loc} u := \mathcal{A}_j^{loc}(\cdot, D)u \quad (u \in D(\mathbb{A}_j^{loc}) := D(\mathbb{A})).$$

For $j = 0, \ldots, N$ and each $\phi > \varphi_{(\mathcal{A}, \mathcal{B})}$ this yields the existence of $\delta = \delta(\phi) > 0$ such that $\mathbb{A}_j^{loc} + \delta \in \mathcal{RS}(X)$ and

$$(8.18) \quad \mathcal{R}\Big(\Big\{\lambda^\rho D^\alpha(\lambda + \mathbb{A}_j^{loc} + \delta)^{-1}; \\ \lambda \in \Sigma_{\pi-\phi},\ \alpha \in \mathbb{N}_0^n,\ \rho \in [0,1],\ 0 \le \rho + \frac{|\alpha^1|}{2m_1} + \frac{|\alpha^2|}{2m_2} \le 1\Big\}\Big) < \infty.$$

Next we choose a partition of unity $(\varphi_j)_{j=0}^N \subset C^\infty(\mathbb{R}^{n_1})$ of $\mathbb{R}^{n_1}$ subordinate to the open covering $(U_j)_{j=0}^N$ such that $0 \le \varphi_j \le 1$. In addition, we fix $\psi_j \in C^\infty(\mathbb{R}^{n_1})$ such that $\psi_j \equiv 1$ on $\operatorname{supp}\varphi_j$ and $\operatorname{supp}\psi_j \subset U_j$. We further set

$$A_{low}(x, D) := \mathcal{A}(x, y, D) - \mathcal{A}^{\#}(x, y, D)$$

which in fact acts on $\mathbb{R}^{n_1}$ only. Pick $\lambda \in \Sigma_{\pi-\phi}$. Then

$$\lambda u + \mathcal{A}(\cdot, D)u = f$$

holds if and only if

$$\lambda u + \mathcal{A}^{\#}(\cdot, D)u = f - A_{low}(\cdot, D)u.$$

Multiplying the line above by φ_j we obtain

$$\lambda \varphi_j u + \mathcal{A}^{\#}(\cdot, D)\varphi_j u = \varphi_j f + [\mathcal{A}^{\#}(\cdot, D), \varphi_j]u - \varphi_j A_{low}(\cdot, D)u,$$

where the commutators

$$[\mathcal{A}^{\#}(\cdot, D), \varphi_j] := \mathcal{A}^{\#}(\cdot, D)\varphi_j - \varphi_j \mathcal{A}^{\#}(\cdot, D) = [A_1^{\#}(\cdot, D), \varphi_j]$$

in fact do only depend on $A_1^{\#}(\cdot, D)$. Applying the resolvent of $\mathbb{A}_j^{loc}$ to the localized equations we deduce

$$\varphi_j u = (\lambda + \mathbb{A}_j^{loc} + \delta)^{-1} \varphi_j f + (\lambda + \mathbb{A}_j^{loc} + \delta)^{-1}([A_1^{\#}(\cdot, D), \varphi_j]u - \varphi_j A_{low}(\cdot, D)u).$$

By multiplying with ψ_j and by summing up over j we gain the representation

$$u = \sum_{j=0}^{N} \psi_j (\lambda + \mathbb{A}_j^{loc} + \delta)^{-1} \varphi_j f$$

$$+ \sum_{j=0}^{N} \psi_j (\lambda + \mathbb{A}_j^{loc} + \delta)^{-1} ([A_1^{\#}(\cdot, D), \varphi_j]u - \varphi_j A_{low}(\cdot, D))u.$$

Hence, we obtain

$$(I - \sum_{j=0}^{N} \psi_j (\lambda + \mathbb{A}_j^{loc} + \delta)^{-1} \mathcal{C}_j(\cdot, D))u = \sum_{j=0}^{N} \psi_j (\lambda + \mathbb{A}_j^{loc} + \delta)^{-1} \varphi_j f,$$

where

$$\mathcal{C}_j(\cdot, D) := [A_1^{\#}(\cdot, D), \varphi_j] - \varphi_j A_{low}(\cdot, D)$$

is a differential operator of lower order, acting on $\mathbb{R}^{n_1}$ only, whose coefficients fulfill the assumptions of Lemma 8.14. We set

(8.19) $$R_0(\lambda, \delta) := \sum_{j=0}^{N} \psi_j (\lambda + \mathbb{A}_j^{loc} + \delta)^{-1} \varphi_j$$

and

$$R_1(\lambda, \delta) := \sum_{j=0}^{N} \psi_j (\lambda + \mathbb{A}_j^{loc} + \delta)^{-1} \mathcal{C}_j(\cdot, D).$$

Both $R_0(\lambda, \delta)$ and $R_1(\lambda, \delta)$ map to

$$\mathbb{E} := L^p(\mathbb{R}^{n_1}, D(A_2)) \cap \bigcap_{\frac{\ell_1}{2m_1} + \frac{\ell_2}{2m_2} \leq 1} W^{\ell_1, p}(\mathbb{R}^{n_1}, W^{\ell_2, p}(V)).$$

Given $u \in \mathbb{E}$, relation (8.18) and Lemma 8.14a) further imply that

$$\|R_1(\lambda, \delta' + \delta_0)u\|_{\mathbb{E}} + \delta_0\|R_1(\lambda, \delta' + \delta_0)u\|_p$$
$$\leq C\left(\|R_1(\lambda + \delta_0, \delta')u\|_{\mathbb{E}} + |\lambda + \delta_0|\|R_1(\lambda + \delta_0, \delta')u\|_p\right)$$
$$\leq C\|\mathcal{C}_j(\cdot, D)u\|_p \leq C\left(\varepsilon\|u\|_{2m_1, p, \mathbb{R}^{n_1}, \mathbb{E}} + C(\varepsilon)\|u\|_p\right)$$
$$\leq \tfrac{1}{2}\left(\|u\|_{2m_1, p, \mathbb{R}^{n_1}, \mathbb{E}} + \delta_0\|u\|_p\right) \leq \tfrac{1}{2}\left(\|u\|_{\mathbb{E}} + \delta_0\|u\|_p\right) \quad (\lambda \in \Sigma_{\pi-\phi})$$

for some $\delta' > 0$ and provided that $\delta_0 > 0$ is sufficiently large. Setting $\delta := \delta' + \delta_0$ we see that then

$$L_\lambda := (I - R_1(\lambda, \delta))^{-1}R_0(\lambda, \delta) \colon L^p(\mathbb{R}^{n_1}, E) \to D(A)$$

is a left inverse of $\lambda + \mathbb{A} + \delta$ which admits an estimate

$$\|\lambda L_\lambda f\|_p \leq C\|f\|_p \quad (\lambda \in \Sigma_{\pi-\phi}).$$

Therefore, if we can show that there exists a right inverse as well we have proved $\mathbb{A} + \delta \in S(X)$ and $\phi_{\mathbb{A}+\delta} \leq \phi$.

To this end, let $f \in X$ be arbitrary and consider

$$(\lambda + \mathcal{A}(\cdot, D) + \delta)R_0(\lambda, \delta)f = (\lambda + \mathcal{A}^{\#}(\cdot, D) + \delta)R_0(\lambda, \delta)f + A_{low}(\cdot, D)R_0(\lambda, \delta)f.$$

The sum on the right can be rewritten as

$$\sum_{j=0}^N \psi_j(\lambda + \mathcal{A}^{\#}(\cdot, D) + \delta)(\lambda + \mathbb{A}_j^{loc} + \delta)^{-1}\varphi_j f + \sum_{j=0}^N \mathcal{D}_j(\cdot, D)(\lambda + \mathbb{A}_j^{loc} + \delta)^{-1}\varphi_j f,$$

where

$$\mathcal{D}_j(\cdot, D) := [A_1^{\#}(\cdot, D), \psi_j] + A_{low}(\cdot, D)\psi_j$$

is again a differential operator of lower order, acting on $\mathbb{R}^{n_1}$ only, whose coefficients fulfill the assumptions of Lemma 8.14. Since supp $\psi_j \subset U_j$ and $\psi \equiv 1$ on supp φ_j, we obtain

$$(\lambda + \mathcal{A}(\cdot, D) + \delta)R_0(\lambda, \delta)f = f + R_2(\lambda, \delta)f$$

with

$$R_2(\lambda, \delta) := \sum_{j=0}^N \mathcal{D}_j(\cdot, D)(\lambda + \mathbb{A}_j^{loc} + \delta)^{-1}\varphi_j.$$

Lemma 8.14b) implies $\|R_2(\lambda,\delta)\|_{\mathcal{L}(X)} \leq \frac{1}{2}$ and thus existence of $(I + R_2(\lambda,\delta))^{-1}$ for large enough $\delta > 0$. Consequently, $R_\lambda := R_0(\lambda,\delta)(I + R_2(\lambda,\delta))^{-1}$ is a right inverse of $\lambda + \mathbb{A} + \delta$.

With the help of the Leibniz rule and the contraction principle of Kahane, from representation (8.19) and relation (8.18) we obtain that

$$\mathcal{R}(\{\lambda^\rho D^\alpha R_0(\lambda,\delta)\}) \leq C(N+1).$$

Note that commutators $[D^\alpha, \psi_j]$ are involved in this calculation. It is here that we make use of the more general assertion that the $\mathcal{R}$-boundedness condition (8.18) is valid for each $\alpha \in \mathbb{N}_0^n$ and all $\ell \in \mathbb{N}_0$ subject to $0 \leq \ell + 2m_2|\alpha^1| + 2m_1|\alpha^2| \leq 4m_1 m_2$ and $\rho := \frac{\ell}{4m_1 m_2}$ instead of for $\rho := 1 - \left(\frac{|\alpha^1|}{2m_1} + \frac{|\alpha^2|}{2m_2}\right)$ only.

In view of Lemma 8.14b) and Lemma 3.2 the representation

$$(\lambda + \mathbb{A} + \delta)^{-1} = R_0(\lambda,\delta) \sum_{i=0}^{\infty} R_2(\lambda,\delta)^i$$

as a Neumann series finally gives us

$$\mathcal{R}(\{\lambda^\rho D^\alpha(\lambda + \mathbb{A} + \delta)^{-1}; \ \lambda \in \Sigma_{\pi-\phi}, \ 0 \leq \ell + |\alpha| \leq 2m\})$$
$$\leq \mathcal{R}(\{\lambda^\rho D^\alpha R_0(\lambda,\delta)\})\mathcal{R}\left(\left\{\sum_{i=0}^{\infty} R_2(\lambda,\delta)^i\right\}\right)$$
$$\leq (N+1)C \sum_{i=0}^{\infty} (N+1)^i (C\varepsilon)^i = \frac{(N+1)C}{1-(N+1)C\varepsilon} < \infty.$$

Hence, the proof of Theorem 8.10 within step 1 is complete.

8.2.2 Step 2:

As in the previous step we take into account that Ω is given as the Cartesian product of two domains only. More precisely, we shift the indices 2 and 3 to 1 and 2, respectively. In particular, we consider $\Omega := \mathcal{Q}_{n_1} \times V \subset \mathbb{R}^n$, where V is a standard domain in $\mathbb{R}^{n_2}$. Within this step, we investigate the ν-periodic boundary value problem $(\mathcal{A}, \mathcal{B}_\nu)$ given through

$$(8.20) \quad \begin{aligned} \lambda u + A_1(x, D)u + A_2(y, D)u &= f \quad \text{in } \Omega, \\ B_j(y, D)u &= 0 \quad \text{on } \mathcal{Q}_{n_1} \times \partial V, \ j = 1, \ldots, m_2, \\ (D^\beta u)|_{x_j=2\pi} - e^{2\pi\nu_j}(D^\beta u)|_{x_j=0} &= 0 \quad (j = 1, \ldots, n_1, \ |\beta| < m_1). \end{aligned}$$

The induced boundary value problem on V and its $L^p(V)$-realization remain unchanged. As the $L^p(\Omega)$-realization of the cylindrical ν-periodic boundary value problem (8.20) we define

$$D(\mathbb{A}_\nu) := L^p(\mathcal{Q}_{n_1}, D(A_2)) \cap \bigcap_{\frac{\ell_1}{2m_1} + \frac{\ell_2}{2m_2} \leq 1} W_{\nu,per}^{\ell_1,p}(\mathcal{Q}_{n_1}, W^{\ell_2,p}(V)),$$

$$\mathbb{A}_\nu u := \mathcal{A}(\cdot, D)u \quad (u \in D(\mathbb{A}_\nu)).$$

Similar to the preceding step, we first consider the model problem for the cylindrical boundary value problem (8.20), that is, we assume $A_1(x, D)$ on $\mathcal{Q}_{n_1}$ to be given as homogeneous differential operator

$$A_1(D) := \sum_{|\alpha|=2m} a_\alpha^1 D^\alpha$$

with constant coefficients $a_\alpha^1 \in \mathbb{C}$. We set

$$\mathcal{A}_0(\cdot, D) := \mathcal{A}_0(y, D) := A_1(D) + A_2(y, D)$$

and

$$\mathbb{A}_{\nu,0}u := \mathcal{A}_0(\cdot, D)u \quad (u \in D(\mathbb{A}_{\nu,0}) := D(\mathbb{A}_\nu)).$$

Again no restrictions on $A_2(y, D)$ have to be assumed.

Proposition 8.15. *For each $\phi > \varphi_{(A_0,B,\nu)}$ we have $\mathbb{A}_{\nu,0} + \delta_2 \in \mathcal{RS}(X)$ with $\delta_2 = \delta_2(\phi)$ as in Proposition 8.3. Moreover, $\phi_{\mathbb{A}_{\nu,0}+\delta_2}^{\mathcal{RS}} \leq \phi$ and it holds that*

$$(8.21) \qquad \mathcal{R}\Big(\Big\{\lambda^{1-\left(\frac{|\alpha^1|}{2m_1} + \frac{|\alpha^2|}{2m_2}\right)} D^\alpha (\lambda + \mathbb{A}_{\nu,0} + \delta_2)^{-1};$$
$$\lambda \in \Sigma_{\pi-\phi}, \ \alpha \in \mathbb{N}_0^n, \ 0 \leq \frac{|\alpha^1|}{2m_1} + \frac{|\alpha^2|}{2m_2} \leq 1\Big\}\Big) < \infty.$$

Proof. Let $\phi > \varphi_{(A_0,B)}$ and consider

$$M_\lambda(k) := \lambda^{1-\left(\frac{|\alpha^1|}{2m_1} + \frac{|\alpha^2|}{2m_2}\right)} k^{\alpha^1} D^{\alpha^2} (\lambda + a_1(k_\nu) + A_2 + \delta_2)^{-1}$$

for $\lambda \in \Sigma_{\pi-\phi}$, $k \in \mathbb{Z}^{n_1}$ and $\alpha \in \mathbb{N}_0^n$ such that $0 \leq \frac{|\alpha^1|}{2m_1} + \frac{|\alpha^2|}{2m_2} \leq 1$.

Distinguishing the cases $\nu = 0$ and $\nu \neq 0$, we combine the proofs of Proposition 7.20 and Proposition 8.11 to prove the assertion. For $\beta \leq \alpha$ observe the existence of $C > 0$ such that

$$\frac{\lambda^{1-\left(\frac{|\alpha^1|}{2m_1} + \frac{|\alpha^2|}{2m_2}\right)} |k|^{|\omega|} |\Delta^\omega N(k)|}{|\lambda + P(k_\nu)|^{1-\frac{|\beta^2|}{2m_2}}} \leq C$$

holds true for $N(k) := k^{\beta^1}$, $0 < \omega \leq \gamma$, all $k \in \mathbb{Z}^{n_1} \setminus \{0\}$ and all $\lambda \in \Sigma_{\pi-\phi}$ due to parameter-ellipticity of P. The validity of $B_j(y, D)u = 0$ follows from density of $\mathbb{T}(\mathcal{Q}_{n_1}, L^p(V))$ in $L^p(\Omega)$ (see Proposition 2.4). $\qquad \square$

From Lemma 8.12 we obtain as in the previous step the following result on small variations.

Corollary 8.16. *Let $R(x^1, D) := \sum_{|\alpha^1|=2m_1} r_{\alpha^1}(x^1) D^{\alpha^1}$ be given such that the condition* $\sum_{|\alpha^1|=2m_1} \|r_{\alpha^1}\|_\infty < \eta$ *is satisfied. Set*

$$(8.22) \qquad \mathcal{A}^{va}(\cdot, D) := \mathcal{A}^{va}(x, y, D) := A_0(y, D) + R(x, D) \quad ((x, y) \in \Omega)$$

and denote its X-realization by

$$\mathbb{A}_\nu^{va} u := \mathcal{A}^{va}(\cdot, D) u \quad (u \in D(\mathbb{A}_\nu^{va}) := D(\mathbb{A}_0)).$$

Then there exists a $\delta > 0$ such that $\mathbb{A}_\nu^{va} + \delta \in \mathcal{RS}(X)$ and $\phi_{\mathbb{A}_\nu^{va}+\delta}^{\mathcal{RS}} \leq \phi_{A_0+\delta_2}^{\mathcal{RS}}$ provided that η is sufficiently small. In this case, for $\phi > \varphi_{(A_0, B, \nu)}$ we have

$$(8.23) \qquad \begin{aligned} &\mathcal{R}\Big(\Big\{\lambda^\rho D^\alpha (\lambda + \mathbb{A}_\nu^{va} + \delta)^{-1}; \\ &\lambda \in \Sigma_{\pi - \phi}, \ \alpha \in \mathbb{N}_0^n, \ \rho \in [0,1], \ 0 \leq \rho + \tfrac{|\alpha^1|}{2m_1} + \tfrac{|\alpha^2|}{2m_2} \leq 1\Big\}\Big) < \infty. \end{aligned}$$

Within this step we use the same strategy of localization as employed in step 1. Again local operators are defined by reflection in order to achieve smallness of variation for all top order coefficients. However, there is the boundary $\partial \mathcal{Q}_{n_1}$ which has to be taken into account.

We denote by

$$A_1^\#(x, D) := \sum_{|\alpha|=2m} a_\alpha^1(x) D^\alpha$$

the principal part of $A_1(x, D)$. Freezing the coefficients at some arbitrary point $x_0 \in \overline{\mathcal{Q}}_{n_1}$, the operator $A_1^\#(x_0, D)$ fulfills the assumptions of Proposition 8.15 and we denote by $\eta = \eta(x_0)$ the constant from Corollary 8.16 applied to $A_1^\#(x_0, D)$.

For the given non-constant top order coefficients $a_{\alpha^1}^1$ we consider the periodic extensions $a_{\alpha^1, per}^1$ to $\mathbb{R}^{n_1}$. Note that we have $a_{\alpha^1, per}^1 \in BUC(\mathbb{R}^{n_1})$ for all α^1 such that $|\alpha^1| = 2m_1$ due to the periodicity assumptions. In particular, there exists a length $d = d(x_0)$ such that

$$|a_{\alpha^1}^1(x) - a_{\alpha^1}^1(x_0)| \leq \eta(x_0)/M_\alpha \quad (|x - x_0|_\infty \leq d(x_0), \ |\alpha^1| = 2m_1).$$

Without loss of generality, we may assume $d(x_0) < \tfrac{\pi}{2}$. Set

$$\mathcal{Q}_{d(x_0)}(x_0) := \{x \in \mathbb{R}^{n_1}; \ |x - x_0|_\infty \leq d(x_0) < \tfrac{\pi}{2}\}.$$

Obviously, the collection $\{\mathcal{Q}_{d(x_0)}(x_0)); \ x_0 \in \overline{\mathcal{Q}}_{n_1}\}$ represents an open covering of $\overline{\mathcal{Q}}_{n_1}$. Thus, by compactness we have

$$\overline{\mathcal{Q}}_{n_1} \subset \bigcup_{j=1}^N \mathcal{Q}_{d(x_j)}(x_j)$$

for a certain finite set $(x_j)_{j=1}^N$. In what follows we may assume every open set of the covering which intersects with $\mathbb{R}^{n_1} \setminus \mathcal{Q}_{n_1}$ to be cut at the boundary of $\mathcal{Q}_{n_1}$ and continued within $\mathcal{Q}_{n_1}$ on the opposite side (cf. Figure 8.2).

For simplicity we set $d_j := d(x_j)$ and $U_j := \mathcal{Q}_{d_j}(x_j)$ for $j = 1, \ldots, N$. For each $j = 1, \ldots, N$ we further define coefficients of $A_1^{\#}(x, D)$-localizations

$$A_j^{1,loc}(x, D) := \sum_{|\alpha|=2m} a_{j,\alpha}^1(x) D^\alpha$$

by reflection as introduced below. While performing the reflection we drop the indication for the coefficients of $A_1(x, D)$ and $A_j^{1,loc}(x, D)$ under consideration and merely write a and a_j instead of a_α^1 and $a_{j,\alpha}^1$ for the sake of readability. Let further e_i, $i = 1, \ldots, n_1$ denote the unit vectors in $\mathbb{R}^{n_1}$.

Initially we set $a_j(x) := a(x)$ for $x \in \mathcal{Q}_{d_j}(x_j)$. In a first step we extend $\mathcal{Q}_{d_j}(x_j)$ to $\mathcal{Q}_{d_j}^1(x_j) := \mathcal{Q}_{d_j}(x_j) \cup \mathcal{Q}_{d_j}(x_j + 2d_j e_1)$ and define $a_j(x)$ on $\mathcal{Q}_{d_j}(x_j + 2d_j e_1)$ by means of even reflection at the hyperplane given through $x_1 = (x_j)_1 + d_j$, i.e.

$$a_j(x) := a(2(x_j)_1 + 2d_j - x_1, x') \quad \text{for} \quad x = (x_1, x') \in \mathcal{Q}_{d_j}(x_j + 2d_j e_1).$$

Again $\mathcal{Q}_{d_j}^1(x_j)$ is cut at the boundary of $\mathcal{Q}_{n_1}$ and continued within $\mathcal{Q}_{n_1}$ on the opposite side if it intersects with the boundary of $\mathcal{Q}_{n_1}$. Note that $d(x_0) < \frac{\pi}{2}$ ensures $\mathcal{Q}_{d_j}^1(x_j)$ to be non-selfintersecting in that sense. In a second step we extend $\mathcal{Q}_{d_j}^1(x_j)$ to the hyperstrip

$$\mathcal{S}_{d_j}^1(x_j) := [0, 2\pi] \times \left(\mathcal{Q}_{d_j}^1(x_j) \right)' := \{(x_1, x');\ x_1 \in [0, 2\pi],\ ((x_j)_1, x') \in \mathcal{Q}_{d_j}^1(x_j)\}$$

and define $a_j(x)$ on $\mathcal{S}_{d_j}^1(x_j) \setminus \mathcal{Q}_{d_j}^1(x_j)$ by constant extension, i.e.

$$a_j(x) := a|_{x_1=(x_j)_1+d_j}(x') \quad \text{for} \quad x = (x_1, x') \in \mathcal{S}_{d_j}^1(x_j) \setminus \mathcal{Q}_{d_j}^1(x_j).$$

Now we proceed with the second coordinate direction. We extend $\mathcal{S}_{d_j}^1(x_j)$ to

$$\mathcal{Q}_{d_j}^2(x_j) := \mathcal{S}_{d_j}^1(x_j) \cup \mathcal{S}_{d_j}^1(x_j + 2d_j e_2)$$

and define $a_j(x)$ on $\mathcal{S}_{d_j}^1(x_j + 2d_j e_2)$ by means of even reflection. Afterwards we extend $\mathcal{Q}_{d_j}^2(x_j)$ to

$$\mathcal{S}_{d_j}^2(x_j) := [0, 2\pi] \times \left(\mathcal{Q}_{d_j}^2(x_j) \right)' := \{(x_2, x');\ x_2 \in [0, 2\pi],\ ((x_j)_2, x') \in \mathcal{Q}_{d_j}^2(x_j)\}$$

and define $a(x)$ on $\mathcal{S}_{d_j}^2(x_j) \setminus \mathcal{Q}_{d_j}^2(x_j)$ by constant extension.

Carrying on throughout all coordinate directions, for each $j = 1, \ldots, N$ we end up with a continuous function $a_{j,\alpha}^1 \in C_{per}(\mathcal{Q}_{n_1})$. Moreover, by definition we have

$$\sum_{|\alpha^1|=2m_1} |a_{j,\alpha^1}^1(x) - a_{\alpha^1}^1(x_j)| \le \eta(x_j)$$

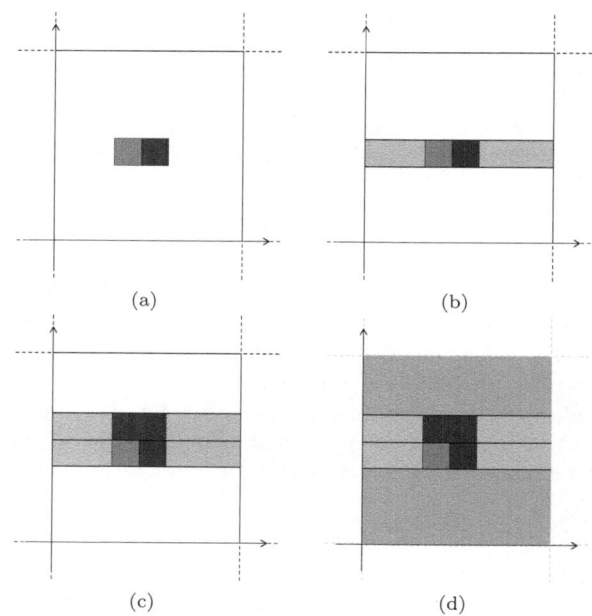

Figure 8.1: The single steps during the reflection procedure ($n_1 = 2$):
(a): Extension of $\mathcal{Q}_{d_j}(x_j)$ to $\mathcal{Q}^1_{d_j}(x_j)$. (b): Extension of $\mathcal{Q}^1_{d_j}(x_j)$ to $\mathcal{S}^1_{d_j}(x_j)$.
(c): Extension of $\mathcal{S}^1_{d_j}(x_j)$ to $\mathcal{Q}^2_{d_j}(x_j)$. (d): Extension of $\mathcal{Q}^2_{d_j}(x_j)$ to $\mathcal{S}^2_{d_j}(x_j)$.

for $x \in \overline{\mathcal{Q}}_{n_1}$ and $j = 1, \ldots, N$, that is,

$$\mathcal{A}^{loc}_j(x, y, D) := A^{1,loc}_j(x, D) + A_2(y, D)$$

is a small variation of

$$\mathcal{A}^{j,\#}(y, D) := A^{\#}_1(x_j, D) + A_2(y, D)$$

in the sense of (8.22). Hence, Corollary 8.16 applies to

$$\mathbb{A}^{loc}_{\nu,j} u := \mathcal{A}^{loc}_j(\cdot, D)u \quad (u \in D(\mathbb{A}^{loc}_{\nu,j}) := D(\mathbb{A}_\nu)).$$

Thus, for each $j = 1, \ldots, N$ and each $\phi > \varphi_{(\mathcal{A}, \mathcal{B}_\nu)}$ there exists $\delta = \delta(\phi) > 0$ such that $\mathbb{A}^{loc}_{\nu,j} + \delta \in \mathcal{RS}(X)$ and we have

$$(8.24) \quad \mathcal{R}\Big(\Big\{\lambda^\rho D^\alpha (\lambda + \mathbb{A}^{loc}_{\nu,j} + \delta)^{-1};$$
$$\lambda \in \Sigma_{\pi - \phi},\ \alpha \in \mathbb{N}^n_0,\ \rho \in [0, 1],\ 0 \le \rho + \frac{|\alpha^1|}{2m_1} + \frac{|\alpha^2|}{2m_2} \le 1\Big\}\Big) < \infty.$$

We choose a partition of unity $(\varphi_j)^N_{j=1} \subset C^\infty(\mathbb{R}^{n_1})$ subordinate to the open covering $(U_j)^N_{j=1}$ such that $0 \le \varphi_j \le 1$. In addition, we fix $\psi_j \in C^\infty(\mathbb{R}^{n_1})$ such that $\psi_j \equiv 1$ on $\operatorname{supp}\varphi_j$ and $\operatorname{supp}\psi_j \subset U_j$. Again we think of φ_j and ψ_j as continued on the opposite side of $\mathcal{Q}_{n_1}$ in case U_j intersects with the boundary of $\mathcal{Q}_{n_1}$.

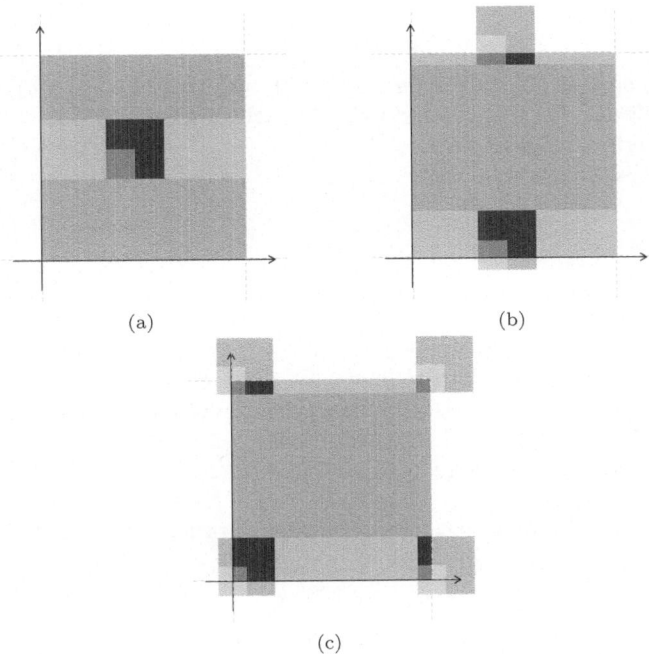

(a)   (b)

(c)

Figure 8.2: Reflections depending on intersections with the boundary.

With $\mathcal{C}_j(\cdot, D)$ and $\mathcal{D}_j(\cdot, D)$ defined similar to the preceding step we set

(8.25)
$$R_0(\lambda, \delta) := \sum_{j=0}^{N} \psi_j (\lambda + \mathbb{A}_{\nu,j}^{loc} + \delta)^{-1} \varphi_j,$$

$$R_1(\lambda, \delta) := \sum_{j=0}^{N} \psi_j (\lambda + \mathbb{A}_{\nu,j}^{loc} + \delta)^{-1} \mathcal{C}_j(\cdot, D),$$

and

$$R_2(\lambda, \delta) := \sum_{j=0}^{N} \mathcal{D}_j(\cdot, D)(\lambda + \mathbb{A}_{\nu,j}^{loc} + \delta)^{-1} \varphi_j.$$

Thanks to the definition of ψ_j for the case that U_j intersects with the boundary of $\mathcal{Q}_{n_1}$, both $R_0(\lambda, \delta)$ and $R_1(\lambda, \delta)$ map to

$$\mathbb{E} := L^p(\mathcal{Q}_{n_1}, D(A_2)) \cap \bigcap_{\ell=1}^{2m} W_{\nu,per}^{\ell,p}(\mathcal{Q}_{n_1}, W^{2m-\ell,p}(V)).$$

Just as in step 1 we deduce existence of $\delta > 0$, such that $(I - R_1(\lambda, \delta))^{-1} R_0(\lambda, \delta)$ defines a left and $R_0(\lambda, \delta)(I + R_2(\lambda, \delta))^{-1}$ a right inverse of $\lambda + \mathbb{A}_\nu + \delta$. Again a Neumann series argument shows (8.11).

This proves the assertion within step 2.

8.2.3 Step 3:

We can now combine the results from step 1 and step 2 to prove Theorem 8.10.

Proof of Theorem 8.10. We apply step 2 to the ν-periodic boundary value problem $(\mathcal{A}, \mathcal{B}_\nu)$ given through $\mathcal{A}_{23}(x^2, x^3, D) := A_2(x^2, D) + A_3(x^3, D)$ supplemented with $B_j(x^3, D)$, $j = 1, \ldots, m_3$ and observe validity of the $\mathcal{R}$-boundedness condition (8.11) for its $L^p(\mathcal{Q}_{n_2} \times V, F)$-realization $\mathbb{A}_\nu$. Hence, the claim follows from step 1 replacing A_2 on V there by $\mathcal{A}_{23}$ on $\mathcal{Q}_{n_2} \times V$.   □

In the subsequent lines we comment on generalizations of Theorem 8.10.

Remark 8.17. We point out that these proofs easily generalize to the case of different p-integrability with respect to $\mathbb{R}^{n_1}$, $\mathcal{Q}_{n_2}$ and V (cf. Remark 10.7).

Remark 8.18. As frequently mentioned in the proof of Theorem 8.10, it is a strength of the Fourier multiplier approach that the induced boundary value problem on V does not have to be investigated during the localization process performed above. Indeed, it can be treated separately for our proof only relies on the $\mathcal{R}$-boundedness estimates (8.4) established in [DHP03]. Therefore, with a Banach space F of class $\mathcal{HT}$ that enjoys property (α), we can also treat F-valued boundary value problems with $\mathcal{L}(F)$-valued coefficients in A_3 and $B_{3,j}$, $j = 1, \ldots, m_3$.

Remark 8.19. Using our results from Chapters 6 and 7 on the transference of an $\mathcal{RH}^\infty$-calculus and the results from [DDH$^+$04] stated in Proposition 8.4, we can similarly prove an $\mathcal{RH}^\infty$-calculus for the model problem of our partially ν-periodic cylindrical boundary value problem (8.7). Then, the coefficients of A_3 and $B_{3,j}$, $j = 1, \ldots, m_3$, have to match the assumptions of Proposition 8.4. In that case, non-constant coefficients of A_1 and A_2 can be treated by means of a localization procedure again. However, perturbation arguments for the $\mathcal{H}^\infty$-calculus are known to be much more involved (see e.g. [DDH$^+$04] or [KKW06]). We thus restrict ourselves to what has been presented so far and refer to Chapter 10 for results on the $\mathcal{RH}^\infty$-calculus for cylindrical boundary value problems.

Remark 8.20. Also observe that to some extent we can weaken the assumption that $(\mathcal{A}, \mathcal{B}_\nu)$ is cylindrical. Indeed, consider homogeneous differential operators with constant coefficients $A_1(D_{x_1})$, $Q_1(D_{x_1})$, $A_2(D_{x^2})$, and $Q_2(D_{x^2})$. Furthermore, let (A_3, B_3) define a boundary value problem as described in Remark 8.18. Then with the aid of the results from Chapter 6 and Chapter 7, we can treat differential operators

$$\mathcal{A}(x^3, D) := A_1(D_{x_1}) + Q_1(D_{x_1})\big(A_2(D_{x^2}) + Q_2(D_{x^2})A_3(x^3, D_{x^3})\big).$$

Remark 8.21. Assume $A(x, D)$ to have constant coefficients with respect to one of its parts A_i, $i = 1, 2$. As in the previous chapters, additional results by means of reflection techniques can be obtained provided $A_i(D)$ is even. To some extent, the reflection techniques can be applied successfully even in case of non-constant coefficients. For instance, consider a second order parameter-elliptic differential operator in divergence form

$$A(\boldsymbol{x}, D) := -\sum_{i,j=1}^{1+n_2} \partial_i a_{ij}(\boldsymbol{x})\partial_j$$

in $\Omega := \mathbb{R}_+ \times V$, $V \subset \mathbb{R}^{n_2}$.

On the lateral surface we impose Dirichlet or Neumann boundary conditions. On $\{0\} \times V$ in what follows we impose Neumann boundary conditions which are obtained by an even extension of the data. Dirichlet boundary conditions with respect to $\{0\} \times V$ can be obtained similarly using an odd extension instead. Set $a(\boldsymbol{x}) := \big(a_{ij}(\boldsymbol{x})\big)_{i,j=1,\dots,n}$, where $n := 1 + n_2$. In order to enter the context of cylindrical boundary value problems we assume that

$$a(\boldsymbol{x}) = \begin{pmatrix} a_{11}(x^1) & 0 \\ 0 & \tilde{a}(x^2) \end{pmatrix} := \begin{pmatrix} a_{11}(x^1) & 0 & \cdots & 0 \\ 0 & a_{22}(x^2) & \cdots & a_{2n}(x^2) \\ \vdots & \vdots & & \vdots \\ 0 & a_{n2}(x^2) & \cdots & a_{nn}(x^2) \end{pmatrix}.$$

Let $a_{ij} \in BUC^1(\overline{\Omega})$ and assume that $a_{11}(\infty) := \lim_{x^1 \to \infty} a_{11}(x^1)$ exists. This yields

$$A(\boldsymbol{x}, D) := -a_{11}(x^1)\Delta_1 - \sum_{i,j=2}^{n} a_{ij}(x^2)\partial_i\partial_j - (\partial_1 a_{11})(x^1)\partial_1 - \sum_{i,j=2}^{n} (\partial_i a_{ij})(x^2)\partial_j$$

and so

$$A(\boldsymbol{x}, D) := -a_{11}(x^1)\Delta_1 - (\partial_1 a_{11})(x^1)\partial_1 + A_2(x^2, D_2),$$

where $A_2(x^2, D)$, acting on V, is defined by

$$A_2(x^2, D) := -\sum_{i,j=1}^{n_2} \tilde{a}_{ij}(x^2)\partial_i\partial_j - \sum_{j=1}^{n_2} (\partial_j \tilde{a}_{ij})(x^2)\partial_i = -\sum_{i,j=1}^{n_2} \partial_i \tilde{a}_{ij}(x^2)\partial_j.$$

Note that A_2 is of divergence form again. The operator

$$A_1(x^1, D_1) := -a_{11}(x^1)\Delta_1 - (\partial_1 a_{11})(x^1)\partial_1 = -\partial_1 a_{11}(x^1)\partial_1$$

is of even structure in the following sense: consider the Dirichlet-Neumann type boundary value problem for $A(\boldsymbol{x}, D)$ on Ω as explained above and let $f \in L^p(\Omega)$ be any given right-hand side. We extend the boundary value problem to the whole of $\mathbb{R} \times V$ by means of even extensions of f to F as well as of a_{11} to A_{11}. This yields $A_{11} \in C(\mathbb{R})$ as well as $\lim_{x^1 \to \pm\infty} A_{11}(x^1) = a_{11}(\infty)$. Moreover, $\partial_1 A_{11}$ exists classically on $\mathbb{R} \setminus \{0\}$, is odd with respect to x^1, and belongs to $L^\infty(\mathbb{R} \times V)$.

Let U define the unique solution to the (possibly shifted) extended problem on $\mathbb{R} \times V$ and set $V(\boldsymbol{x}) := U(-x^1, x^2)$. Then

$$A_{11}(x^1)(\Delta_1 V)(\boldsymbol{x}) = A_{11}(-x^1)(\Delta_1 U)(-x^1, x^2)$$

and

$$
\begin{aligned}
(\partial_1 A_{11})(x^1)\partial_1 V(\boldsymbol{x}) &= -(\partial_1 A_{11})(-x^1)\big(-\partial_1 U\big)(-x^1, x^2) \\
&= (\partial_1 A_{11})(-x^1)(\partial_1 U)(-x^1, x^2).
\end{aligned}
$$

Therefore,

$$A_\delta(\boldsymbol{x}, D)V(\boldsymbol{x}) = A_\delta(\boldsymbol{x}, D)U(-x^1, x^2) = F(-x^1, x^2) = F(\boldsymbol{x})$$

and by uniqueness $U = V$, i.e. $U \in W^{2,p}(\mathbb{R} \times V)$ is even with respect to x^1. Hence, $u := U|_\Omega$ is a solution to the original (possibly shifted) boundary value problem subject to Neumann boundary conditions with respect to $\{0\} \times V$.

Along the same lines we can treat the according boundary value problem in $\Omega := (0, \pi) \times V$ subject to mixed Dirichlet-Neumann boundary conditions. Here we make use of our previous results on periodic respectively antiperiodic boundary value problems applied to the extended problem in $(0, 2\pi) \times V$.

We do not specify all possible boundary value problems which are covered by the results obtained by reflection so far but rather focus on the Laplacian at the end of this chapter.

8.3 A focus on the Laplacian

In what follows let $n_i \in \mathbb{N}_0$, $i = 1, \ldots, 5$, $n := \sum_{i=1}^5 n_i$, $V \subset \mathbb{R}^{n_5}$, and

$$\Omega := \mathbb{R}^{n_1} \times \mathcal{Q}_{n_2} \times \mathbb{K}^{n_3} \times \tilde{\mathcal{Q}}_{n_4} \times V.$$

Given $\nu \in i(-1, 1)^{n_2}$, we consider the boundary value problem for the Laplacian

$$
\begin{array}{rll}
\lambda u - \Delta u &=& f \quad \text{in } \Omega, \\
\mathcal{B}_3 u &=& 0 \quad \text{on } \mathbb{R}^{n_1} \times \mathcal{Q}_{n_2} \times \partial\mathbb{K}^{n_3} \times \tilde{\mathcal{Q}}_{n_4} \times V, \\
\mathcal{B}_4 u &=& 0 \quad \text{on } \mathbb{R}^{n_1} \times \mathcal{Q}_{n_2} \times \mathbb{K}^{n_3} \times \partial\tilde{\mathcal{Q}}_{n_4} \times V, \\
\mathcal{B}_5 u &=& 0 \quad \text{on } \mathbb{R}^{n_1} \times \mathcal{Q}_{n_2} \times \mathbb{K}^{n_3} \times \tilde{\mathcal{Q}}_{n_4} \times \partial V, \\
u|_{x_j = 2\pi} - e^{2\pi\nu_j} u|_{x_j = 0} &=& 0 \quad j = n_1 + 1, \ldots, n_1 + n_2, \\
\partial_j u|_{x_j = 2\pi} - e^{2\pi\nu_j} \partial_j u|_{x_j = 0} &=& 0 \quad j = n_1 + 1, \ldots, n_1 + n_2.
\end{array}
$$

(8.26)

Here the boundary operator $\mathcal{B} := \{\mathcal{B}_i; \ i = 3, 4, 5\}$ endows the boundary value problem for the Laplacian with mixed Dirichlet-Neumann boundary conditions on

$$\mathbb{R}^{n_1} \times \mathcal{Q}_{n_2} \times \partial(\mathbb{K}^{n_3} \times \tilde{\mathcal{Q}}_{n_4} \times V).$$

We point out that on each part of this boundary the types of boundary conditions might be different. On

$$\mathbb{R}^{n_1} \times \mathcal{Q}_{n_2} \times \mathbb{K}^{n_3} \times \partial(\tilde{\mathcal{Q}}_{n_4}) \times V,$$

in particular, we can even distinguish between opposite sides of the boundary. The Laplacians with respect to each of the five components of Ω are denoted by Δ_i, $i = 1, \ldots, 5$. Here we consider different L^p-realizations of these operators. For $i = 1, \ldots, 4$ the domains of definition are defined as

$$D(\Delta_{p,1}) := W^{2,p}(\mathbb{R}^{n_1}, L^p(\mathcal{Q}_{n_2} \times \mathbb{K}^{n_3} \times \tilde{\mathcal{Q}}_{n_4} \times V)),$$

$$D(\Delta_{p,2}) := W^{2,p}_{\nu,per}(\mathcal{Q}_{n_2}, L^p(\mathbb{K}^{n_3} \times \tilde{\mathcal{Q}}_{n_4} \times V)),$$

$$D(\Delta_{p,3}) := \left\{ u \in W^{2,p}(\mathbb{K}^{n_3}, L^p(\tilde{\mathcal{Q}}_{n_4} \times V)); \ \mathcal{B}_3 u = 0 \right\},$$

$$D(\Delta_{p,4}) := \left\{ u \in W^{2,p}(\tilde{\mathcal{Q}}_{n_4}, L^p(V)); \ \mathcal{B}_4 u = 0 \right\}.$$

For $i = 5$ we consider either the strong Dirichlet or Neumann Laplacian $\Delta_{p,s,5}$ or the weak Dirichlet or Neumann Laplacian $\Delta_{p,w,5}$ as defined in Section 8.1. For all operators the same notation will be used for their canonical extensions to $L^p(\Omega)$. Without further explanations we also employ the notation

$$\Delta u = \Delta_1 u + \ldots + \Delta_5 u,$$

where Δ_i acts on the according component of Ω. We define the strong Laplacian subject to the boundary conditions $(\mathcal{B}, \nu)$ to be

$$D(\Delta_{p,s,\mathcal{B},\nu}) := \bigcap_{i=1}^{4} D(\Delta_{p,i}) \cap D(\Delta_{p,s,5})$$

$$\cap \bigcap_{m=0}^{2} W^{m,p}\left(\mathbb{R}^{n_1}, \bigcap_{\ell=0}^{2-m} W^{\ell,p}_{\nu,per}(\mathcal{Q}_{n_2}, W^{2-m-\ell,p}(\mathbb{K}^{n_3} \times \tilde{\mathcal{Q}}_{n_4} \times V))\right),$$

$$\Delta_{p,s,\mathcal{B},\nu} \, u := \Delta u \quad (u \in D(\Delta_{p,s,\mathcal{B},\nu})).$$

Additionally, we will consider the weak Laplacian subject to the boundary conditions $(\mathcal{B}, \nu)$ defined as

$$D(\Delta_{p,w,\mathcal{B},\nu}) := \bigcap_{i=1}^{4} D(\Delta_{p,i}) \cap D(\Delta_{p,w,5})$$

$$\cap \bigcap_{m=0}^{2} W^{m,p}\left(\mathbb{R}^{n_1}, \bigcap_{\ell=0}^{2-m} W^{\ell,p}_{\nu,per}(\mathcal{Q}_{n_2}, W^{2-m-\ell,p}(\mathbb{K}^{n_3} \times \tilde{\mathcal{Q}}_{n_4}, L^p(V)))\right),$$

$$\Delta_{p,w,\mathcal{B},\nu} \, u := \Delta u \quad (u \in D(\Delta_{p,w,\mathcal{B},\nu})).$$

Note that $D(\Delta_{p,w,\mathcal{B},\nu}) \subset W^{2,p}(\mathbb{R}^{n_1} \times \mathcal{Q}_{n_2} \times \mathbb{K}^{n_3} \times \tilde{\mathcal{Q}}_{n_4}, L^p(V)) \cap W^{1,p}(\Omega)$ only, whereas $D(\Delta_{p,s,\mathcal{B},\nu}) \subset W^{2,p}(\Omega)$.

Theorem 8.22. a) *Let $1 < p < \infty$ and let $V \subset \mathbb{R}^{n_5}$ be a C^2 standard domain. Then the strong Laplacian subject to the boundary conditions $(\mathcal{B}, \nu)$ fulfills*

$$-\Delta_{p,s,\mathcal{B},\nu} \in \Psi\mathcal{R}\mathcal{H}^\infty(L^p(\Omega)) \quad and \quad \phi^{\mathcal{R}\infty}_{-\Delta_{p,s,\mathcal{B},\nu}} < \frac{\pi}{2}.$$

b) *Let $V \subset \mathbb{R}^{n_5}$ be a Lipschitz domain. If $n_5 \geq 3$, then there exists $\varepsilon > 0$ depending only on the Lipschitz character of V such that the weak Laplacian subject to the boundary conditions $(\mathcal{B}, \nu)$ fulfills*

$$-\Delta_{p,w,\mathcal{B},\nu} \in \Psi\mathcal{R}\mathcal{H}^\infty(L^p(\Omega)) \quad and \quad \phi^{\mathcal{R}\infty}_{-\Delta_{p,w,\mathcal{B},\nu}} < \frac{\pi}{2}$$

for all p subject to $(3 + \varepsilon)' < p < 3 + \varepsilon$.
If $n_5 = 2$ and if $\mathcal{B}_5$ represents Dirichlet boundary conditions, then the assertion remains true for all p subject to $(4 + \varepsilon)' < p < 4 + \varepsilon$.

Proof. We first mimic the proofs of Proposition 8.11 and 8.15 presented to deal with the model problems in steps 1 and 2 of Theorem 8.10. The only difference is that we employ Proposition 8.7 instead of Proposition 8.3. This already yields $-\Delta_{p,s,\mathcal{B},\nu} \in \Psi\mathcal{R}S(L^p(\Omega))$ and $-\Delta_{p,w,\mathcal{B},\nu} \in \Psi\mathcal{R}S(L^p(\Omega))$ with angles less than $\frac{\pi}{2}$ and the domains of definition as introduced above. Recall Propositions 6.13 and 7.20 as well as Propositions 6.16 and 7.23. To prove the statements on the pseudo-$\mathcal{R}\mathcal{H}^\infty$-calculus we make use of Propositions 6.20 and 7.25 as well as of Propositions 6.21 and 7.26. □

A few remarks on the outcome as well as on possible extensions of Theorem 8.22 are in order.

Remark 8.23. Given the assumptions of Theorem 8.22 for p as described in Theorem 8.22 let further $n_1 + n_3 \neq 0$, or $\nu \neq 0$, or let $\mathcal{B}$ represent non-pure Neumann boundary conditions only. Then

$$-\Delta_{p,s,\mathcal{B},\nu} \in \mathcal{R}\mathcal{H}^\infty(L^p(\Omega)) \quad and \quad \phi^{\mathcal{R}\infty}_{-\Delta_{p,s,\mathcal{B},\nu}} < \frac{\pi}{2}$$

as well as

$$-\Delta_{p,w,\mathcal{B},\nu} \in \mathcal{R}\mathcal{H}^\infty(L^p(\Omega)) \quad and \quad \phi^{\mathcal{R}\infty}_{-\Delta_{p,w,\mathcal{B},\nu}} < \frac{\pi}{2}.$$

Moreover, if $\nu \neq 0$ or if Dirichlet conditions with respect to V are imposed where V is a bounded domain, then $0 \in \rho(\Delta_{p,s,\mathcal{B},\nu})$ and $0 \in \rho(\Delta_{p,w,\mathcal{B},\nu})$, respectively.

Remark 8.24. Theorem 8.22 and Remark 8.23 remain valid in the case of different p-integrability with respect to the five components of Ω (cf. Remark 8.17).

Remark 8.25. Compared to existing literature, it is worthwhile to highlight two facts concerning the outcome of Theorem 8.22: the result simultaneously includes classes of unbounded Lipschitz domains and mixed boundary conditions. Furthermore, recall that mixed derivatives of u with respect to all of the first four components of Ω belong to $L^p(\Omega)$. Thus, for the special situation of cylindrical domains this improves previous results in the literature, since for Lipschitz domains second order derivatives in general are not expected to belong to $L^p(\Omega)$ (see e.g. [JK95]). Also observe that the larger range of p known for two-dimensional Lipschitz domains is extended to a class of higher-dimensional Lipschitz domains (see Figure 8.3).

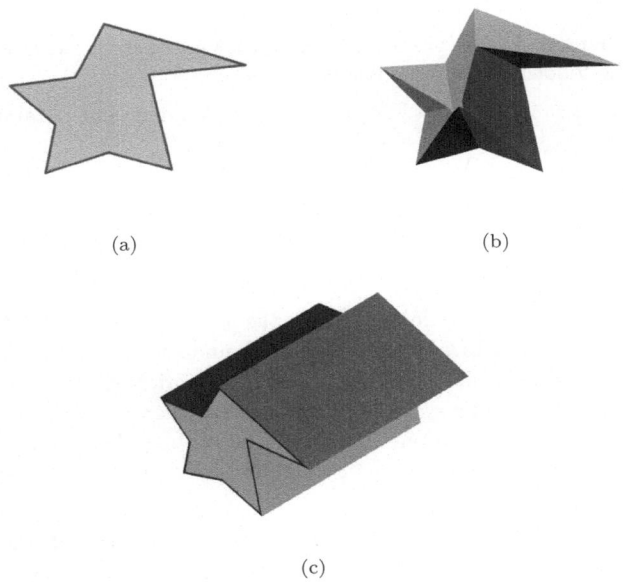

(a)   (b)

(c)

Figure 8.3: Lipschitz domains with different ranges of p. The range of p in case of a three-dimensional Lipschitz cylinder is inherited from the two-dimensional cross-section:
(a): Two-dimensional Lipschitz domain with range $(4+\varepsilon_1)' < p < 4+\varepsilon_1$.
(b): Three-dimensional Lipschitz domain with range $(3+\varepsilon_2)' < p < 3+\varepsilon_2$.
(c): Three-dimensional Lipschitz domain with range $(4+\varepsilon_1)' < p < 4+\varepsilon_1$.

Remark 8.26. Note that for every bounded Lipschitz domain G we have the usual relation $W_0^{1,p}(G) = \{u \in W^{1,p}(G); \gamma_{\partial G} u = 0\}$ (cf. [MM08]). Let $n_2 = 0$, i.e. $\Delta_{p,w,\mathcal{B},\nu} = \Delta_{p,w,\mathcal{B}}$. If $\mathcal{B}$ represents Dirichlet boundary conditions only then

$$D(\Delta_{p,w,\mathcal{B}}) \subset \{u \in W_0^{1,p}(\Omega); \ \Delta u \in L^p(\Omega)\}.$$

Accordingly, Fubini's theorem implies

$$D(\Delta_{p,w,\mathcal{B}}) \subset \left\{ u \in W^{1,p}(\Omega); \ \exists v \in L^p(\Omega) \ \forall \varphi \in W^{1,p'}(\Omega) : -\int_\Omega \nabla u \nabla \varphi = \int_\Omega v\varphi \right\}$$

in case of pure Neumann conditions. Hence, in both cases $\Delta_{p,w,\mathcal{B}}$ coincides with the usual weak realizations of the Laplacian defined on Lipschitz domains (cf. [JK95] and [Woo07]).

Remark 8.27. The precise angles in Theorem 8.22 and the subsequent remarks do only depend on $\Delta_{p,s,5}$ and $\Delta_{p,w,5}$, respectively. In many cases they are known to be zero. In fact, we have $\phi_{-\Delta_{p,s,\mathcal{B},\nu}}^{\mathcal{R}\infty} = 0$, for instance, if $V = \mathbb{R}^{n_5}$ is the whole space or, given Dirichlet boundary conditions with respect to V, if V is a bounded domain (see e.g. [PS93, Theorem D]).

Problems handled by Theorem 8.22 might arise in concrete technological applications as the following example demonstrates.

Example 8.28. Let $\Omega := \tilde{\mathcal{Q}}_{n_4} \times V$ with $V \subset \mathbb{R}^{n_5}$ bounded. The parabolic problem

$$\begin{aligned}
u_t - \Delta u &= f & &\text{in } (0,T) \times \Omega, \\
\mathcal{B}_4 u &= 0 & &\text{on } (0,T) \times \partial\tilde{\mathcal{Q}}_{n_4} \times V, \\
\mathcal{B}_5 u &= 0 & &\text{on } (0,T) \times \tilde{\mathcal{Q}}_{n_4} \times \partial V, \\
u|_{t=0} &= u_0 & &\text{in } \Omega
\end{aligned}$$

corresponding to equation (8.26) represents the heat-equation in a bounded cylindrical domain with mixed Dirichlet-Neumann boundary conditions. This equation serves, for instance, as a model for the cooling of cylindrical electronic components. For such devices the cooling system is often located on opposite sides with respect to one space dimension or on one side only. The cooling of a graphic board with heatpipes placed on bottom and top or the cooling of a processor represent typical practical examples.

Recall that mixed Dirichlet-Neumann boundary conditions with respect to $\tilde{\mathcal{Q}}_{n_4}$ as in Example 8.28 were treated by means of ν-periodic boundary conditions. Apart from that, periodic or mixed periodic and Dirichlet-Neumann boundary conditions are of their own interest. They occur for instance in the modeling of formation of keratin networks which are one component of the cytoskeleton of biological cells. In [ABF$^+$08] the authors use the evolution equation

$$\begin{aligned}
u_t - \Delta u &= f & &\text{in } (0,T) \times \Omega, \\
u|_{x_j=0} - u|_{x_j=2\pi} &= 0 & &\text{for } j = 1,2,3, \\
u|_{t=0} &= u_0 & &\text{in } \Omega
\end{aligned}$$

to investigate concentration of a pool of soluble polymers in a small, cubical section $\Omega := \mathcal{Q}_3$ of a biological cell. Here Ω represents an observation window within the cell in which network growth depends on the locally available amount of soluble polymers. The authors point out that their model is designed to simulate network formation within a small compartment of the cytoplasm which lies in the very inner of the cell. Thus, no bound such as a membrane is considered. Consequently, periodic boundary conditions are chosen to make allowance for the interaction of the observation window Ω with a structurally similar environment.

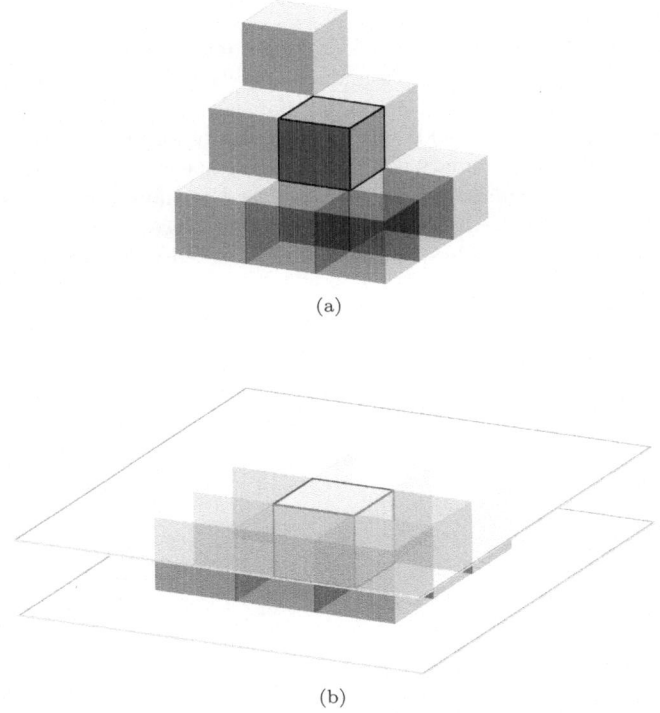

(a)

(b)

Figure 8.4: Different observation windows in a biological cell. The cubical blue domain represents the observation window in a structurally similar environment.
(a): Observation window in the inner.
(b): Observation window with membranous boundary at bottom and top.

In [ABF$^+$08] the underlying space is chosen to be the Hilbert space $L^2(\Omega)$ which makes a definition of the periodic Laplacian $\Delta^{2\pi}$ via an appropriate bilinear form available. The precise domain $D(\Delta^{2\pi})$ is defined in an abstract manner but is known to be a subset of the set

$$\left\{ u \in W_{per}^{1,2}(\Omega); \; \Delta u \in L^2(\Omega) \right\}.$$

Besides many other results concerning different components of the model, the main result on $\Delta^{2\pi}$ is that it is the generator of a strictly positive, analytic semigroup.

Our approach improves this result in two directions. Firstly, it makes L^p-theory possible. More precisely, maximal regularity is deduced which implies the generation of an analytic semigroup. Secondly, regularity and periodicity of solutions are improved since the domain $D(\Delta_{p,s,per})$ of the (strong) periodic Laplacian $\Delta_{p,s,per}$ in $L^p(\Omega)$ as investigated with our technique equals $W_{per}^{2,p}(\Omega)$. Thus, mixed second order derivatives belong to $L^p(\Omega)$ and not only the function itself but also its first partial derivatives are periodic.

Moreover, the observation window is no longer forced to be located in the very inner of the cell. Purely periodic boundary conditions have required absence of membranes so far. In particular, no mainly flat, layer-like cell could be modeled. Our approach, in contrast, yields maximal regularity for the negative Laplacian endowed with periodic boundary conditions with respect to two coordinate directions and Dirichlet, Neumann, or mixed Dirichlet-Neumann boundary conditions with respect to the third coordinate direction. This, however, can additionally be used to model different membranous boundaries located at bottom and top of the cell.

9 Application to the Stokes equation

In this chapter we consider the Stokes problem

(9.1)
$$\begin{aligned}
u_t - \Delta u + \nabla p &= f & \text{in } J \times \Omega, \\
\operatorname{div} u &= 0 & \text{in } J \times \Omega, \\
\mathcal{B}(u, p) &= 0 & \text{on } J \times \partial\Omega, \\
u(0) &= 0 & \text{in } \Omega.
\end{aligned}$$

Here $J := [0, T)$ with $T \in (0, \infty]$ denotes the time interval and Ω is given as a layer or a rectangular cylinder. The boundary operator $\mathcal{B}$ represents different boundary conditions on $\partial\Omega$. It allows for Neumann-type boundary conditions for u as well as for periodic or antiperiodic boundary conditions with respect to opposite sides for u and p.

Let $\mathcal{Q}_n := (0, 2\pi)^n$. We start with the resolvent problem associated with (9.1) via Laplace transform in $\Omega := \mathcal{Q}_n \times \mathbb{R}^m$ subject to ν-periodic boundary conditions, i.e.

(9.2)
$$\begin{aligned}
\lambda u - \Delta u + \nabla p &= f & \text{in } \Omega, \\
\operatorname{div} u &= 0 & \text{in } \Omega, \\
(u, p)|_{x_j = 2\pi} - e^{2\pi\nu_j}(u, p)|_{x_j = 0} &= 0 & (j = 1, \dots, n), \\
(\partial_j u_\ell)|_{x_j = 2\pi} - e^{2\pi\nu_j}(\partial_j u_\ell)|_{x_j = 0} &= 0 & (j = 1, \dots, n; \ell = 1, \dots, n + m; \ell \neq j).
\end{aligned}$$

Here p is scalar-valued, $u = (u_1, \dots, u_{n+m})$ and $f = (f_1, \dots, f_{n+m})$, respectively. Again $\nu \in i(-1, 1)^n$ is a given vector of zeros and purely imaginary numbers with absolute values less than one. Recall that $\nu_j = 0$ corresponds to periodic boundary conditions and $\nu_j = \frac{i}{2}$ corresponds to antiperiodic boundary conditions with respect to the j-th coordinate. Observe that the boundary conditions in (9.2) can be replaced by

$$\begin{aligned}
p|_{x_j = 2\pi} - e^{2\pi\nu_j} p|_{x_j = 0} &= 0 & (j = 1, \dots, n), \\
(\partial^\beta u)|_{x_j = 2\pi} - e^{2\pi\nu_j}(\partial^\beta u)|_{x_j = 0} &= 0 & (j = 1, \dots, n; |\beta| < 2)
\end{aligned}$$

due the condition $\operatorname{div} u = 0$ in Ω.

The investigation of (9.2) will provide structural knowledge which in turn will be used to identify the Stokes operator subject to periodic boundary conditions as a part of the periodic negative Laplacian. Moreover, we will be able to define a periodic Helmholtz projection as a Fourier multiplier operator. By reflection techniques the Helmholtz projection itself can be deduced from this rather non-physical periodic analogue.

Given $f \in L^p(\Omega)^{n+m}$ we assume for the moment that

$$u \in \bigcap_{\ell=0}^{2} W_{\nu,per}^{\ell,p}(\mathcal{Q}_n, W^{2-\ell,p}(\mathbb{R}^m))^{n+m}$$

and

$$p \in W_{\nu,per}^{1,p}(\mathcal{Q}_n, L^p(\mathbb{R}^m)) \cap L^p(\mathcal{Q}_n, W^{1,p}(\mathbb{R}^m))$$

solves (9.2). Note that the regularity of p is higher than the regularity we can expect, in general. In order to enter a setting which fits into the context of discrete operator-valued Fourier multipliers, we make use of $L^p(\Omega) \cong L^p(\mathcal{Q}_n, E)$, where $E := L^p(\mathbb{R}^m)$. Since $u_j \in W_{\nu,per}^{2,p}(\mathcal{Q}_n, E)$ if and only if $e^{-\nu \cdot} u_j \in W_{per}^{2,p}(\mathcal{Q}_n, E)$, we further multiply the first two lines of (9.2) with

$$e^{-\nu \cdot} : \mathcal{Q}_n \to \mathbb{C}; \ x \mapsto e^{-\nu x}.$$

In what follows we write $u = (u_1, \ldots, u_n, u')$, respectively $f = (f_1, \ldots, f_n, f')$, and abbreviate as before $k_\nu := k - i\nu$ and $k_{\nu,j} := k_j - i\nu_j$. Recall from Lemma 2.16 the equality

$$(e^{-\nu \cdot} \partial^\alpha u)\hat{\ }(k) = (ik_\nu)^\alpha (e^{-\nu \cdot} u)\hat{\ }(k) \quad (|\alpha| \leq 2, \ k \in \mathbb{Z}^n)$$

which yields

$$
\begin{aligned}
\left(\lambda + |k_\nu|^2 - \Delta'\right)(e^{-\nu \cdot} u_1)\hat{\ }(k) + i\, k_{\nu,1}(e^{-\nu \cdot} p)\hat{\ }(k) &= (e^{-\nu \cdot} f_1)\hat{\ }(k), \\
\vdots\ \ &= \ \ \vdots \\
\left(\lambda + |k_\nu|^2 - \Delta'\right)(e^{-\nu \cdot} u_n)\hat{\ }(k) + i\, k_{\nu,n}(e^{-\nu \cdot} p)\hat{\ }(k) &= (e^{-\nu \cdot} f_n)\hat{\ }(k), \\
\left(\lambda + |k_\nu|^2 - \Delta'\right)(e^{-\nu \cdot} u')\hat{\ }(k) + \nabla'(e^{-\nu \cdot} p)\hat{\ }(k) &= (e^{-\nu \cdot} f')\hat{\ }(k), \\
i \sum_{j=1}^{n} k_{\nu,j}(e^{-\nu \cdot} u_j)\hat{\ }(k) + \mathrm{div}'(e^{-\nu \cdot} u')\hat{\ }(k) &= 0
\end{aligned}
$$

(9.3)

for all $k \in \mathbb{Z}^n$. As multiplication with $e^{-\nu \cdot}$ defines an isomorphism in $L^p(\mathcal{Q}_n, E)$, we neglect the weight function $e^{-\nu \cdot}$ in the upcoming calculations. Hence, for each $k \in \mathbb{Z}^n$ we are faced with the multiple parameter problem in the whole space $\mathbb{R}^m$ given through

$$
\begin{aligned}
\left(\lambda + |k_\nu|^2 - \Delta'\right)\hat{u}_1 + i\, k_{\nu,1}\hat{p} &= \hat{f}_1 & \text{in } \mathbb{R}^m, \\
\vdots\ \ &= \ \ \vdots\ \ \vdots \\
\left(\lambda + |k_\nu|^2 - \Delta'\right)\hat{u}_n + i\, k_{\nu,n}\hat{p} &= \hat{f}_n & \text{in } \mathbb{R}^m, \\
\left(\lambda + |k_\nu|^2 - \Delta'\right)\hat{u}' + \nabla'\hat{p} &= \hat{f}' & \text{in } \mathbb{R}^m, \\
i \sum_{j=1}^{n} k_{\nu,j}\hat{u}_j + \mathrm{div}'\hat{u}' &= 0 & \text{in } \mathbb{R}^m.
\end{aligned}
$$

(9.4)

Note that $k_\nu = 0$ if and only if $k = \nu = 0$. In that case the system (9.4) decomposes into n resolvent problems for the heat equation in $L^p(\mathbb{R}^m)$ and the Stokes resolvent problem in $L^p(\mathbb{R}^m)^m$. It is well-known that the Helmholtz decomposition $\mathbb{P}_{\mathbb{R}^m}$ in $L^p(\mathbb{R}^m)^m$ can be defined by means of Fourier transform. In fact,

$$\mathbb{P}_{\mathbb{R}^m} := \mathcal{F}^{-1}\big(I + Q(\xi)\big)\mathcal{F},$$

where $Q(\xi) := -\frac{\xi\xi^T}{|\xi|^2}$. Let $\big(\lambda - \Delta'\big)^{-1}$ denote the resolvent of Δ' in $L^p(\mathbb{R}^m)^{(m)}$ with $D(\Delta') = W^{2,p}(\mathbb{R}^m)^{(m)}$. Then the unique solution

$$(\hat{u}, \hat{p}) \in W^{2,p}(\mathbb{R}^m)^{n+m} \times \hat{W}^{1,p}(\mathbb{R}^m)$$

is defined by $\hat{u}_j := (\lambda - \Delta')^{-1}\hat{f}_j$ for each $j = 1, \ldots, n$, $\hat{u}' := (\lambda - \Delta')^{-1}\mathbb{P}_{\mathbb{R}^m}\hat{f}'$, and $\nabla\hat{p} := \big(I - \mathbb{P}_{\mathbb{R}^m}\big)\hat{f}'$. Here

$$\hat{W}^{1,p}(\mathbb{R}^m) := \big\{v \in L^1_{loc}(\mathbb{R}^m)/\mathbb{R}; \; \nabla v \in L^p(\mathbb{R}^m)^m\big\}$$

endowed with $\|v\|_{\hat{W}^{1,p}(\mathbb{R}^m)} := \|\nabla v\|_p$ denotes the first order homogeneous Sobolev space.

The following lemma shows more regularity for p to be given in case $k_\nu \neq 0$.

Lemma 9.1. *For each $k \in \mathbb{Z}^n$ such that $k_\nu \neq 0$ there exists a unique solution*

$$(\hat{u}, \hat{p}) \in W^{2,p}(\mathbb{R}^m)^{n+m} \times W^{1,p}(\mathbb{R}^m)$$

of (9.4).

Proof. We apply $-ik_{\nu,j}$ to the j-th line for $j = 1, \ldots, n$ as well as $-\operatorname{div}'$ to the line next to the last. Adding up yields

$$-\big(\lambda + |k_\nu|^2 - \Delta'\big)\big(i\sum_{j=1}^n k_{\nu,j}\hat{u}_j + \operatorname{div}'\hat{u}'\big) + \big(|k_\nu|^2 - \Delta'\big)\hat{p} = -\big(i\sum_{j=1}^n k_{\nu,j}\hat{f}_j + \operatorname{div}'\hat{f}'\big).$$

We make use of the last line of (9.4) to get

$$(9.5) \qquad \big(|k_\nu|^2 - \Delta'\big)\hat{p} = -i\sum_{j=1}^n k_{\nu,j}\hat{f}_j - \operatorname{div}'\hat{f}'.$$

In view of the right-hand side, we formally decompose the pressure $p = \sum_{\ell=1}^{n+m} p_\ell$ and consider the problems

$$\big(|k_\nu|^2 - \Delta'\big)\hat{p}_\ell = -ik_{\nu,\ell}\hat{f}_\ell$$

for $\ell = 1, \ldots, n$ and

$$\big(|k_\nu|^2 - \Delta'\big)\hat{p}_\ell = -\partial_\ell\hat{f}_\ell$$

for $\ell = n+1, \ldots, n+m$.

Recall that we do not consider $k = \nu = 0$ and that there exists $\varepsilon > 0$ such that $\{k_\nu; \; (k,\nu) \neq (0,0)\} \subset \mathbb{R}^n \setminus B_\varepsilon(0)$ by our choice of ν. Thus, by means of Fourier transformation with respect to $x' \in \mathbb{R}^m$ and Michlin's multiplier theorem, for each $k_\nu \neq 0$ these problems are uniquely solvable for $\hat{f} \in L^p(\mathbb{R}^m)^{n+m}$ with

$$\hat{p}_\ell = -ik_{\nu,\ell}(|k_\nu|^2 - \Delta')^{-1}\hat{f}_\ell = -(|k_\nu|^2 - \Delta')^{-1}ik_{\nu,\ell}\hat{f}_\ell,$$

respectively

$$\hat{p}_\ell = -(|k_\nu|^2 - \Delta')^{-1}\partial_\ell \hat{f}_\ell := -\partial_\ell(|k_\nu|^2 - \Delta')^{-1}\hat{f}_\ell.$$

Therefore,

$$\hat{p} = \sum_{\ell=1}^{n+m} \hat{p}_\ell = -i\sum_{\ell=1}^{n}(|k_\nu|^2 - \Delta')^{-1}k_{\nu,\ell}\hat{f}_\ell - \sum_{\ell=n+1}^{n+m}(|k_\nu|^2 - \Delta')^{-1}\partial_\ell \hat{f}_\ell$$

is the unique solution of (9.5).

We set $\nabla'_{k_\nu} := \left(ik_{\nu,1}, \ldots, ik_{\nu,n}, \nabla'^T\right)^T$ and

$$Q_{k_\nu} := \nabla'_{k_\nu}(|k_\nu|^2 - \Delta')^{-1}\nabla'^T_{k_\nu}.$$

Then $\hat{p} = -(|k_\nu|^2 - \Delta')^{-1}\nabla'^T_{k_\nu}\hat{f}$ and for (9.4) we find that

$$\hat{u} = U_{k_\nu,\lambda}(I + Q_{k_\nu})\hat{f},$$

where $I := \mathrm{id}_{L^p(\mathbb{R}^m)^{n+m}}$ denotes the identity on $L^p(\mathbb{R}^m)^{n+m}$ and

$$U_{k_\nu,\lambda} := (\lambda + |k_\nu|^2 - \Delta')^{-1} \in \mathcal{L}(L^p(\mathbb{R}^m)^{n+m}).$$

Moreover, since $\hat{p}$ solves (9.5), we have

(9.6) $(ik_{\nu,1}, \ldots, ik_{\nu,n}, \nabla'^T)(I + Q_{k_\nu})\hat{f} = 0$

and therefore

$$(\lambda + |k_\nu|^2 - \Delta')(i\sum_{j=1}^{n} k_{\nu,j}\hat{u}_j + \mathrm{div}'\hat{u}') = 0.$$

Hence, $i\sum_{j=1}^{n} k_{\nu,j}\hat{u}_j + \mathrm{div}'\hat{u}' = 0$ and

(9.7) $R_{k_\nu,\lambda} := U_{k_\nu,\lambda}(I + Q_{k_\nu})$

defines the solution operator for (9.4).   □

Remark 9.2. Note that the representation formula (9.7) remains valid for $k_\nu = 0$ since $I + Q_0 = (I_{\mathbb{R}^n}, \mathbb{P}_{\mathbb{R}^m})^T$ and $U_{0,\lambda} = (\lambda - \Delta')^{-1} \in \mathcal{L}(L^p(\mathbb{R}^m)^{n+m})$.

In view of the discrete multiplier theorem, a representation formula for discrete derivatives is in order. Recall the set $\mathcal{Z}_\vartheta$ of all additive decompositions of the multi-index ϑ from (6.1).

Lemma 9.3. *Let $\nu \in i(-1,1)^n$, $\beta \in \mathbb{N}_0^n$, and $\lambda \in \mathbb{C} \setminus (-\infty, 0)$ be arbitrary. Define $S(k) := k_\nu^\beta$, $T(k) := |k_\nu|^2$, and $M(k) := S(k)(\lambda + T(k) - \Delta')^{-1}$. Then for each $\gamma \in \mathbb{N}_0^n$ we formally have*

$$\Delta^\gamma M(k) = \sum_{\vartheta \leq \gamma} \sum_{\mathcal{W} \in \mathcal{Z}_\vartheta} c_{\mathcal{W}} \left(\Delta^{\gamma - \vartheta} S(k - \vartheta) \right) \left(\lambda + T(k - \vartheta) - \Delta' \right)^{-1}$$
$$\cdot \prod_{j=1}^{r_{\mathcal{W}}} \left(\Delta^{\omega_j} T(k - \omega_j^*) \right) \left(\lambda + T(k - \omega_j^*) - \Delta' \right)^{-1}.$$

Proof. This follows from the more general formula deduced from Lemma 7.1 in the proof of Proposition 7.8. $\qquad\square$

Lemma 9.4. *Let $\nu \in i(-1,1)^n$, $\alpha \in \mathbb{N}_0^m$, $\beta \in \mathbb{N}_0^n$, and $\phi \in (0, \pi]$ be arbitrary. Define $M_\lambda^{\alpha,\beta}(k) := \lambda^{1 - \frac{|\alpha| + |\beta|}{2}} k^\beta \partial'^\alpha U_{k_\nu, \lambda}$. Then the sets*

$$\left\{ M_\lambda^{\alpha,\beta}(k); \ k \in \mathbb{Z}^n, \ \lambda \in \Sigma_{\pi - \phi}, \ 0 \leq |\alpha| + |\beta| \leq 2 \right\}$$

and

$$\left\{ k^\gamma \Delta^\gamma M_\lambda^{\alpha,\beta}(k); \ k \in \mathbb{Z}^n \setminus [-1,1], \ \lambda \in \Sigma_{\pi - \phi}, \ 0 \leq |\alpha| + |\beta| \leq 2, \ 0 \leq \gamma \leq 1 \right\}$$

are $\mathcal{R}$-bounded. The $\mathcal{R}$-boundedness remains valid for $\lambda \in \Sigma_{\pi - \phi} \cup \{0\}$ if $\nu \neq 0$ or if α and β fulfill $|\alpha| + |\beta| = 2$.

Proof. Let $\phi \in (0, \pi]$. Due to Michlin's theorem there exists $C_\phi > 0$ such that

$$\mathcal{R}\left(\left\{ \mu^{1 - \frac{|\alpha|}{2}} \partial'^\alpha (\mu - \Delta')^{-1}; \ \mu \in \Sigma_{\pi - \phi}, \ \alpha \in \mathbb{N}_0^m, \ 0 \leq |\alpha| \leq 2 \right\} \right) \leq C_\phi$$

as well as

$$\mathcal{R}\left(\left\{ \partial'^\alpha (-\Delta')^{-1}; \ |\alpha| = 2 \right\} \right) \leq C_\phi.$$

With the help of Kahane's contraction principle, for $\alpha \in \mathbb{N}_0^m$ and $\beta \in \mathbb{N}_0^n$ we easily deduce $\mathcal{R}$-boundedness of

$$\left\{ \lambda^{1 - \frac{|\alpha| + |\beta|}{2}} k_\nu^\beta \partial'^\alpha \left(\lambda + |k_\nu|^2 - \Delta' \right)^{-1}; \ k \in \mathbb{Z}^n, \ \lambda \in \Sigma_{\pi - \phi}, \ 0 \leq |\alpha| + |\beta| \leq 2 \right\}$$

and

$$\left\{ k_\nu^\beta \partial'^\alpha \left(|k_\nu|^2 - \Delta' \right)^{-1}; \ k \in \mathbb{Z}^n, \ |\alpha| + |\beta| = 2 \right\}.$$

Let $0 < \gamma \leq 1$. We have to estimate $\mathcal{R}$-bounds of $\{ k^\gamma \Delta^\gamma M_\lambda^{\alpha,\beta}(k) \}$, where it suffices to consider $k \in \mathbb{Z}^n \setminus [-1,1]$ thanks to our intermediate condition (3.12). In view of Lemma 9.3 and Lemma 3.2 with $S_\beta(k) := k_\nu^\beta$, $T(k) := |k_\nu|^2$ and arbitrary $0 \leq \omega \leq 1$ it is sufficient to prove $\mathcal{R}$-boundedness of the families

$$\left\{ \lambda^{1 - \frac{|\alpha| + |\beta|}{2}} k^\omega \Delta^\omega S_\beta(k) \partial'^\alpha \left(\lambda + T(k) - \Delta' \right)^{-1}; \ k \in \mathbb{Z}^n \setminus \{0\}, \ \lambda \in \Sigma_{\pi - \phi} \right\}$$

and

$$\left\{k^{\omega}\Delta^{\omega}T(k)\big(\lambda+T(k)-\Delta'\big)^{-1};\ k\in\mathbb{Z}^n\setminus\{0\},\ \lambda\in\Sigma_{\pi-\phi}\right\},$$

where $0\le|\alpha|+|\beta|\le 2$ as well as

$$\left\{k^{\omega}\Delta^{\omega}S_{\beta}(k)\partial'^{\alpha}\big(T(k)-\Delta'\big)^{-1};\ k\in\mathbb{Z}^n\setminus\{0\}\right\}$$

and

$$\left\{k^{\omega}\Delta^{\omega}T(k)\big(T(k)-\Delta'\big)^{-1};\ k\in\mathbb{Z}^n\setminus\{0\}\right\},$$

where $|\alpha|+|\beta|=2$. It is here that we can employ the intermediate condition (3.12) in its full strength. Due to the fact that we do not have to consider $k=0$ there exists $\varepsilon>0$ such that

$$\{k_{\nu};\ k\in\mathbb{Z}^n\setminus\{0\}\}\subset\mathbb{R}^n\setminus B_{\varepsilon}(0).$$

Hence, there exists $C>0$ such that

$$\left|\lambda^{1-\frac{|\alpha|+|\beta|}{2}}k^{\omega}\Delta^{\omega}S_{\beta}(k)\right|\le C|\lambda+T(k)|^{1-\frac{|\alpha|+|\beta|}{2}}\quad(k\in\mathbb{Z}^n\setminus\{0\},\ \lambda\in\Sigma_{\pi-\phi})$$

due to parameter-ellipticity of T. By Kahane's contraction principle the claim follows. $\qquad\square$

We agree to simply write E for the set $\mathrm{Const}(\mathcal{Q}_n,E)$ of constant, E-valued functions defined on $\mathcal{Q}_n$. Let further $W^{1,p}_{(0),per}(\mathcal{Q}_n,E):=W^{1,p}_{per}(\mathcal{Q}_n,E)\cap L^p_{(0)}(\mathcal{Q}_n,E)$ denote the subset of functions of mean value zero, that is, the set of functions $f\in W^{1,p}_{per}(\mathcal{Q}_n,E)$ such that $\hat{f}(0)=0$.

Theorem 9.5. *For the Stokes resolvent problem* (9.2) *there exists a unique solution*

$$u\ \in\ \bigcap_{\ell=0}^{2}W^{\ell,p}_{\nu,per}\big(\mathcal{Q}_n,W^{2-\ell,p}(\mathbb{R}^m)\big)^{n+m},$$

$$p\ \in\ \begin{cases}W^{1,p}(\Omega)\cap W^{1,p}_{\nu,per}(\mathcal{Q}_n,L^p(\mathbb{R}^m)),&\nu\ne 0\ or\ \hat{f}(0)=0,\\ \big(W^{1,p}(\Omega)\cap W^{1,p}_{(0),per}(\mathcal{Q}_n,L^p(\mathbb{R}^m))\big)+\hat{W}^{1,p}(\mathbb{R}^m),&else.\end{cases}$$

We have

$$\|\nabla^2 u\|_p+\sqrt{\lambda}\|\nabla u\|_p+\lambda\|u\|_p+\|\nabla p\|_p\le C\|f\|_p$$

and additionally $\|p\|_p\le C\|f\|_p$ *if* $\nu\ne 0$ *or* $\hat{f}(0)=0$. *Moreover, for each* $\phi>0$ *the family of solution operators* $R_{\lambda}f:=u$ *fulfills*

$$\mathcal{R}\Big(\big\{\lambda^{1-\frac{|\alpha|}{2}}\partial^{\alpha}R_{\lambda};\ \lambda\in\Sigma_{\pi-\phi},\ 0\le|\alpha|\le 2\big\}\Big)<\infty.$$

Proof. Lemma 9.4 and Theorem 3.24 prove $M_\lambda^{\alpha,\beta}$ to define a Fourier multiplier for $\lambda \in \Sigma_{\pi-\phi}$ and $0 \le |\alpha| + |\beta| \le 2$.

Let $\nu \ne 0$. Then by the last assertion of Lemma 9.4 we see that

$$Q^* : \mathbb{Z}^n \to \mathcal{L}\big(L^p(\mathbb{R}^m)^{n+m}, L^p(\mathbb{R}^m)\big); \quad k \mapsto Q_{k_\nu}^* := \big(|k_\nu|^2 - \Delta'\big)^{-1} \nabla_{k_\nu}'^T$$

and

$$Q : \mathbb{Z}^n \to \mathcal{L}\big(L^p(\mathbb{R}^m)^{n+m}\big); \quad k \mapsto \nabla_{k_\nu} Q_{k_\nu}^* = \nabla_{k_\nu}\big(|k_\nu|^2 - \Delta'\big)^{-1} \nabla_{k_\nu}'^T$$

define Fourier multipliers. Hence,

$$T_{Q^*} \in \mathcal{L}\big(L^p(\Omega)^{m+n}, \ L^p(\mathcal{Q}_n, W^{1,p}(\mathbb{R}^m)) \cap W_{\nu,per}^{1,p}(\mathcal{Q}_n, L^p(\mathbb{R}^m))\big)$$

and $u := e^{\nu\cdot} T_{\lambda^{-1} M_\lambda^{0,0}}\big(e^{-\nu\cdot} f\big)$ and $p := -e^{\nu\cdot} T_{Q^*}\big(e^{-\nu\cdot} f\big)$ define the solution of (9.2) for fixed $\lambda \in \Sigma_{\pi-\phi}$. Recall that we dropped the factor $e^{-\nu\cdot}$ right after (9.3).

If $\nu = 0$, we still have

$$Q(k) = \nabla_k Q_k^* \in \mathcal{L}\big(L^p(\mathbb{R}^m)^{n+m}\big) \quad (k \in \mathbb{Z}^n)$$

and

$$Q^*(k) = Q_k^* \in \mathcal{L}\big(L^p(\mathbb{R}^m)^{n+m}, L^p(\mathbb{R}^m)\big) \quad (k \in \mathbb{Z}^n \setminus \{0\}).$$

However,

$$Q^*(0) = Q_0^* \in \mathcal{L}(L^p(\mathbb{R}^m)^{n+m}, \hat{W}^{1,p}(\mathbb{R}^m)).$$

We therefore define

$$Q_I^*(k) := \begin{cases} 0, & k = 0, \\ Q_k^*, & k \ne 0 \end{cases} \quad \text{and} \quad Q_{II}^*(k) := \begin{cases} Q_0^*, & k = 0, \\ 0, & k \ne 0. \end{cases}$$

Then

$$Q_I^* : \mathbb{Z}^n \to \mathcal{L}(L^p(\mathbb{R}^m)^{n+m}, L^p(\mathbb{R}^m)) \quad \text{and} \quad Q_{II}^* : \mathbb{Z}^n \to \mathcal{L}(L^p(\mathbb{R}^m)^{n+m}, \hat{W}^{1,p}(\mathbb{R}^m))$$

define Fourier multipliers. Note that by their structure $T_{Q_I^*}$ maps to functions of mean value zero, whereas $T_{Q_{II}^*}$ maps to constant functions which of course are periodic. Thus, we have

$$T_{Q_I^*} \in \mathcal{L}\big(L^p(\Omega)^{m+n}, L^p(\mathcal{Q}_n, W^{1,p}(\mathbb{R}^m)) \cap W_{(0),per}^{1,p}(\mathcal{Q}_n, L^p(\mathbb{R}^m))\big)$$

as well as

$$T_{Q_{II}^*} \in \mathcal{L}\big(L^p(\Omega)^{m+n}, \hat{W}^{1,p}(\mathbb{R}^m)\big).$$

Consequently, for each $\lambda \in \Sigma_{\pi-\phi}$ we find as above that $u := e^{\nu\cdot} T_{\lambda^{-1} M_\lambda^{0,0}}\big(e^{-\nu\cdot} f\big)$ and $p := -e^{\nu\cdot}\big(T_{Q_I^*} f - T_{Q_{II}^*}\big)\big(e^{-\nu\cdot} f\big)$ solve problem (9.2). Uniqueness finally follows from the uniqueness theorem for Fourier series. $\qquad\square$

In what follows we aim for the definition of a ν-periodic Helmholtz decomposition. To this end, we set

$$
G^p_{\nu,per}(\Omega) := \begin{cases} \{\nabla p;\ p \in W^{1,p}(\Omega) \cap W^{1,p}_{\nu,per}(\mathcal{Q}_n, L^p(\mathbb{R}^m))\}, & \nu \neq 0, \\ \{\nabla p;\ p \in \left(W^{1,p}(\Omega) \cap W^{1,p}_{(0),per}(\mathcal{Q}_n, L^p(\mathbb{R}^m))\right) + \hat{W}^{1,p}(\mathbb{R}^m)\}, & \text{else.} \end{cases}
$$

Recall the class $\mathcal{D}'_{\nu,per,n}(\mathbb{R}^{n+m})$ of partially ν-periodic distributions and its properties from Lemma 2.21. In particular, $\operatorname{div} T \in \mathcal{D}'_{\nu,per}(\mathbb{R}^{n+m})$ if $T \in \mathcal{D}'_{\nu,per}(\mathbb{R}^{n+m})$. For the sake of convenience in this chapter we agree to write $\mathcal{D}'_{\nu,per}(\Omega)$ instead of $\mathcal{D}'_{\nu,per,n}(\mathbb{R}^{n+m})$. Furthermore, we drop the argument k and write $\hat{f}$ only instead of $\hat{f}(k)$ if an assertion is valid for all $k \in \mathbb{Z}^n$.

Lemma 9.6. *For $\nu \in i(-1,1)^n$ define*

$$
\mathbb{P}_{\Omega,\nu} := e^{\nu\cdot} T_{I+Q_{k_\nu}} e^{-\nu\cdot}.
$$

Then $\mathbb{P}_{\Omega,\nu} \in \mathcal{L}(L^p(\Omega)^{n+m})$ defines a projection in $L^p(\Omega)^{n+m}$ such that

$$
N(\mathbb{P}_{\Omega,\nu}) = G^p_{\nu,per}(\Omega) \quad and \quad R(\mathbb{P}_{\Omega,\nu}) = \{f \in L^p(\Omega)^{n+m};\ \operatorname{div} f = 0 \text{ in } \mathcal{D}'_{\nu,per}(\Omega)\}.
$$

Proof. Due to Lemma 9.4 $I + Q_{k_\nu}$ defines a discrete Fourier multiplier. In particular, $\mathbb{P}_{\Omega,\nu} \in \mathcal{L}(L^p(\Omega)^{n+m})$ is well-defined. Since

$$
\nabla'^T_{k_\nu} \nabla'_{k_\nu} \left(|k_\nu|^2 - \Delta'\right)^{-1} = -I,
$$

we have $(I + Q_{k_\nu})(I + Q_{k_\nu}) = I + 2Q_{k_\nu} + Q_{k_\nu} Q_{k_\nu} = I + Q_{k_\nu}$ and $\mathbb{P}_{\Omega,\nu}$ defines a projection.

In what follows we consider partial Fourier coefficients to prove the equalities for the kernel and range of $\mathbb{P}_{\Omega,\nu}$. Due to Lemma 2.21 we have

$$
\begin{aligned}
\left(e^{-\nu\cdot} \operatorname{div}(e^{\nu\cdot} T_{I+Q_{k_\nu}} e^{-\nu\cdot} f))\right)^{\wedge} &= ik^T_\nu \left(T_{I+Q_{k_\nu}} e^{-\nu\cdot} f\right)^{\wedge} + \nabla'^T \left(T_{I+Q_{k_\nu}} e^{-\nu\cdot} f\right)^{\wedge} \\
&= \nabla'^T_{k_\nu}(I + Q_{k_\nu})\left(e^{-\nu\cdot} f\right)^{\wedge} = 0,
\end{aligned}
$$

where we have used (9.6) in the last line. Hence, $\operatorname{div}(e^{\nu\cdot} T_{I+Q_{k_\nu}} e^{-\nu\cdot} f) = 0$ in $\mathcal{D}'_{\nu,per}(\Omega)$. On the other hand let $\operatorname{div} f = 0$ in $\mathcal{D}'_{\nu,per}(\Omega)$. Then $e^{-\nu\cdot} \operatorname{div} f = 0$ in $\mathcal{D}'_{per}(\Omega)$ and therefore $\nabla'^T_{k_\nu}\left(e^{-\nu\cdot} f\right)^{\wedge} = 0$ which implies

$$
Q_{k_\nu}\left(e^{-\nu\cdot} f\right)^{\wedge} = 0.
$$

Thus,

$$
\left(T_{I+Q_{k_\nu}} e^{-\nu\cdot} f\right)^{\wedge} = (I + Q_{k_\nu})\left(e^{-\nu\cdot} f\right)^{\wedge} = \left(e^{-\nu\cdot} f\right)^{\wedge}
$$

and

$$
e^{\nu\cdot} T_{I+Q_{k_\nu}} e^{-\nu\cdot} f = f
$$

by means of the representation formula from Theorem 2.19. Altogether we have shown $R(\mathbb{P}_{\Omega,\nu}) = \{f \in L^p(\Omega);\ \mathrm{div} f = 0 \text{ in } \mathcal{D}'_{\nu,per}(\Omega)\}$.

Now let $f \in G^p_{\nu,per}(\Omega)$, i.e. $f = \nabla p \in L^p(\Omega)$. This implies

$$\big(T_{I+Q_{k_\nu}} e^{-\nu\cdot} f\big)\hat{} = \big(T_{I+Q_{k_\nu}} e^{-\nu\cdot}\nabla p\big)\hat{} = (I+Q_{k_\nu})\big(e^{-\nu\cdot}\nabla p\big)\hat{}$$
$$= (I+Q_{k_\nu})\nabla'_{k_\nu}\big(e^{-\nu\cdot} p\big)\hat{} = \big(I+\nabla'_{k_\nu}\big(|k_\nu|^2 - \Delta'\big)^{-1}\nabla'^T_{k_\nu}\big)\nabla'_{k_\nu}\big(e^{-\nu\cdot} p\big)\hat{} = 0.$$

In turn consider $f \in L^p(\Omega)^{n+m}$ such that $\mathbb{P}_{\Omega,\nu} f = 0$. Then $(I+Q_{k_\nu})\big(e^{-\nu\cdot} f\big)\hat{} = 0$ and

$$\big(e^{-\nu\cdot} f\big)\hat{} = -Q_{k_\nu}\big(e^{-\nu\cdot} f\big)\hat{}$$
$$= \nabla'_{k_\nu}\big(|k_\nu|^2 - \Delta'\big)^{-1}\nabla'^T_{k_\nu}\big(-e^{-\nu\cdot} f\big)\hat{} = \big(-e^{-\nu\cdot}\nabla T_{U_{k_\nu,0}}\nabla'^T f\big)\hat{}.$$

Hence, $f = \nabla p$, where $p := -T_{Q^*_{k_\nu}} f \in G^p_{\nu,per}(\Omega)$. Here just as in the proof of Theorem 9.5 we decompose $Q^* = Q^*_I + Q^*_{II}$ in case $\nu = 0$. Thus, the kernel of $\mathbb{P}_{\Omega,\nu}$ is equal to $G^p_{\nu,per}(\Omega)$ which finishes the proof. $\qquad\square$

We refer to the projection from Lemma 9.6 as the ν-periodic Helmholtz projection. It implies the decomposition

$$L^p(\Omega) = L^p_{\sigma,\nu,per}(\Omega) \oplus G^p_{\nu,per},$$

where

(9.8) $$L^p_{\sigma,\nu,per}(\Omega) := R(\mathbb{P}_{\Omega,\nu}).$$

With the aid of this ν-periodic Helmholtz decomposition, we are now in the position to define the ν-periodic Stokes operator in $L^p_{\sigma,\nu,per}(\Omega)$.

Definition 9.7. For $\nu \in i(-1,1)^n$ we define the ν-periodic Stokes operator $A_{S,\nu}$ in $L^p_{\sigma,\nu,per}(\Omega)$ to be

$$D(A_{S,\nu}) := \bigcap_{\ell=0}^{2} W^{\ell,p}_{\nu,per}(\mathcal{Q}_n, W^{2-\ell,p}(\mathbb{R}^m))^{n+m} \cap L^p_{\sigma,\nu,per}(\Omega)$$
$$A_{S,\nu} u := -\mathbb{P}_{\Omega,\nu}\Delta u \quad (u \in D(A_{S,\nu})).$$

As a consequence of the following lemma, just as in $\mathbb{R}^m$, the ν-periodic Stokes operator $A_{S,\nu}$ is given as a part of the $L^p(\Omega)^{n+m}$-realization of the ν-periodic negative Laplacian $-\Delta_\nu$ with $D(\Delta_\nu) := \bigcap_{\ell=0}^{2} W^{\ell,p}_{\nu,per}(\mathcal{Q}_n, W^{2-\ell,p}(\mathbb{R}^m))^{n+m}$.

Lemma 9.8. Let $\nu \in i(-1,1)^n$ and $u \in \bigcap_{\ell=0}^{2} W^{\ell,p}_{\nu,per}(\mathcal{Q}_n, W^{2-\ell,p}(\mathbb{R}^m))^{n+m}$. Then

$$\partial^\alpha \mathbb{P}_{\Omega,\nu} u = \mathbb{P}_{\Omega,\nu}\partial^\alpha u \quad (\alpha \in \mathbb{N}_0^{n+m},\ |\alpha| \le 2).$$

Proof. Let $\alpha = (\overline{\alpha}, \alpha') \in \mathbb{N}_0^{n+m}$. In what follows we write $e^{-\nu x}$ instead of $e^{-\nu \cdot}$ to indicate the dependency on $x \in \mathbb{R}^n$ only. Then Lemma 2.21 shows

$$\left(e^{-\nu x}\partial^\alpha \mathbb{P}_{\Omega,\nu} u\right)\hat{} = \left(e^{-\nu x}\partial^\alpha e^{\nu x}T_{I+Q_{k_\nu}}e^{-\nu x}u\right)\hat{} = (ik_\nu)^{\overline{\alpha}}\partial^{\alpha'}(I+Q_{k_\nu})\left(e^{-\nu x}u\right)\hat{}$$

as well as

$$\left(e^{-\nu x}\mathbb{P}_{\Omega,\nu}\partial^\alpha u\right)\hat{} = (I+Q_{k_\nu})\left(e^{-\nu x}\partial^\alpha u\right)\hat{} = (I+Q_{k_\nu})(ik_\nu)^{\overline{\alpha}}\partial^{\alpha'}\left(e^{-\nu x}u\right)\hat{}.$$

Since $\partial^{\alpha'}$ and Q_{k_ν} commute, the assertion follows. $\square$

For the ν-periodic Stokes operator $A_{S,\nu}$ we therefore have

$$(9.9) \qquad A_{S,\nu} = -\Delta_\nu|_{D(A_{S,\nu})} \quad \text{and} \quad (\lambda - A_{S,\nu})^{-1} = (\lambda + \Delta_\nu)^{-1}|_{L^p_{\sigma,\nu,per}(\Omega)}.$$

From that we deduce the following result on the $\mathcal{RH}^\infty$-calculus of the ν-periodic Stokes operator.

Theorem 9.9. *Let $\nu \in i(-1,1)^n$. Then we have $A_{S,\nu} \in \mathcal{RH}^\infty(L^p_{\sigma,\nu,per}(\Omega))$ and $\phi^{\mathcal{R}\infty}_{A_{S,\nu}} = 0$.*

Proof. Thanks to the observation in (9.9) this follows from $-\Delta_\nu \in \mathcal{RH}^\infty(L^p(\Omega))$ with $\phi^{\mathcal{R}\infty}_{-\Delta_\nu} = 0$ by Theorem 8.22 and the subsequent remarks. $\square$

In the sequel we prove existence of the standard Helmholtz decomposition for the space $L^p(\tilde{\Omega})$. For technical reasons we choose $\tilde{\Omega} := \tilde{\mathcal{Q}}_n \times \mathbb{R}^m$, where $\tilde{\mathcal{Q}}_n := (0,\pi)^n$. To this end, we make use of the periodic Helmholtz projection $\mathbb{P}_{\Omega,per} \in \mathcal{L}(L^p(\Omega))$ to prove existence of the Helmholtz projection $\mathbb{P}_{\tilde{\Omega}} \in \mathcal{L}(L^p(\tilde{\Omega}))$. Moreover, with suitable extension and restriction operators $\mathfrak{E}$ and $\mathfrak{R}$ we establish the representation formula

$$\mathbb{P}_{\tilde{\Omega}} = \mathfrak{R}\mathbb{P}_{\Omega,per}\mathfrak{E}.$$

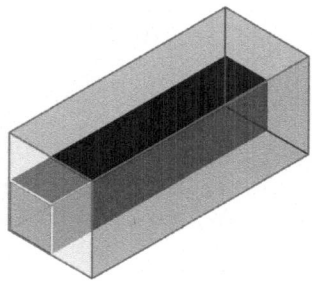

Figure 9.1: The domains $\tilde{\Omega}$ (blue) and Ω (gray).

First we give an equivalent representation of $L^p_{\sigma,per}(\Omega)$. Given $f \in L^p(\Omega)$ such that $\mathrm{div}\, f \in L^p(\Omega)$, the expression

$$\gamma_{\mathbf{n}\cdot f}(\omega) := \int_\Omega f \cdot \nabla\varphi + \int_\Omega \varphi\, \mathrm{div} f \quad (\varphi \in W^{1,p'}(\Omega))$$

is well-defined. Here $\mathbf{n}$ denotes the outer normal, $\omega = \gamma(\varphi)$ denotes the trace of φ on $\partial\Omega$ (see e.g. [Gal94, Section III.2]), and p' is the Hölder conjugate of p, i.e. $\frac{1}{p} + \frac{1}{p'} = 1$. If $\mathrm{div}\, f = 0$, we can extend $\gamma_{\mathbf{n}\cdot f}(\omega)$ formally to $\varphi \in \hat{W}^{1,p'}(\Omega)$ as the integral $\int_\Omega f \cdot \nabla\varphi$ is still well-defined. We abbreviate

$$W^{1,p}_{per}(\Omega) := \left(W^{1,p}(\Omega) \cap W^{1,p}_{per}(\mathcal{Q}_n, L^p(\mathbb{R}^m))\right) \subset W^{1,p}(\Omega)$$

and

$$W^{1,p}_{per,\hat{+}}(\Omega) := \left(W^{1,p}(\Omega) \cap W^{1,p}_{(0),per}(\mathcal{Q}_n, L^p(\mathbb{R}^m))\right) + \hat{W}^{1,p}(\mathbb{R}^m) \subset \hat{W}^{1,p}(\Omega).$$

For f subject to $\mathrm{div}\, f = 0$ in Ω we define periodic traces by means of

$$(9.10) \qquad \gamma_{per} f = 0 \quad :\Longleftrightarrow \quad \gamma_{\mathbf{n}\cdot f}(\omega) = 0 \quad (\varphi \in W^{1,p'}_{per,\hat{+}}(\Omega)).$$

Note that the trace $\omega = \gamma(\varphi)$ of $\varphi \in W^{1,p'}_{per,\hat{+}}(\Omega)$ is a well-defined element in the space $W^{1-1/p'}_{p'}(\partial\Omega) + \hat{W}^{1,p'}(\partial\Omega)$. This extends the notion of periodic boundary conditions as known from smooth functions. Precisely, consider the following class

$$\mathcal{D}_{\sigma,per}(\Omega) := \{\varphi \in C^\infty(\Omega);\ \mathrm{div}\,\varphi = 0,$$
$$\forall y \in \mathbb{R}^m : D_y^\alpha \varphi(\cdot, y)_{per} \in C^\infty_{per}(\mathcal{Q}_n) \quad (\alpha \in \mathbb{N}_0^m),$$
$$\exists U = U_\varphi \subset \mathbb{R}^m \text{ bounded} : \forall x \in \overline{\mathcal{Q}}_n :\ \mathrm{supp}\,\varphi(x,\cdot) \subset U\}$$

of smooth, divergence-free, partially periodic functions with tangentially bounded support. Recall the 2π-periodic extension g_{per} of a function $g \in L^p(\mathcal{Q}_n)$ from Chapter 2. Let $f \in \mathcal{D}_{\sigma,per}(\Omega)$, $U = U_f \subset \mathbb{R}^m$ and $\Omega_U = \mathcal{Q}_n \times U$. Then

$$\gamma_{\mathbf{n}\cdot f}(\omega) = \int_\Omega f \cdot \nabla\varphi + \int_\Omega \varphi\, \mathrm{div} f = \int_{\Omega_U} f \cdot \nabla\varphi + \int_{\Omega_U} \varphi\, \mathrm{div} f = \int_{\partial\Omega_U} \omega f \cdot \mathbf{n}$$

by the Gauß identity for bounded Lipschitz domains (see [Gal94, Section II.3]). We decompose the boundary of Ω_U into $\partial\Omega_U = \partial\mathcal{Q}_n \times U \cup \mathcal{Q}_n \times \partial U$. By our choice of f and U it follows that $\omega f \cdot \mathbf{n}$ vanishes on $\mathcal{Q}_n \times \partial U$. The remaining part can be decomposed further using

$$\partial\mathcal{Q}_n = \bigcup_{j=1}^n \mathcal{Q}^+_{n,j} \cup \mathcal{Q}^-_{n,j},$$

where

$$\mathcal{Q}^+_{n,j} := \mathcal{Q}_{j-1} \times \{2\pi\} \times \mathcal{Q}_{n-j} \quad \text{and} \quad \mathcal{Q}^-_{n,j} := \mathcal{Q}_{j-1} \times \{0\} \times \mathcal{Q}_{n-j}.$$

Then

$$\int_{\partial\Omega_U} \omega f \cdot \mathbf{n} = \int_{\partial\mathcal{Q}_n \times U} \omega f \cdot \mathbf{n} = \sum_{j=1}^{n} \int_{\mathcal{Q}_{n,j}^+ \times U} \omega f \cdot \mathbf{n} + \int_{\mathcal{Q}_{n,j}^- \times U} \omega f \cdot \mathbf{n}$$

$$= \sum_{j=1}^{n} \int_{\mathcal{Q}_{n,j}^+ \times U} \omega f \cdot e_j - \int_{\mathcal{Q}_{n,j}^- \times U} \omega f \cdot e_j = \sum_{j=1}^{n} \int_{\mathcal{Q}_{n,j}^+ \times U} \omega f_j - \int_{\mathcal{Q}_{n,j}^- \times U} \omega f_j$$

and so

$$\int_{\partial\Omega_U} \omega f \cdot \mathbf{n} = \sum_{j=1}^{n} \int_{\mathcal{Q}_{n-1} \times U} \omega|_{\mathcal{Q}_{n,j}^- \times U} \left(f_j|_{\mathcal{Q}_{n,j}^+ \times U} - f_j|_{\mathcal{Q}_{n,j}^- \times U} \right),$$

where we have used $\omega|_{\mathcal{Q}_{n,j}^+ \times U} = \omega|_{\mathcal{Q}_{n,j}^- \times U}$ for $\omega = \gamma(\varphi)$, $\varphi \in W_{per}^{1,p'}(\Omega)$. Since $\varphi \in W_{per}^{1,p'}(\Omega)$ can be chosen arbitrarily, we conclude

$$\gamma_{per} f = 0 \quad \Longleftrightarrow \quad f_j|_{\mathcal{Q}_{n,j}^+ \times U} = f_j|_{\mathcal{Q}_{n,j}^- \times U} \quad (j = 1, \ldots, n).$$

In the sequel we frequently denote the periodic Helmholtz projection on $L^p(\Omega)$ by $\mathbb{P}_{\Omega,per,p} = T_{I+Q_{k,p}}$. This is done in order to emphasize the underlying L^p-space.

Lemma 9.10. *The periodic Helmholtz projection fulfills* $\mathbb{P}'_{\Omega,per,p} = \mathbb{P}_{\Omega,per,p'}$.

Proof. Let $u \in \mathbb{T}(\mathcal{Q}_n, L^p(\mathbb{R}^m))$ and $v \in \mathbb{T}(\mathcal{Q}_n, L^{p'}(\mathbb{R}^m))$ denote two arbitrary trigonometric polynomials, say $u = \sum_{k\in[-\alpha,\alpha]} e_k \otimes u_k$ and $v = \sum_{k\in[-\alpha,\alpha]} e_k \otimes v_k$ with $u_k \in L^p(\mathbb{R}^m)$ and $v_k \in L^{p'}(\mathbb{R}^m)$, respectively.

Then $u(x,y) := \sum_{k\in[-\alpha,\alpha]} e^{ikx} u_k(y)$ and $v(x,y) := \sum_{k\in[-\alpha,\alpha]} e^{ikx} v_k(y)$ define elements of $L^p(\Omega)$ and $L^{p'}(\Omega)$, respectively. Moreover, we have

$$\mathbb{P}_{\Omega,per,p} u = \sum_{k\in[-\alpha,\alpha]} e_k \otimes (I+Q_{k,p})u_k \quad \text{and} \quad \mathbb{P}_{\Omega,per,p'} v = \sum_{k\in[-\alpha,\alpha]} e_k \otimes (I+Q_{k,p'})v_k.$$

This allows for the calculation

$$\left(\mathbb{P}_{\Omega,per,p} u, v \right)_{\Omega} := \int_{\Omega} (\mathbb{P}_{\Omega,per,p} u)(x,y) v(x,y) d(x,y)$$

$$= \int_{\Omega} \sum_{k\in[-\alpha,\alpha]} e^{ikx} ((I+Q_{k,p})u_k)(y) \sum_{k\in[-\alpha,\alpha]} e^{ikx} v_k(y) d(x,y)$$

$$= \int_{\Omega} \sum_{k\in[-\alpha,\alpha]} e^{ikx} \sum_{j\in[-\alpha,\alpha]} e^{ijx} ((I+Q_{k,p})u_k)(y) v_j(y) d(x,y)$$

$$= \sum_{k,j\in[-\alpha,\alpha]} \int_{\mathcal{Q}_n} e^{i(k+j)x} dx \int_{\mathbb{R}^m} ((I+Q_{k,p})u_k)(y) v_j(y) dy$$

$$= (2\pi)^n \sum_{k\in[-\alpha,\alpha]} ((I+Q_{k,p})u_k, v_{-k})_{\mathbb{R}^m}.$$

Since $Q'_{k,p} = Q_{-k,p'}$ holds true for all $k \in \mathbb{Z}^n$, we deduce

$$\left(\mathbb{P}_{\Omega,per,p}u, v\right)_\Omega = (2\pi)^n \sum_{k\in[-\alpha,\alpha]} \left((I + Q_{k,p})u_k, v_{-k}\right)_{\mathbb{R}^m}$$

$$= (2\pi)^n \sum_{k\in[-\alpha,\alpha]} \left(u_k, (I + Q_{-k,p'})v_{-k}\right)_{\mathbb{R}^m} = \left(u, \mathbb{P}_{\Omega,per,p'}v\right)_\Omega.$$

Thanks to density of trigonometric polynomials (see Proposition 2.4) the claim follows for arbitrary $u \in L^p(\Omega)$ and $v \in L^{p'}(\Omega)$. $\qquad\square$

Proposition 9.11. *For $L^p_{\sigma,per}(\Omega)$ defined as in (9.8) we have*

$$L^p_{\sigma,per}(\Omega) = \{f \in L^p(\Omega); \; \operatorname{div} f = 0 \text{ in } \Omega, \; \gamma_{per}f = 0\}.$$

Proof. First let $f \in L^p_{\sigma,per}(\Omega)$. Trivially $\operatorname{div} f = 0$ in Ω. Moreover, for arbitrary $\varphi \in W^{1,p'}_{per,\hat{+}}(\Omega)$ we have

$$\left(f, \nabla\varphi\right)_\Omega = \left(\mathbb{P}_{\Omega,per,p}f, \nabla\varphi\right)_\Omega = \left(f, \mathbb{P}_{\Omega,per,p'}\nabla\varphi\right)_\Omega = 0.$$

Thus, $\gamma_{per}f = 0$ by (9.10).

Now let $f \in L^p(\Omega)$ fulfill $\operatorname{div} f = 0$ in Ω and $\gamma_{per}f = 0$. Then

$$0 = \int_\Omega f \cdot \nabla\varphi + \int_\Omega \varphi \operatorname{div} f = \int_\Omega f \cdot \nabla\varphi = \left(f, \nabla\varphi\right)_\Omega \quad \left(\varphi \in W^{1,p'}_{per,\hat{+}}(\Omega)\right).$$

Since $N(\mathbb{P}_{\Omega,per,p'}) = R(I - \mathbb{P}_{\Omega,per,p'})$, for each $v \in L^{p'}(\Omega)$ it holds that

$$0 = \left(f, (I - \mathbb{P}_{\Omega,per,p'})v\right)_\Omega = \left((I - \mathbb{P}_{\Omega,per,p})f, v\right)_\Omega.$$

Thus, $f = \mathbb{P}_{\Omega,per,p}f \in R(\mathbb{P}_{\Omega,per,p}) = L^p_{\sigma,per}(\Omega)$. $\qquad\square$

Recall the standard Helmholtz decomposition in $L^p(\Omega)$ given by

$$L^p(\Omega) = L^p_\sigma(\Omega) \oplus G^p(\Omega),$$

where

$$L^p_\sigma(\Omega) = \{f \in L^p(\Omega)^{n+m}; \; \operatorname{div} f = 0 \text{ in } \Omega, \; \gamma_{f\cdot n} = 0\}$$

and

$$G^p(\Omega) = \{\nabla p; \; p \in L^1_{loc}(\overline{\Omega}), \; \nabla p \in L^p(\Omega)^{n+m}\},$$

as well as the standard Helmholtz projection $\mathbb{P}_\Omega \in \mathcal{L}(L^p(\Omega))$ which fulfills

$$\mathbb{P}_\Omega(L^p(\Omega)) = L^p_\sigma(\Omega) \quad \text{and} \quad N(\mathbb{P}_\Omega) = G^p(\Omega).$$

Our aim is to derive the standard Helmholtz projection in $\mathcal{L}(L^p(\tilde{\Omega}))$ from the periodic Helmholtz projection in $\mathcal{L}(L^p(\Omega))$. As mentioned, we scale the underlying rectangular domain and consider $\tilde{\Omega} := \tilde{\mathcal{Q}}_n \times \mathbb{R}^m$ with $\tilde{\mathcal{Q}}_n := (0,\pi)^n$.

A few preparations are in order. Consider $u \in L^p((0,\pi) \times G)^{m+1}$ with an arbitrary domain $G \subset \mathbb{R}^m$. Define $U \in L^p((0,2\pi) \times G)^{m+1}$ by

$$U(x,y) := (\mathfrak{E}_1 u)(x,y) := \begin{cases} u(x,y), & (x,y) \in (0,\pi) \times G, \\ \begin{pmatrix} -u_1 \\ u' \end{pmatrix} (2\pi - x, y), & (x,y) \in (\pi, 2\pi) \times G. \end{cases}$$

If $u \in C^1((0,\pi) \times G)^{m+1}$, then

$$\operatorname{div} U(x,y) = \operatorname{div} U(2\pi - x, y) \quad ((x,y) \in (0,\pi) \times G).$$

Iteratively, for $u \in L^p((0,\pi)^n \times G)^{m+n}$ we define $U \in L^p((0,2\pi)^n \times G)^{m+n}$ by

$$U := \mathfrak{E}u := \mathfrak{E}_n \cdots \mathfrak{E}_1 u,$$

where the $\mathfrak{E}_j$ are defined to extend the j-th variable accordingly.

Lemma 9.12. *Let* $u \in L^p(\tilde{\Omega})^{m+n}$. *Then we have* $u \in L^p_\sigma(\tilde{\Omega})$ *if and only if* $U := \mathfrak{E}u \in L^p_{\sigma,per}(\Omega)$.

Proof. It is sufficient to consider the case $n = 1$ as the assertion for arbitrary $n \in \mathbb{N}$ follows with minor adjustments by iteration.

First let $u \in L^p_\sigma(\tilde{\Omega})$. We show $\operatorname{div} U_{per} = 0$ in $\mathcal{D}'_{per}(\Omega)$. To this end, let $\varphi \in \mathcal{D}(\mathbb{R}^{1+m})$ be arbitrary and $\ell \in \mathbb{N}$ such that $\operatorname{supp}\varphi \subset (-2\ell\pi, 2\ell\pi) \times \mathbb{R}^m$. Then

$$\int_\Omega U_{per}(x,y)\nabla\varphi(x,y)d(x,y) = \int_{\mathbb{R}^m} \int_{(-2\ell\pi, 2\ell\pi)} \sum_{j=1}^{m+1} U_{per,j}(x,y)\partial_j\varphi(x,y)dxdy$$

$$= \sum_{i=-\ell}^{\ell-1} \int_{\mathbb{R}^m} \int_{(0,2\pi)} \sum_{j=1}^{m+1} U_j(x,y)(\partial_j\varphi)(x + 2i\pi, y)dxdy.$$

It therefore suffices to prove

$$\int_{\mathbb{R}^m} \int_{(0,2\pi)} \sum_{j=1}^{m+1} U_j(x,y)\partial_j\varphi(x,y)dxdy = 0$$

for arbitrary $\varphi \in W^{1,p'}(\Omega)$. Obviously

$$\int_{(0,2\pi)} \sum_{j=1}^{m+1} U_j(x,y)\partial_j\varphi(x,y)dx = \int_{(0,\pi)} \sum_{j=1}^{m+1} u_j(x,y)\partial_j\varphi(x,y)dx$$

$$+ \int_{(\pi,2\pi)} -u_1(2\pi - x, y)\partial_1\varphi(x,y) + \sum_{j=2}^{m+1} u_j(2\pi - x, y)\partial_j\varphi(x,y)dx$$

holds true for $U = \mathfrak{E}_1 u$. A substitution of variables and the relation

(9.11) $$(\partial_1 \varphi)(2\pi - x, y) = -\partial_x \varphi(2\pi - x, y)$$

yields

$$\int_{(0,2\pi)} \sum_{j=1}^{m+1} U_j(x,y)\partial_j \varphi(x,y)dx = \int_{(0,\pi)} \sum_{j=1}^{m+1} u_j(x,y)\partial_j \varphi(x,y)dx$$

$$+ \int_{(0,\pi)} -u_1(x,y)(\partial_1\varphi)(2\pi - x, y) + \sum_{j=2}^{m+1} u_j(x,y)\partial_j \varphi(2\pi - x, y)dx.$$

We define $\psi \in W^{1,p'}(\tilde{\Omega})$ by

$$\psi(x,y) := \varphi(x,y) + \varphi(2\pi - x, y).$$

Then

$$\int_{\mathbb{R}^m} \int_{(0,2\pi)} \sum_{j=1}^{m+1} U_j(x,y)\partial_j \varphi(x,y)dxdy = \int_{\tilde{\Omega}} u(x,y)\nabla\psi(x,y)d(x,y) = 0$$

due to $u \in L^p_\sigma(\tilde{\Omega})$.

In turn let $U \in L^p_{\sigma,per}(\Omega)$ which implies $\operatorname{div} u = 0$ in $\tilde{\Omega}$. Given $\varphi \in W^{1,p'}(\tilde{\Omega})$ we set

$$\Phi(x,y) := \begin{cases} \varphi(x,y), & (x,y) \in (0,\pi) \times \mathbb{R}^m, \\ \varphi(2\pi - x, y), & (x,y) \in (\pi, 2\pi) \times \mathbb{R}^m. \end{cases}$$

Due to the fact that this extension is even it holds that $\Phi \in W^{1,p'}_{per}(\Omega)$. Hence,

$$0 = \int_\Omega U(x,y)\nabla\Phi(x,y)d(x,y) = \int_{\mathbb{R}^m} \int_{(0,2\pi)} \sum_{j=1}^{m+1} U_j(x,y)\partial_j \Phi(x,y)dxdy$$

by the fact that $U \in L^p_{\sigma,per}(\Omega)$, Proposition 9.11, and the definition of γ_{per} in (9.10). The structures of U and Φ further allow for the calculation

$$\int_{(0,2\pi)} \sum_{j=1}^{m+1} U_j(x,y)\partial_j \Phi(x,y)dx = \int_{(0,\pi)} \sum_{j=1}^{m+1} u_j(x,y)\partial_j \varphi(x,y)dx$$

$$+ \int_{(\pi,2\pi)} -u_1(2\pi - x, y)\partial_x \varphi(2\pi - x, y) + \sum_{j=2}^{m+1} u_j(2\pi - x, y)\partial_j \varphi(2\pi - x, y)dx.$$

In the second integral on the right-hand side we employ relation (9.11) to arrive at

$$\int_{(0,2\pi)} \sum_{j=1}^{m+1} U_j(x,y)\partial_j\Phi(x,y)dx = \int_{(0,\pi)} \sum_{j=1}^{m+1} u_j(x,y)\partial_j\varphi(x,y)dx$$

$$+ \int_{(\pi,2\pi)} u_1(2\pi-x,y)\big(\partial_1\varphi\big)(2\pi-x,y) + \sum_{j=2}^{m+1} u_j(2\pi-x,y)\partial_j\varphi(2\pi-x,y)dx.$$

Altogether this yields

$$0 = \int_{\mathbb{R}^m}\int_{(0,\pi)} \sum_{j=1}^{m+1} u_j(x,y)\partial_j\varphi(x,y)dxdy$$

$$+ \int_{\mathbb{R}^m}\int_{(0,\pi)} u_1(x,y)\partial_1\varphi(x,y) + \sum_{j=2}^{m+1} u_j(x,y)\partial_j\varphi(x,y)dxdy$$

$$= 2\int_{\tilde{\Omega}} u(x,y)\nabla\varphi(x,y)d(x,y)$$

and the proof is complete. $\qquad\square$

Lemma 9.13. *Let* $u \in L^p(\tilde{\Omega})^{m+n}$. *Then* $u \in G^p(\tilde{\Omega})$ *if and only if* $\mathfrak{E}u \in G^p_{per}(\Omega)$.

Proof. First let $\mathfrak{E}u \in G^p_{per}(\Omega)$, that is, $\mathfrak{E}u = \nabla p$ with $p \in W^{1,p}_{per,\hat{+}}(\Omega)$. Thus, $u = \nabla p|_{\tilde{\Omega}}$ with $p|_{\tilde{\Omega}} \in L^1_{loc}(\overline{\tilde{\Omega}})$.

On the other hand, let $u \in G^p(\tilde{\Omega})$, that is, $u = \nabla p$ with $p \in L^1_{loc}(\overline{\tilde{\Omega}})$. Again it is sufficient to consider the case $n = 1$. Then $\mathfrak{E}u = \mathfrak{E}_1\nabla p$, where the first component $(\mathfrak{E}_1\nabla p)_1$ is odd and all other components $(\mathfrak{E}_1\nabla p)_\ell$ for $\ell = 2,\ldots,m+1$ are even with respect to $x_{1,0} = \pi$. Let p_e denote the even extension of p to Ω. From $\partial_1 p_e = (\mathfrak{E}_1\nabla p)_1$, Proposition 2.15, and $\mathfrak{E}_1\nabla p \in L^p(\Omega)$ we deduce $p_e = p_{(0)} + p_+$ with $p_{(0)} \in W^{1,p}(\Omega) \cap W^{1,p}_{(0),per}(\mathcal{Q}_n, L^p(\mathbb{R}^m))$ and $p_+ \in \hat{W}^{1,p}(\mathbb{R}^m)$. $\qquad\square$

The following lemma shows that $\mathbb{P}_{\Omega,per}$ preserves the structure endowed by $\mathfrak{E}$.

Lemma 9.14. *Let* $u \in L^p(\tilde{\Omega})^{m+n}$ *and* $U := \mathfrak{E}u$. *Then there exists* $v \in L^p_\sigma(\tilde{\Omega})^{m+n}$ *such that* $\mathfrak{E}v = \mathbb{P}_{\Omega,per}U$.

Proof. By definition of $\mathfrak{E}$ for each $j = 1,\ldots,n$ the j-th component U_j is odd with respect to $x_{j,0} := \pi$, whereas the components U_ℓ are even with respect to $x_{j,0}$ if $\ell \neq j$. For the sake of convenience in the sequel we regard U as a vector-valued function defined on $\mathcal{Q}_n$, i.e. $U \in L^p(\mathcal{Q}_n, L^p(\mathbb{R}^m))^{m+n}$. The calculation of Fourier coefficients then yields

$$\hat{U}_\ell(k_j,k') = \begin{cases} -\hat{U}_\ell(-k_j,k'), & \ell = j \\ \hat{U}_\ell(-k_j,k'), & \ell \neq j \end{cases} \quad (k \in \mathbb{Z}^n,\ \ell = 1,\ldots n+m,\ j = 1,\ldots,n).$$

Let $V := \mathbb{P}_{\Omega,per}U$, that is, $V = T_{I+Q_k}U$. Hence,

$$\hat{V}_\ell(k) = \begin{cases} \hat{U}_\ell(k) + ik_\ell Q_k^*\hat{U}(k), & \ell = 1,\ldots,n \\ \hat{U}_\ell(k) + \partial_\ell Q_k^*\hat{U}(k), & \ell = n+1,\ldots,n+m \end{cases} \quad (k \in \mathbb{Z}^n),$$

where

$$Q_k^*\hat{U}(k) = i\sum_{\ell=1}^{n} \left(|k|^2 - \Delta'\right)^{-1} k_\ell \hat{U}_\ell(k) + \sum_{\ell=n+1}^{n+m} \left(|k|^2 - \Delta'\right)^{-1} \partial_\ell \hat{U}_\ell(k)$$

fulfills

$$Q_{(k_j,k')}^*\hat{U}(k_j,k') = Q_{(-k_j,k')}^*\hat{U}(-k_j,k') \quad (k \in \mathbb{Z}^n, \; j = 1,\ldots,n).$$

Thus,

$$\hat{V}_\ell(k_j,k') = \begin{cases} -\hat{V}_\ell(-k_j,k'), & \ell = j \\ \hat{V}_\ell(-k_j,k'), & \ell \neq j \end{cases} \quad (k \in \mathbb{Z}^n, \; \ell = 1,\ldots n+m, \; j = 1,\ldots,n).$$

This shows existence of $v \in L^p(\tilde{\Omega})^{m+n}$ such that $\mathfrak{E}v = V = \mathbb{P}_{\Omega,per}U$. Since $V \in L_\sigma^p(\Omega)^{m+n}$ implies $v \in L_\sigma^p(\tilde{\Omega})^{m+n}$ by Lemma 9.12 the assertion is proved. $\qquad \square$

We are now in the position to prove a representation formula for the Helmholtz projection $\mathbb{P}_{\tilde{\Omega}}$ in $\mathcal{L}(L^p(\tilde{\Omega}))$. To this end, let $\mathfrak{R}$ define the restriction operator from Ω to $\tilde{\Omega}$. In particular,

$$\mathfrak{R}\mathfrak{E} = I \in \mathcal{L}(L^p(\tilde{\Omega})).$$

Theorem 9.15. *The standard Helmholtz projection $\mathbb{P}_{\tilde{\Omega}} \in \mathcal{L}(L^p(\tilde{\Omega}))$ and the periodic Helmholtz projection $\mathbb{P}_{\Omega,per} \in \mathcal{L}(L^p(\Omega))$ fulfill $\mathbb{P}_{\tilde{\Omega}} = \mathfrak{R}\mathbb{P}_{\Omega,per}\mathfrak{E}$.*

Proof. Since $\mathbb{P}_{\Omega,per}$ defines a bounded projection in $L^p(\Omega)$, it follows immediately that $\mathfrak{R}\mathbb{P}_{\Omega,per}\mathfrak{E}$ defines a bounded projection in $L^p(\tilde{\Omega})$.

We show the range of both projections to coincide. Let $u \in L_\sigma^p(\tilde{\Omega})$ be arbitrary. Then $U := \mathfrak{E}u \in L_{\sigma,per}^p(\Omega)$ by Lemma 9.12 which immediately yields $\mathbb{P}_{\Omega,per}U = U$. This shows $u = \mathfrak{R}\mathbb{P}_{\Omega,per}\mathfrak{E}u$. On the other hand let $v \in R(\mathfrak{R}\mathbb{P}_{\Omega,per}\mathfrak{E})$, that is, there exists $u \in L^p(\tilde{\Omega})$ such that $v = \mathfrak{R}\mathbb{P}_{\Omega,per}\mathfrak{E}u$. Then $V := \mathbb{P}_{\Omega,per}\mathfrak{E}u \in L_{\sigma,per}^p(\Omega)$ and $V = \mathfrak{E}v$ due to Lemma 9.14 which thanks to Lemma 9.12 yields $v \in L_\sigma^p(\tilde{\Omega})$.

Finally assume $u \in N(\mathfrak{R}\mathbb{P}_{\Omega,per}\mathfrak{E})$. Due to Lemma 9.14 there exists $v \in L^p(\tilde{\Omega})$ such that $\mathbb{P}_{\Omega,per}\mathfrak{E}u = \mathfrak{E}v$. By assumption we have $v = \mathfrak{R}\mathbb{P}_{\Omega,per}\mathfrak{E}u = 0$ which implies $\mathbb{P}_{\Omega,per}\mathfrak{E}u = 0$. Hence, $\mathfrak{E}u \in N(\mathbb{P}_{\Omega,per})$ and Lemma 9.13 yields $u \in N(\mathbb{P}_{\tilde{\Omega}})$. In turn $u \in N(\mathbb{P}_{\tilde{\Omega}})$ implies $\mathfrak{E}u \in N(\mathbb{P}_{\Omega,per})$ by Lemma 9.13, thus, $u \in N(\mathfrak{R}\mathbb{P}_{\Omega,per}\mathfrak{E})$.

Altogether we have proved

$$R(\mathbb{P}_{\tilde{\Omega}}) = R(\mathfrak{R}\mathbb{P}_{\Omega,per}\mathfrak{E}) \quad \text{and} \quad N(\mathfrak{R}\mathbb{P}_{\Omega,per}\mathfrak{E}) = N(\mathbb{P}_{\tilde{\Omega}})$$

from which we conclude $\mathbb{P}_{\tilde{\Omega}} = \mathfrak{R}\mathbb{P}_{\Omega,per}\mathfrak{E}$. $\qquad \square$

With the standard Helmholtz projection at hand, we can now easily transfer the result of Theorem 9.9 on the ν-periodic Stokes operator to the Stokes operator subject to pure-slip boundary conditions. To this end, we first establish useful results on the resolvent problem in $\tilde{\Omega}$ subject to pure-slip boundary conditions given through

(9.12)
$$\begin{aligned}
\lambda u - \Delta u + \nabla p &= f && \text{in } \tilde{\Omega}, \\
\operatorname{div} u &= 0 && \text{in } \tilde{\Omega}, \\
u_{\mathbf{n}} &= 0 && \text{on } \partial\tilde{\Omega}, \\
\partial_{\mathbf{n}} u_\tau &= 0 && \text{on } \partial\tilde{\Omega}.
\end{aligned}$$

The boundary $\partial\tilde{\Omega}$ is understood as

$$\partial\tilde{\Omega} := \bigcup_{j=1}^{n} \tilde{\mathcal{Q}}_{n,j}^{\pm}, \quad \text{where} \quad \tilde{\mathcal{Q}}_{n,j}^{\pm} := \tilde{\mathcal{Q}}_{j-1} \times \{0,\pi\} \times \tilde{\mathcal{Q}}_{n-j} \quad (j = 1,\dots,n).$$

Note that the Lebesgue null set of edges of $\tilde{\mathcal{Q}}_n \times \mathbb{R}^m$ is neglected. Here $\mathbf{n}$ denotes the outer normal, $u_{\mathbf{n}}$ the normal component of u, u_τ the tangential component of u, and $\partial_{\mathbf{n}}$ the normal derivative with respect to the boundary. Hence,

$$u_{\mathbf{n}} = u_j \quad \text{and} \quad \partial_{\mathbf{n}} u_\tau = \left(\partial_j u_1, \dots, \partial_j u_{j-1}, \partial_j u_{j+1}, \dots, \partial_j u_n\right)^T$$

on $\tilde{\mathcal{Q}}_{j-1} \times \{0,\pi\} \times \tilde{\mathcal{Q}}_{n-j}$.

Theorem 9.16. *For the Stokes resolvent problem* (9.12) *there exists a unique solution*

$$(u,p) \quad \in \quad W^{2,p}(\tilde{\Omega})^{n+m} \times \hat{W}^{1,p}(\tilde{\Omega}).$$

It admits the estimate

$$\|\nabla^2 u\|_p + \sqrt{\lambda}\|\nabla u\|_p + \lambda\|u\|_p + \|\nabla p\|_p \leq C\|f\|_p.$$

The solution operator $\tilde{R}_\lambda f := u$ *is given by* $\tilde{R}_\lambda := \mathfrak{R} R_\lambda \mathfrak{E}$ *with* R_λ *as defined in Theorem 9.5. Moreover, for each* $\phi > 0$ *it holds that*

$$\mathcal{R}\left(\left\{\lambda^{1-\frac{|\alpha|}{2}} \partial^\alpha \tilde{R}_\lambda; \ \lambda \in \Sigma_{\pi-\phi}, \ 0 \leq |\alpha| \leq 2\right\}\right) < \infty.$$

Proof. We extend the right-hand side f with the extension operator $\mathfrak{E}$ to F defined on the whole of Ω. Due to Theorem 9.5 there exists a unique solution

$$U \in W^{2,p}(\Omega)^{n+m} \cap W^{2,p}_{per}(\mathcal{Q}_n, L^p(\mathbb{R}^m))^{n+m},$$
$$P \in \left(W^{1,p}(\Omega) \cap W^{1,p}_{(0),per}(\mathcal{Q}_n, L^p(\mathbb{R}^m))\right) + \hat{W}^{1,p}(\mathbb{R}^m)$$

of the periodic Stokes resolvent problem with right-hand side F. Set

$$V^{(\ell)}(x,y) := \begin{cases} U_j(x', -x_\ell, y), & j = \ell \\ -U_j(x', -x_\ell, y), & j \neq \ell \end{cases} \quad (\ell = 1,\dots,n)$$

and

$$Q^{(\ell)}(x,y) := P(x', -x_\ell, y) \quad (j = 1, \ldots, n+m, \ \ell = 1, \ldots, n).$$

Then $(V^{(\ell)}, Q^{(\ell)})$ define solutions of the periodic Stokes resolvent problem with right-hand side F, too. By uniqueness

$$V^{(\ell)} = U \quad \text{and} \quad Q^{(\ell)} = P \qquad (\ell = 1, \ldots, n)$$

follows. Taking into account periodicity of U, these symmetry properties yield

$$\Re U_{\mathbf{n}} = 0 \text{ on } \partial\tilde{\Omega},$$
$$\partial_{\mathbf{n}} \Re U_\tau = 0 \text{ on } \partial\tilde{\Omega}.$$

Hence, $u := \Re U$ and $p := \Re P$ solve (9.12). An indirect argument along the same steps proves uniqueness. $\square$

We define the *Stokes operator* $A_{S,ps}$ *subject to pure-slip boundary conditions in* $L^p_\sigma(\tilde{\Omega})$ to be

$$D(A_{S,ps}) := \left\{ u \in W^{2,p}(\tilde{\Omega})^{m+n} \cap L^p_\sigma(\tilde{\Omega}); \ u_{\mathbf{n}} = \partial_{\mathbf{n}} u_\tau = 0 \text{ on } \partial\tilde{\Omega} \right\},$$
$$A_{S,ps} u := -\mathbb{P}_{\tilde{\Omega}} \Delta u = -\Delta \mathbb{P}_{\tilde{\Omega}} u \quad (u \in D(A_{S,ps})).$$

Note that $\Delta \mathbb{P}_{\tilde{\Omega}} u = \mathbb{P}_{\tilde{\Omega}} \Delta u$ holds true for all $u \in W^{2,p}(\tilde{\Omega})^{m+n}$ which fulfill the boundary condition $u_{\mathbf{n}} = \partial_{\mathbf{n}} u_\tau = 0$ on $\partial\tilde{\Omega}$. This follows immediately from the representation $\mathbb{P}_{\tilde{\Omega}} = \Re \mathbb{P}_{\Omega,per} \mathfrak{E}$ and Lemma 9.8. Consequently, we obtain the following result on the $\mathcal{RH}^\infty$-calculus of $A_{S,ps}$ by Theorem 9.9.

Theorem 9.17. *We have* $A_{S,ps} \in \mathcal{RH}^\infty(L^p_\sigma(\tilde{\Omega}))$ *and* $\phi^{\mathcal{R}\infty}_{A_{S,ps}} = 0$.

10 The functional calculus approach

This chapter extends the results from Chapter 8 in two directions. Firstly, domains given as the Cartesian product of finitely many standard domains are considered. Secondly, with a Banach space F, the cylindrical boundary value problems are F-valued and contain $\mathcal{L}(F)$-valued coefficients. Here we employ the operator-valued Dunford calculus and the Kalton-Weis-Theorem. Again we first consider rather arbitrary cylindrical boundary value problems and focus on the Laplacian at the end. The results of this chapter also appear in [NS11a].

10.1 Operator-valued Dunford calculus

In this section the Dunford calculus from Chapter 4 is extended to the operator-valued context. Let $\mathcal{A} \subset \mathcal{L}(X)$ denote the subalgebra of bounded operators on X which commute with the resolvent $(\mu - A)^{-1}$. For $\sigma \in (0, \pi]$ we denote by $\mathcal{H}^\infty(\Sigma_\sigma, \mathcal{A})$ the commutative algebra of bounded, $\mathcal{A}$-valued, holomorphic functions on Σ_σ, that is,

$$\mathcal{H}^\infty(\Sigma_\sigma, \mathcal{A}) := \{f : \Sigma_\sigma \to \mathcal{A}; \ f \text{ is holomorphic, } |f|_\infty^\sigma < \infty\},$$

where

$$|f|_\infty^\sigma := \sup\{\|f(z)\|_{\mathcal{L}(X)}; \ z \in \Sigma_\sigma\}.$$

Using $\rho(z) := \frac{z}{(1+z)^2}$ we define the subalgebra

$$\mathcal{H}_0^\infty(\Sigma_\sigma, \mathcal{A}) := \{f \in \mathcal{H}^\infty(\Sigma_\sigma, \mathcal{A}); \ \text{there are } C, \varepsilon > 0 \text{ such that}$$
$$\|f(z)\|_{\mathcal{L}(X)} \leq C|\rho(z)|^\varepsilon \text{ for all } z \in \Sigma_\sigma\}.$$

Consider a sectorial operator A in X with spectral angle $\phi_A \in [0, \pi)$. We pick $\sigma \in (\phi_A, \pi]$ and $\psi \in (\phi_A, \sigma)$ and set $\Gamma := (\infty, 0]e^{i\psi} \cup [0, \infty)e^{-i\psi}$. Similar to the scalar valued case, by Cauchy's integral formula and the sectoriality of A, the Bochner integral

$$f(A) := \frac{1}{2\pi i} \int_\Gamma f(\mu)(\mu - A)^{-1} d\mu$$

represents a well-defined element in $\mathcal{L}(X)$ for every $f \in \mathcal{H}_0^\infty(\Sigma_\sigma, \mathcal{A})$. As f is supposed to take values in $\mathcal{A}$, the above formula defines an algebra homomorphism

(10.1) $$\Phi_A : \mathcal{H}_0^\infty(\Sigma_\sigma, \mathcal{A}) \to \mathcal{L}(X); \quad f \mapsto f(A),$$

known as operator-valued Dunford calculus. For arbitrary $f \in \mathcal{H}^\infty(\Sigma_\sigma, \mathcal{A})$ we set

$$f(A) := \rho(A)^{-1}(\rho f)(A).$$

As in the scalar-valued setting, this definition gives rise to a closed, densely defined operator in X. Moreover, by Cauchy's theorem it is consistent with the former one for $f \in \mathcal{H}_0^\infty(\Sigma_\sigma, \mathcal{A})$.

Again convergence properties are important in order to make the functional calculus a useful tool. The following most important result in this direction is a special operator-valued version of the convergence lemma for sectorial operators (see e.g. [DV05, Theorem 4.7] or [HD03]). Recall the approximation sequence $\rho_n \in \mathcal{H}_0^\infty(\Sigma_\sigma)$ from Lemma 4.7 defined by

$$\rho_n(z) := \frac{1}{1 + z/n} - \frac{1}{1 + zn} = \frac{n^2 z - z}{(1 + nz)(n + z)}.$$

Lemma 10.1. *Let $f \in \mathcal{H}^\infty(\Sigma_\sigma, \mathcal{A})$ and $A \in S(X)$. Then $f(A) \in \mathcal{L}(X)$ if and only if $\sup_{n \in \mathbb{N}} \|(\rho_n f)(A)\| < \infty$. In this case, $f(A)x = \lim_{n \to \infty} (\rho_n f)(A)x$ for all $x \in X$.*

Given an operator admitting an $\mathcal{R}$-bounded $\mathcal{H}^\infty$-calculus, the question arises, whether $\mathcal{R}$-boundedness can be extended to some class of $\mathcal{A}$-valued functions. An affirmative answer to this question is given in [KW01, Corollary 5.4]. This result nowadays is known as the Kalton-Weis-Theorem.

Theorem 10.2. *Let X be a Banach space that has property (α) and let $A \in S(X)$. Given an $\mathcal{R}$-bounded subset $\mathcal{T} \subset \mathcal{L}(X)$ set*

$$\mathcal{H}^\infty(\Sigma_\sigma, \mathcal{T}) := \{f \in \mathcal{H}^\infty(\Sigma_\sigma, \mathcal{A}); \ f(z) \in \mathcal{T} \ (z \in \Sigma_\sigma)\}.$$

If A admits a bounded $\mathcal{H}^\infty$-calculus, then for $\sigma > \phi_A^\infty$ we have

$$\mathcal{R}(\{f(A); \ f \in \mathcal{H}^\infty(\Sigma_\sigma, \mathcal{T})\}) < \infty.$$

10.2 Applications to cylindrical boundary value problems

In what follows we consider a domain $\Omega \subset \mathbb{R}^n$ given as product of finitely many domains $V_i \subset \mathbb{R}^{n_i}$ with $n_i \in \mathbb{N}$ and $\sum_{i=1}^N n_i = n$, that is $\Omega = \prod_{i=1}^N V_i$. For $\boldsymbol{x} \in \Omega$ we write $\boldsymbol{x} = (x^1, \ldots, x^N)$ with $x^i \in V_i$ and $i = 1, \ldots, N$, whenever we want to refer to the cylindrical geometry of Ω. Accordingly, for a multi-index $\alpha \in \mathbb{N}^n$ we write $\alpha = (\alpha^1, \ldots, \alpha^N) \in \prod_{i=1}^N \mathbb{N}_0^{n_i}$. Finally we set

$$(10.2) \qquad \partial \mathcal{V}_i := V_1 \times \ldots \times V_{i-1} \times \partial V_i \times V_{i+1} \times \ldots \times V_N$$

and $\partial \Omega := \bigcup_{i=1}^N \partial \mathcal{V}_i$. As $\partial \mathcal{V}_i \cap \partial \mathcal{V}_j = \emptyset$ for $i \neq j$, all points belonging to the Lebesgue null set of edges of Ω are neglected in this definition of $\partial \Omega$.

Let F be a Banach space. In this section we particularly deal with F-valued boundary value problems

(10.3)
$$\begin{aligned} \lambda u + \mathcal{A}(\boldsymbol{x}, D)u &= f \text{ in } \Omega, \\ \mathcal{B}(\boldsymbol{x}, D)u &= 0 \text{ on } \partial\Omega \end{aligned}$$

with operators $\mathcal{A}(\boldsymbol{x}, D)$ and $\mathcal{B}(\boldsymbol{x}, D)$ of cylindrical form. Hence, we first extend our previous definition of cylindrical boundary value problems defined for the Cartesian product of three sets to the present situation of finitely many sets.

Definition 10.3. Let $m_i \in \mathbb{N}$ for $i = 1, \ldots, N$. The boundary value problem (10.3) is called *cylindrical* if Ω is a cylindrical domain as introduced above and if the operator $\mathcal{A}(\cdot, D)$ is represented as

$$\mathcal{A}(\boldsymbol{x}, D) = \sum_{i=1}^{N} A_i(x^i, D) = \sum_{i=1}^{N} \sum_{|\alpha^i| \le 2m_i} a_{\alpha^i}^i(x^i) D_{x^i}^{\alpha^i}$$

and if the boundary operator on $\partial\Omega$ is given as

$$\mathcal{B} := (\mathcal{B}_i; \ i = 1, \ldots, N),$$

where $\mathcal{B}_i$, acting on ∂V_i only, is defined as

$$\mathcal{B}_i(\boldsymbol{x}, D) := \left\{ B_{i,j}(x^i, D); \ j = 1, \ldots, m_i \right\} \quad (i = 1, \ldots, N)$$

with

$$B_{i,j}(x^i, D)u = \sum_{|\beta^i| \le m_{i,j}} b_{j,\beta^i}^i(x^i)(D_{x^i}^{\beta^i}u)|_{\partial V_i} \quad (m_{i,j} < m_i, \ j = 1, \ldots, m_i).$$

In other words, the differential operators $\mathcal{A}(\boldsymbol{x}, D)$ and $\mathcal{B}(\boldsymbol{x}, D)$ resolve completely into parts $A_i(x^i, D)$ and $B_{i,j}(x^i, D)$ of which each one acts merely on V_i. The boundary operator $\mathcal{B}$ acting on the whole of $\partial\Omega$ can as well be represented by means of

$$\mathcal{B}(\boldsymbol{x}, D)u = \sum_{i=1}^{N} \chi_{\partial V_i}(\boldsymbol{x})\mathcal{B}_i(\boldsymbol{x}, D)u,$$

where $\chi_{\partial V_i}$ denotes the characteristic function of the set ∂V_i. With the aid of the definition $m := \max\{m_i; \ i = 1, \ldots, N\}$ and $B_{i,j}(\cdot, D) := 0$ for $j > m_i$ we can further write

$$\mathcal{B}(\boldsymbol{x}, D) := \{B_j(\boldsymbol{x}, D); \ j = 1, \ldots, m\},$$

where

$$B_j(\boldsymbol{x}, D)u = \sum_{i=1}^{N} \chi_{\partial V_i}(\boldsymbol{x})B_{i,j}(x^i, D)u.$$

For $1 < p < \infty$ the $L^p(\Omega, F)$-realization of the boundary value problem $(\mathcal{A}, \mathcal{B})$ given through (10.3) is defined by

$$D(\mathbb{A}) := \left\{ u \in L^p(\Omega); \ D^\alpha u \in L^p(\Omega) \text{ for } \sum_{i=1}^{N} \frac{|\alpha^i|}{2m_i} \leq 1, \right.$$

$$\left. B_j(\cdot, D)u = 0 \quad (j = 1, \ldots, m) \right\},$$

$$\mathbb{A}u := \mathcal{A}(\cdot, D)u \quad (u \in D(\mathbb{A})).$$

In case that $m_i = m$ for all $i = 1, \ldots, N$ we obviously have

$$D(\mathbb{A}) := \left\{ u \in W^{2m,p}(\Omega, F); \ B_j(\cdot, D)u = 0 \quad (j = 1, \ldots, m) \right\}$$

which coincides with the domains of definition of the operators considered in Proposition 8.3 and Proposition 8.4. For $i = 1, \ldots, N$ we consider the induced boundary value problems

$$(A_i, B_i) := (A_i(\cdot, D), B_{i,1}(\cdot, D), \ldots, B_{i,m_i}(\cdot, D))$$

given through

$$(10.4) \qquad \begin{aligned} \lambda u + A_i(x, D)u &= f \text{ in } V_i, \\ B_{i,j}(x, D)u &= 0 \text{ on } \partial V_i \quad (j = 1, \ldots, m_i) \end{aligned}$$

which arise by *cylindrical decomposition* of (A, B).

From now on we assume every $V_i \subset \mathbb{R}^{n_i}$ to be given as a C^{2m_i} standard domain (cf. Definition 8.1). Furthermore, we consider the following *cylindrical parameter-ellipticity*.

Definition 10.4. A cylindrical boundary value problem in the sense of Definition 10.3 is called *parameter-elliptic* in Ω if

(i) each induced boundary value problem (A_i, B_i) is parameter-elliptic in V_i with angle $\varphi_i := \varphi_{(A_i, B_i)} \in [0, \pi)$ in the sense of Definition 8.2, and

(ii) it holds that $\varphi_i + \varphi_j < \pi$ for $i, j = 1, \ldots, N$, $i \neq j$.

We call $\varphi_{(\mathcal{A}, \mathcal{B})} := \max\{\varphi_i; \ i = 1, \ldots, N\}$ the angle of parameter-ellipticity of the cylindrical boundary value problem.

Finally, with arbitrary $\gamma_i \in (0,1)$, $i = 1, \ldots, N$, we impose the following smoothness assumptions on the coefficients of (A_i, B_i):

$$(10.5) \quad \begin{cases} a^i_{\alpha^i} \in BUC^{\gamma_i}(\overline{V_i}, \mathcal{L}(F)) \text{ for all } |\alpha^i| = 2m_i, \\ a^i_{\alpha^i}(\infty) := \lim_{|x^i| \to \infty} a^i_{\alpha^i}(x^i) \text{ exists if } V_i \text{ is unbounded, and} \\ \|a^i_{\alpha^i}(x^i) - a^i_{\alpha^i}(\infty)\| \leq C|x^i|^{-\gamma_i} \quad (x^i \in V_i, |x^i| \geq 1)), \\ a^i_{\alpha^i} \in [L^\infty + L^{r_\nu}](V_i, \mathcal{L}(F)) \text{ for all } |\alpha^i| = \nu < 2m_i, \\ \quad \text{where } r_\nu \geq p, \ \dfrac{2m_i - \nu}{n_i} > \dfrac{1}{r_\nu}, \\ b^i_{j,\beta^i} \in C^{2m_i - m_{i,j}}(\partial V_i, \mathcal{L}(F)) \quad (j = 1, \ldots, m_i; \ |\beta^i| \leq m_{i,j}). \end{cases}$$

Note that the limit behavior of the top order coefficients ensures parameter-ellipticity of the limit operators $A_i(\infty, D)$ in case V_i is an unbounded domain to be well-defined. Thus, we can extend the notion of parameter-ellipticity of a cylindrical boundary value problem to

$$\overline{\Omega} := \prod_{i=1}^{N} \overline{V}_i,$$

where we again agree on $\{\infty\} \subset \overline{V}_i$ in case V_i is unbounded. We are now in the position to present the main theorem of this section.

Theorem 10.5. *Let $1 < p < \infty$ and let F be a Banach space of class $\mathcal{HT}$ enjoying property (α). Let $\Omega := \prod_{i=1}^{N} V_i$, where every V_i is a C^{2m_i} standard domain in $\mathbb{R}^{n_i}$, $n_i \in \mathbb{N}$, $i = 1, \ldots, N$, and let $\sum_{i=1}^{N} n_i = n$. Furthermore, we assume that*

(i) the boundary value problem $(\mathcal{A}, \mathcal{B})$ is cylindrical,

(ii) the coefficients of (A_i, B_i) satisfy (10.5),

(iii) $(\mathcal{A}, \mathcal{B})$ is parameter-elliptic with angle $\varphi_{(\mathcal{A},\mathcal{B})} \in [0, \pi)$ in $\overline{\Omega}$,

(iv) $a^i_{\alpha^i}(x^i) a^j_{\alpha^j}(x^j) = a^j_{\alpha^j}(x^j) a^i_{\alpha^i}(x^i)$ in $\mathcal{L}(F)$ for $i, j = 1, \ldots, N$, $i \neq j$ and a.e. $x \in \Omega$.

Then for every $\phi > \varphi_{(\mathcal{A},\mathcal{B})}$, there exists $\delta = \delta(\phi) > 0$ such that the realization $\mathbb{A}$ of $(\mathcal{A}, \mathcal{B})$ fulfills $\mathbb{A} + \delta \in \mathcal{RH}^\infty(L^p(\Omega, F))$ and $\phi^{\mathcal{R},\infty}_{\mathbb{A}+\delta} \leq \phi$. Moreover, we have

$$(10.6) \quad \mathcal{R}\left(\left\{\lambda^{1 - \sum_{i=1}^{N} \frac{|\alpha^i|}{2m_i}} D^\alpha (\lambda + \mathbb{A} + \delta)^{-1}; \ \lambda \in \Sigma_{\pi - \phi}, \ 0 \leq \sum_{i=1}^{N} \frac{|\alpha^i|}{2m_i} \leq 1\right\}\right) < \infty.$$

Proof. Step 1: First we perform the cylindrical decomposition.

We define $L^p(V_i, F)$-realizations of the boundary value problems (A_i, B_i) by

$$D(A_i) := \{u \in W^{2m_i,p}(V_i, F); \ B_{i,j}(\cdot, D)u = 0 \quad (j = 1, \ldots, m_i)\},$$
$$A_i u := A_i(\cdot, D)u \quad (u \in D(A_i)).$$

Assumptions (ii) and (iii) allow for an application of Propositions 8.3 and 8.4 to each boundary value problem (A_i, B_i). Thus, for each $i = 1, \ldots, N$ and for every $\phi > \varphi_i$ there exists $\delta_i = \delta_i(\phi) \geq 0$ such that $A_i + \delta_i \in \mathcal{RH}^\infty(L^p(V_i, F))$ and $\phi_{A_i+\delta_i}^{\mathcal{R}\infty} \leq \phi$. Moreover,

$$(10.7) \qquad \mathcal{R}\left(\left\{\lambda^{1-\frac{|\alpha|}{2m_i}} D^\alpha (\lambda + A_i + \delta_i)^{-1}; \ \lambda \in \Sigma_{\pi-\phi}, \ 0 \leq |\alpha| \leq 2m_i\right\}\right) < \infty.$$

These statements remain true for the canonical extension of A_i to $L^p(\Omega, F)$ which for simplicity will be denoted by A_i again. Note that the domain of A_i in $L^p(\Omega, F)$ reads as

$$(10.8) \qquad \begin{aligned} D(A_i) := &\Big\{u \in L^p\big(V_1 \times \cdots \times V_{i-1}, W^{2m_i,p}(V_i, L^p(V_{i+1} \times \cdots \times V_N, F)))\big); \\ &B_{i,j}(\cdot, D)u = 0 \quad (j = 1, \ldots, m_i)\Big\}. \end{aligned}$$

Step 2: Within this step, we restrict ourselves to the case $N = 2$.

a) We first show that resolvents of the extensions A_1 and A_2 commute. To this end, we will frequently make use of the following observation. If $T \in \mathcal{L}(E_1, E_2)$ and $u \in W^{k,p}(G, E_1)$ for Banach spaces E_1 and E_2 and an open set $G \subset \mathbb{R}$, then

$$(10.9) \qquad\qquad D^\alpha T u = T D^\alpha u \quad (|\alpha| \leq k)$$

in $L^p(G, E_2)$. This follows easily by the fact that a derivative $\partial_{x_j} u$ represents the limit of a convergent sequence in $L^p(G, E_1)$ and by the continuity of T. For the following argumentation it will be convenient to introduce the notation

$$D(A_1, X) := \{u \in W^{2m_1,p}(V_1, X); \ B_{1,j}(\cdot, D)u = 0, \ j = 1, \ldots, m_1\}$$

for the domain of A_1 in the X-valued space $L^p(V_1, X)$. Here we are particularly interested in the case $X = D(A_2, F)$. According to step 1,

$$\lambda + A_1 : D(A_1, D(A_2, F)) \to L^p(V_1, D(A_2, F))$$

is an isomorphism for $\lambda \in \rho(-A_1)$. Fubini's theorem yields

$$D(A_1, D(A_2, F)) \hookrightarrow W^{2m_1,p}(V_1, W^{2m_2,p}(V_2, F)) \cong W^{2m_2,p}(V_2, W^{2m_1,p}(V_1, F)).$$

First we set $E_1 := W^{2m_1,p}(V_1, F)$ and $E_2 := W_p^{1-1/p}(\partial V_1, F)$ and consider $T = B_{1,j}$, where $j \in \{1, \ldots, m_1\}$ is arbitrary. Here $W_p^{1-1/p}(\partial V_1, F)$ defines the Slobodeckij space, see [Tri78, Section 4.2.1] for the precise definition. Then boundedness of $T : E_1 \to E_2$ (cf. [ADF97]) and relation (10.9) implies

$$D^{\alpha_2} B_{1,j} u = B_{1,j} D^{\alpha_2} u \quad (u \in W^{2m_2,p}(V_2, E_1), \ |\alpha_2| \leq 2m_2).$$

This shows that

$$D(A_1, D(A_2, F)) \hookrightarrow W^{2m_2, p}(V_2, D(A_1, F)).$$

Since $B_{2,j}u = 0$ for $u \in D(A_1, D(A_2, F))$ and $j \in \{1, \ldots, m_2\}$, we even have

$$D(A_1, D(A_2, F)) \hookrightarrow D(A_2, D(A_1, F)).$$

Interchanging the roles of A_1 and A_2, we obtain the converse embedding. Hence, we have

$$D(A_1, D(A_2, F)) \cong D(A_2, D(A_1, F))$$

with equivalent norms. The above arguments also include that

$$L^p(V_1, D(A_2, F)) \cong D(A_2, L^p(V_1, F)).$$

From this we conclude that

(10.10) $$\lambda + A_1 \colon D(A_2, D(A_1, F)) \to D(A_2, L^p(V_1, F))$$

defines an isomorphism. Setting $E_1 = D(A_1, F)$, $E_2 = L^p(V_1, F)$, and $T = \lambda + A_1$, relation (10.9) yields

$$D^{\alpha_2}(\lambda + A_1)u = (\lambda + A_1)D^{\alpha_2}u \quad (u \in D(A_2, D(A_1, F))).$$

Setting $E_1 = E_2 = F$ and $T = a_{\alpha 1}^1$, in view of (10.9) we also see that D^{α_2} and the coefficients $a_{\alpha 1}^1$ commute. By our assumption (iv) on the coefficients this shows

(10.11) $$(\mu + A_2)(\lambda + A_1)u = (\mu + A_1)(\lambda + A_2)u \quad (u \in D(A_2, D(A_1, F))).$$

For $f \in L^p(V_2, L^p(V_1, F))$ we have $(\mu + A_2)^{-1}f \in D(A_2, L^p(V_1, F))$, provided $\mu \in \rho(-A_2)$. Since (10.10) is an isomorphism, we obtain

$$(\lambda + A_1)^{-1}(\mu + A_2)^{-1}f \in D(A_2, D(A_1, F)).$$

Hence, the application of $(\lambda + A_1)(\mu + A_2)$ to this expression makes sense and we obtain by virtue of (10.11) that

$$(\lambda + A_1)^{-1}(\mu + A_2)^{-1}f = (\lambda + A_2)^{-1}(\mu + A_1)^{-1}f.$$

b) Let $\phi > \max\{\varphi_1, \varphi_2\}$. Due to assumption (iii) and step 1 of the proof for $i = 1, 2$ there exist $\phi > \phi_i > \varphi_i$ with $\phi_1 + \phi_2 < \pi$ and $\delta_i = \delta_i(\phi_i) \geq 0$ such that $A_i + \delta_i \in \mathcal{RH}^\infty(L^p(\Omega, F))$ and $\phi_{A_i + \delta_i}^{\mathcal{R}\infty} < \phi_i$. Since $\phi_{A_1 + \delta_1}^{\mathcal{R}\infty} + \phi_{A_2 + \delta_2}^{\mathcal{R}\infty} < \pi$ we can employ Proposition 4.19a). Setting $D(\tilde{\mathbb{A}}) := D(A_1) \cap D(A_2)$ and $\tilde{\mathbb{A}} := A_1 + A_2$, this yields $\tilde{\mathbb{A}} + \delta \in \mathcal{RH}^\infty(L^p(\Omega, F))$ and $\phi_{\tilde{\mathbb{A}} + \delta}^{\mathcal{R}, \infty} < \max\{\phi_1, \phi_2\} < \phi$, where $\delta := \delta_1 + \delta_2$.

c) It remains to show the $\mathcal{R}$-boundedness statement (10.6) which implies that $D(\tilde{\mathbb{A}}) \subset D(\mathbb{A})$. For $\frac{|\alpha^1|}{2m_1} + \frac{|\alpha^2|}{2m_2} \leq 1$ we consider the family of operators

$$\left\{ \lambda^{1 - \left(\frac{|\alpha^1|}{2m_1} + \frac{|\alpha^2|}{2m_2}\right)} D^{(\alpha^1, \alpha^2)}(\lambda + \tilde{\mathbb{A}})^{-1}; \; \lambda \in \Sigma_{\pi - \phi} \right\}.$$

By the fact that $A_1 + \delta_1 \in \mathcal{R}\mathcal{H}^\infty(L^p(\Omega, F))$ has bounded imaginary powers we obtain $D(A_1^\nu) = [L^p(\Omega, F), D(A_1)]_\nu$ for $\nu \in (0, 1)$ (see Proposition 4.10), where $[L^p(\Omega, F), D(A_1)]_\nu$ denotes the complex interpolation space between $L^p(\Omega, F)$ and $D(A_1)$ of order ν. From this we deduce

$$D(A_1^\nu) = [L^p(\Omega, F), D(A_1)]_\nu \hookrightarrow H_p^{2m_1\nu}(V_1, L^p(V_2, F)),$$

where $H_p^{2m_1\nu}(V_1, L^p(V_2, F))$ denotes the Bessel-potential space of order $2m_1\nu$ (cf. [Tri78, Section 4.3.1]). This shows that $D^{(\alpha^1, 0)}(A_1 + \delta_1)^{-|\alpha^1|/2m_1}$ is bounded in $L^p(V_1, L^p(V_2, F))$ for $|\alpha^1| \leq 2m_1$, provided δ_1 is suitably large. Thus, thanks to Lemma 3.2a) it suffices to show that the family

$$\left\{\lambda^{1 - \left(\frac{|\alpha^1|}{2m_1} + \frac{|\alpha^2|}{2m_2}\right)}(A_1 + \delta_1)^{\frac{|\alpha^1|}{2m_1}} D^{(0, \alpha^2)}(\lambda + \tilde{\mathbb{A}})^{-1}; \ \lambda \in \Sigma_{\pi - \phi}, \ \frac{|\alpha^1|}{2m_1} + \frac{|\alpha^2|}{2m_2} \leq 1\right\}$$

is $\mathcal{R}$-bounded. To this end, pick $\sigma \in (\phi_1, \min\{\phi, \pi - \phi_2\})$. For $\lambda \in \Sigma_{\pi - \phi}$ we define the holomorphic functions

$$G_\lambda(z) := \lambda^{1 - \left(\frac{|\alpha^1|}{2m_1} + \frac{|\alpha^2|}{2m_2}\right)} z^{\frac{|\alpha^1|}{2m_1}} D^{(0, \alpha^2)}(\lambda + z + A_2 + \delta_2)^{-1} \quad (z \in \Sigma_\sigma).$$

A homogeneity argument yields the existence of $C = C(\phi, \sigma) > 0$ such that

$$\left| \lambda^{1 - \left(\frac{|\alpha^1|}{2m_1} + \frac{|\alpha^2|}{2m_2}\right)} z^{\frac{|\alpha^1|}{2m_1}} \right| \leq C |\lambda + z|^{1 - \frac{|\alpha^2|}{2m_2}}.$$

By virtue of Lemma 3.2b) and relation (10.7) we conclude

$$\mathcal{R}(\{G_\lambda(z); \ z \in \Sigma_\sigma, \ \lambda \in \Sigma_{\pi - \phi}\}) < \infty.$$

From step 2b) we also know that

$$D^{(0, \alpha^2)}(\lambda + z + A_2 + \delta_2)^{-1}(\mu - A_1)^{-1}$$
$$= (\mu - A_1)^{-1} D^{(0, \alpha^2)}(\lambda + z + A_2 + \delta_2)^{-1}.$$

Hence, we may apply Theorem 10.2 to the result that

$$\mathcal{R}(\{G_\lambda(A_1); \ \lambda \in \Sigma_{\pi - \phi}\}) < \infty.$$

By an approximation argument and Lemma 10.1 we therefore see that

$$G_\lambda(A_1) = \lambda^{1 - \left(\frac{|\alpha^1|}{2m_1} + \frac{|\alpha^2|}{2m_2}\right)}(A_1 + \delta_1)^{\frac{|\alpha^1|}{2m_1}} D^{(0, \alpha^2)}(\lambda + \tilde{\mathbb{A}})^{-1}.$$

Consequently, relation (10.6) follows. This yields $D(\tilde{\mathbb{A}}) \subset D(\mathbb{A})$, hence $\mathbb{A} = \tilde{\mathbb{A}}$.

Step 3: In this step, we prove the assertion for $N > 2$.

Given $f \in L^p(\Omega, F)$, Lemma 10.1 and step 2a) of the proof imply

$$(\zeta - A_l)^{-1}(\lambda - (A_i + A_j))^{-1}f$$
$$= \lim_{n \to \infty} \frac{1}{2\pi i} \int_\Gamma \rho_n(\mu)(\zeta - A_l)^{-1}(\lambda - \mu - A_i)^{-1}(\mu - A_j)^{-1}d\mu \ f$$
$$= \lim_{n \to \infty} \frac{1}{2\pi i} \int_\Gamma \rho_n(\mu)(\lambda - \mu - A_i)^{-1}(\mu - A_j)^{-1}(\zeta - A_l)^{-1}d\mu \ f$$
$$= (\lambda - (A_i + A_j))^{-1}(\zeta - A_l)^{-1}f.$$

Hence, the resolvent of an extension commutes with the resolvent of finite sums of extensions. As the bounded $\mathcal{H}^\infty$-calculus as well as the estimate (10.6) are preserved in each iteration step, the claim for arbitrary N follows by induction. $\qquad\square$

Remark 10.6. Note that no continuity of the boundary conditions at the edges of Ω has to be assumed.

Remark 10.7. It is worthwhile to mention that another advantage of this approach lies in the fact that it easily generalizes to the case of different p-integrability in the single cross-sections V_i. In fact, if $p = (p_1, \ldots, p_k) \in (1, \infty)^k$ we set

$$L^p(\Omega, F) := L^{p_1}(V_1, L^{p_2}(V_2, \ldots L^{p_k}(V_k, F) \ldots)).$$

In the smoothness assumptions (10.5) then we have to replace p by p_i. The remaining definitions, such as the domain of A, remain exactly the same. Also the statement of Theorem 10.5 holds without any change. Observe that the operators $(\lambda + A_1)^{-1}(\mu + A_2)^{-1}$ and $(\mu + A_2)^{-1}(\lambda + A_1)^{-1}$ are well-defined elements of $\mathcal{L}(L^{p_1}(V_1, L^{p_2}(V_2)))$ for $\lambda \in \rho(-A_1)$ and $\mu \in \rho(-A_2)$ due to step 1 of the proof. In step 2a) it is further shown that both operators coincide in $\mathcal{L}(L^p(V_1, L^p(V_2)))$, i.e., for $p_1 = p_2$. This implies equality of these operators being restricted to $C_0^\infty(\Omega)$. Since this space is dense in $L^p(\Omega)$ for $p = (p_1, \ldots, p_k)$, A_1 and A_2 are resolvent commuting in the case of different p_i as well. The remaining parts of the proof then copy verbatim.

We close this section with the treatment of a situation where operators on cross-sections do not necessarily commute. We emphasize that we do not aim for the greatest generality. The purpose is just to demonstrate that the approach is not restricted to the commuting situation. Improvements and generalizations in one or the other direction are certainly possible. In particular, we restrict ourselves to the case of two domains and of two operators. Hence, let $\Omega = V_1 \times V_2$ and differential operators $A_1(x^1, D)$ and $A_2(x^2, D)$ such as in Theorem 10.5 be given. Then, for $\phi_i > \varphi_i$ there exist $\delta_i \geq 0$ such that the canonical extensions of A_i fulfill $A_i + \delta_i \in \mathcal{RH}^\infty(L^p(\Omega, F))$ and $\phi_{A_i + \delta_i}^{\mathcal{R}, \infty} \leq \phi_i$. For the sake of simplicity we will assume $\delta_i = 0$ and the operators A_i to be subject to generalized Dirichlet boundary conditions for $i = 1, 2$.

We assume the cylindrical structure to be disturbed in the following way: given a function r on V_1, we consider the differential operator

$$A_1(x^1, D) + r(x^1)A_2(x^2, D)$$

in Ω subject to Dirichlet boundary conditions. Associated to r we define an operator of pointwise multiplication in $L^p(\Omega, F)$ by

$$D(M_r) := \{u \in L^p(\Omega, F); \ ru \in L^p(\Omega, F)\},$$
$$M_r u := ru \quad (u \in D(M_r)).$$

In the sequel we will investigate the operator

$$D(\mathbb{A}_r) := D(A_1) \cap D(M_r A_2),$$
$$\mathbb{A}_r := A_1 + M_r A_2 \quad (u \in D(\mathbb{A}_r)).$$

The main difference to previous boundary value problems is that the operators A_1 and $M_r A_2$ are no longer resolvent commuting on $L^p(\Omega, F)$. Therefore, we have to impose conditions on r that allow for an application of Proposition 4.21.

Theorem 10.8. *Let $1 < p < \infty$ and let F, V_1, V_2, as well as A_1 and A_2 fulfill the assumptions of Theorem 10.5 subject to Dirichlet boundary conditions. Let $\vartheta > 0$ with $\varphi_1 + \varphi_2 + \vartheta < \pi$ and assume that*

(i) $r \in [W^{2m_1, p} + W^{2m_1, \infty}](V_1)$ if $2m_1 p > n_1$ and $r \in W^{2m_1, \infty}(V_1)$ else,

(ii) $r(x^1) \in \Sigma_\vartheta$ for all $x^1 \in V_1$, and

(iii) $r^{-1} D^\eta r \in L^\infty(V_1)$ for all $|\eta| \le 2m_1$,

(iv) $r^{-1} \in L^\infty(V_1)$ or $0 \in \rho(A_2)$.

Then for every $\phi > \max\{\varphi_1, \varphi_2 + \vartheta\}$ there exists $\delta = \delta(\phi) \ge 0$ such that $\mathbb{A}_r$ fulfills $\mathbb{A}_r + \delta \in \mathcal{R}\mathcal{H}^\infty(L^p(\Omega, F))$ and $\phi_{\mathbb{A}_r + \delta}^{\mathcal{R}, \infty} \le \phi$.

Proof. For r subject to assumption (i) we have $M_r \in \mathcal{L}(L^p(\Omega, F))$, in particular $D(M_r) = L^p(\Omega, F)$. This implies

$$D(M_r A_2) := \{u \in D(A_2) : \ A_2 u \in D(M_r)\} = D(A_2),$$

hence $D(\mathbb{A}_r) = D(A_1) \cap D(A_2)$. In addition, $M_r \in \mathcal{H}^\infty(L^p(\Omega, F))$ with $\phi_{M_r}^\infty \le \vartheta$ by assumption (ii) and from assumption (iv) we deduce $0 \in \rho(M_r)$ or $0 \in \rho(A_2)$. Since $M_r A_2 = A_2 M_r$ due to the special form of the two operators in each case we can apply Proposition 4.19(b) to the result that $M_r A_2 \in \mathcal{H}^\infty(L^p(\Omega, F))$ and $\phi_{M_r A_2}^\infty \le \varphi_2 + \phi_{M_r}^\infty$.

We show that A_1 and $M_r A_2$ satisfy the Labbas-Terreni condition (4.8). To this end, we may assume $0 \in \rho(M_r A_2)$ since this can always be derived by a shift which we can compensate at the end by choosing $\delta \ge 0$ a bit larger if necessary. By the fact that we assume Dirichlet boundary conditions and in view of assumption (i) we obtain $M_r(D(A_1)) \subset D(A_1)$. This implies

$$(10.12) \qquad D(M_r A_2 A_1) = D(A_2 A_1) = D(A_1 A_2) \subset D(A_1 M_r A_2).$$

For $u \in D(A_1 A_2)$ therefore the equality

$$(10.13) \qquad M_r A_2(\mu + A_1)u = (\mu + A_1 - R)M_r A_2 u$$

with $R := [A_1, M_r] M_{r^{-1}}$ makes sense in $L^p(\Omega, F)$. Thanks to (10.12) we may also identify R as

$$Ru = [A_1(x^1, D), r(x^1)] r(x^1)^{-1} u =: R(x^1, D)u$$

for all $u \in M_r A_2 D(A_1 A_2) \subset D(A_1)$. Due to assumptions on r the differential operator $R(x^1, D)$, and hence also R, is well-defined on all of $D(A_1)$ and represented as a linear combination of differential operators of the form

$$R_\gamma(x^1, D) = a_{\alpha^1}(x^1) \prod_{\eta \in \mathcal{M}_\gamma} \left(r^{-1} D^\eta r\right)^{l_\eta}(x^1) D^\gamma,$$

with $\gamma < \alpha^1$, some $\mathcal{M}_\gamma \subset \{\eta \in \mathbb{N}_0^{n_1}; \ \eta \neq 0, \ 0 \leq \eta_j \leq \alpha_j^1 \ \text{for } j = 1, \dots, n_1\}$ and integers $l_\eta \in \mathbb{N}$ such that $\sum_{\eta \in \mathcal{M}_\gamma} l_\eta \eta = \alpha^1 - \gamma$. This shows that $R(x^1, D)$ is of lower order with respect to A_1. In view of assumption (iii) we also see that the coefficients of $R(x^1, D)$ satisfy condition (10.5). Hence, there is a $\delta_1 \geq 0$ such that $A_1 - R + \delta_1 \in \mathcal{S}(L^p(\Omega, F))$ and $\phi_{A_1 - R + \delta_1} \leq \phi_{A_1}$.

Let $\phi_1 > \phi_{A_1}$, $\mu \in \Sigma_{\pi - \phi_1}$, and $v \in D(A_2)$. With $u = (\mu + A_1)^{-1} v \in D(A_1 A_2)$ inserted into (10.13) and $(\mu + A_1 - R + \delta_1)^{-1}$ applied to the resulting equation, we deduce

$$M_r A_2(\mu + A_1)^{-1} v = (\mu + A_1 - R + \delta_1)^{-1} M_r A_2 v.$$

From this for $v \in D(A_2)$ and $\mu \in \Sigma_{\pi - \phi_1}$ we infer that

$$[M_r A_2, (\mu + A_1)^{-1}] v = (\mu + A_1)^{-1} (R + \delta_1)(\mu + A_1 - R + \delta_1)^{-1} M_r A_2 v.$$

Let $\phi_2 > \phi_{M_r A_2}$ and $\lambda \in \Sigma_{\pi - \phi_2}$. With the relation given above, the expression appearing in the Labbas-Terreni commutator condition turns into

$$M_r A_2(\lambda + M_r A_2)^{-1} [(M_r A_2)^{-1}, (\mu + A_1)^{-1}]$$
$$= -(\lambda + M_r A_2)^{-1} (\mu + A_1)^{-1} (R + \delta_1)(\mu + A_1 - R + \delta_1)^{-1}.$$

This formula can easily be estimated to the result

$$\|M_r A_2(\lambda + M_r A_2)^{-1} [(M_r A_2)^{-1}, (\mu + A_1)^{-1}]\|$$
$$\leq \frac{C}{(1 + |\lambda|)|\mu|^{1 + \frac{1}{2m_2}}} \qquad (\mu \in \Sigma_{\pi - \phi_1}, \ \lambda \in \Sigma_{\pi - \phi_2}),$$

where we employed $0 \in \rho(M_r A_2)$ and the fact that R is relatively bounded by A_1. The assertion now follows from Proposition 4.21. $\qquad \square$

10.3 A focus on the Laplacian

In this section we consider the Laplacian on domains $\Omega := \prod_{i=1}^{N} V_i$, however, with the difference that $V_i \subset \mathbb{R}^{n_i}$ now each may be a bounded Lipschitz domain (cf. Definition 8.6). More precisely, we consider the resolvent problem for the Laplacian with mixed Dirichlet-Neumann boundary conditions

$$
(10.14) \qquad
\begin{aligned}
\lambda u - \Delta u &= f \text{ in } \Omega, \\
u &= 0 \text{ on } \Gamma_0, \\
\partial_{\mathbf{n}} u &= 0 \text{ on } \Gamma_1,
\end{aligned}
$$

where $\partial_\nu u$ denotes the outer normal derivative of u. Here $\Gamma_0 := \bigcup_{i \in N_0} \partial V_i$ and $\Gamma_1 := \bigcup_{i \in N_1} \partial V_i$ with $N_0 \cup N_1 \subset \{1, \ldots, N\}$, and $N_0 \cap N_1 = \emptyset$. The sets ∂V_i are defined as in (10.2) and $\partial V_i = \emptyset$ if and only if $i \notin N_0 \cup N_1$. The boundary value problem (10.14) decomposes into

$$
(10.15) \qquad
\begin{aligned}
\lambda u - \Delta u &= f \text{ in } V_i, \\
B_i u &= 0 \text{ on } \partial V_i,
\end{aligned}
$$

where $B_i u := u$ for $i \in \mathcal{N}_0$ and $B_i u := \partial_{\mathbf{n}} u$ for $i \in \mathcal{N}_1$. In case $\partial V_i = \emptyset$, of course, the boundary conditions drop out.

For our purposes weak versions of the Dirichlet and Neumann Laplacian have to be considered in Banach space-valued spaces. Their definitions read essentially the same as in scalar-valued context: given a Banach space E, the L^p-realizations are defined as

$$
(10.16) \qquad
\begin{aligned}
D(\Delta_{p,w,i}^D, E) &:= \left\{ u \in W_0^{1,p}(V_i, E); \ \Delta u \in L^p(V_i, E) \right\}, \\
\Delta_{p,w,i}^D u &:= \Delta u \quad (u \in D(\Delta_{p,w,i}^D, E))
\end{aligned}
$$

for the E-valued Dirichlet Laplacian on V_i, i.e. in case $i \in \mathcal{N}_0$, and

$$
D(\Delta_{p,w,i}^N, E) := \Big\{ u \in W^{1,p}(V_i, E); \ \exists v \in L^p(V_i, E) \ \forall \varphi \in W^{1,p'}(V_i) :
$$

$$
- \int_{V_i} \nabla u \nabla \varphi = \int_{V_i} v \varphi \Big\},
$$

$$
\Delta_{p,w,i}^N u := v \quad (u \in D(\Delta_{p,w,i}^N, E))
$$

for the E-valued Neumann Laplacian on V_i, i.e. in case $i \in \mathcal{N}_1$. If $E = \mathbb{C}$ these definitions coincide with the former ones. From now on we just write $D(\Delta_{p,w,i})$ if it is clear from the context or the subindex i which boundary conditions are assumed or if investigations in progress do not require to distinguish with respect to the boundary conditions. On the other hand, to emphasize the boundary conditions in special situations we occasionally come back to the notation $\Delta_{p,w,i}^D$ or $\Delta_{p,w,i}^N$, respectively. As before, we use the same symbol for the canonical extensions of

$\Delta_{p,w,i}$ to $L^p(\Omega)$. We define the L^p-realization of the weak Laplacian with mixed boundary conditions on Ω by

(10.17)
$$D(\Delta_{p,w}) := \bigcap_{i=1}^{N} D(\Delta_{p,w,i}),$$

$$\Delta_{p,w} u := \sum_{i=1}^{N} \Delta_{p,w,i} u \quad (u \in D(\Delta_{p,w})).$$

Theorem 10.9. *For $i = 1, \ldots, N$ let V_i be a C^2 standard domain in $\mathbb{R}^{n_i}$, $n_i \in \mathbb{N}$, or a bounded Lipschitz domain in $\mathbb{R}^{n_i}$, $n_i \geq 2$. On two-dimensional Lipschitz cross-sections V_i we assume $\Delta_{p,w,i}$ to be the Dirichlet Laplacian. Then there exists $\varepsilon > 0$ depending only on the Lipschitz character of the different V_i such that for all $(3 + \varepsilon)' < p < 3 + \varepsilon$ and all $\delta > 0$ we have $-\Delta_{p,w} + \delta \in \mathcal{RH}^{\infty}(L^p(\Omega))$ and $\phi_{-\Delta_{p,w}+\delta}^{\mathcal{R},\infty} < \frac{\pi}{2}$.*

Proof. Step 1: As in Theorem 10.5, we decompose the problem first.

Observe that the assumptions imposed on the Laplacian with Dirichlet or with Neumann conditions on the cross-sections V_i are exactly those which make the results obtained in Proposition 8.7 available, i.e. there exist $\delta_i \geq 0$ such that $-\Delta_{p,w,i} + \delta_i \in \mathcal{RH}^{\infty}(L^p(V_i))$ and $\phi_{-\Delta_{p,w,i}+\delta_i}^{\mathcal{R},\infty} < \frac{\pi}{2}$. Recall that the shift δ_i is inserted to assure injectivity in case of Neumann boundary conditions. Again these results remain true for the canonical extensions of the operators to $L^p(\Omega)$.

Step 2: As above we first consider the case $N = 2$.

Unlike in the proof of Theorem 10.5 we have $-\Delta_{p,w,i} + \delta_i \in \mathcal{RH}^{\infty}(L^p(V_i, E))$ a priori only for $E := \mathbb{C}$ instead of more general Banach spaces E. Moreover, $D(\Delta_{p,w,i}, E)$ is in general no longer a subset of $W^{2,p}(V_i, E)$. By these facts we first have to show that

(10.18) $$\lambda - \Delta_{p,w,1} \colon D(\Delta_{p,w,1}, D(\Delta_{p,w,2})) \to L^p(V_1, D(\Delta_{p,w,2}))$$

defines an isomorphism. This before was guaranteed by known results (see step 2 of the proof of Theorem 10.5).

Let $\Delta_{p,w,1}$ be either the Dirichlet or the Neumann Laplacian in $L^p(V_1)$ and let $\lambda \in \rho(\Delta_{p,w,1})$. By Fubini's theorem we see that

(10.19) $$-\Delta_{p,w,1} + \delta \in \mathcal{RH}^{\infty}(L^p(V_1, W^{k,p}(V_2))) \quad \text{and} \quad \phi_{-\Delta_{p,w,1}+\delta}^{\mathcal{R},\infty} < \frac{\pi}{2}$$

for $k = 0, 1$. In particular, $\lambda - \Delta_{p,w,1} + \delta \colon D(\Delta_{p,w,1}, W^{1,p}(V_2)) \to L^p(V_1, W^{1,p}(V_2))$ defines an isomorphism. For the sake of readability in what follows we assume the shift $\delta \geq 0$ to be included in λ. In order to show that (10.18) is an isomorphism as well, it remains to prove surjectivity. To this end, pick $f \in L^p(V_1, D(\Delta_{p,w,2}))$. In view of (10.19) there exists $u \in D(\Delta_{p,w,1}, W^{1,p}(V_2))$ such that $(\lambda - \Delta_{p,w,1})u = f$.

First assume $\Delta_{p,w,2}$ to be the Neumann Laplacian. Then by definition there exists $v \in L^p(V_1, L^p(V_2))$ such that for all $\varphi \in W^{1,p'}(V_2)$ it holds that

$$-\int_{V_2} \nabla_2(\lambda - \Delta_{p,w,1})u\nabla_2\varphi = -\int_{V_2} \nabla_2 f\nabla_2\varphi = \int_{V_2} v\varphi.$$

Since $(\lambda - \Delta_{p,w,1})^{-1} \in \mathcal{L}(L^p(V_1, L^p(V_2)))$, we deduce

$$-\int_{V_2} \nabla_2 u\nabla_2\varphi = -(\lambda - \Delta_{p,w,1})^{-1}\int_{V_2} \nabla_2(\lambda - \Delta_{p,w,1})u\nabla_2\varphi$$

$$= (\lambda - \Delta_{p,w,1})^{-1}\int_{V_2} v\varphi = \int_{V_2} (\lambda - \Delta_{p,w,1})^{-1}v\varphi = \int_{V_2} w\varphi$$

where $w := (\lambda - \Delta_{p,w,1})^{-1}v \in D(\Delta_{p,w,1}, L^p(V_2))$. Observe that $(\lambda - \Delta_{p,w,1})^{-1}$ can be replaced by $\nabla_1(\lambda - \Delta_{p,w,1})^{-1}$ or $\Delta_{p,w,1}(\lambda - \Delta_{p,w,1})^{-1}$ and u by $\nabla_1 u$ or $\Delta_{p,w,1}u$, respectively. Hence, we obtain $u \in D(\Delta_{p,w,1}, D(\Delta_{p,w,2}))$. In a very similar way, surjectivity of (10.18) can be proved if both $\Delta_{p,w,1}$ and $\Delta_{p,w,1}$ are given as Dirichlet Laplacians.

Now we continue as in step 2a) of the proof of Theorem 10.5. Indeed, Fubini's theorem yields

$$D(\Delta_{p,w,1}, D(\Delta_{p,w,2})) \hookrightarrow W^{1,p}(V_1, W^{1,p}(V_2)) \cong W^{1,p}(V_2, W^{1,p}(V_1))$$

and by very similar calculations as above we obtain

$$u, \nabla_2 u, \Delta_{p,w,2} u \in L^p(V_2, D(\Delta_{p,w,1}))$$

for $u \in D(\Delta_{p,w,1}, D(\Delta_{p,w,2}))$. Thus, we arrive at

$$D(\Delta_{p,w,1}, D(\Delta_{p,w,2})) \cong D(\Delta_{p,w,2}, D(\Delta_{p,w,1})).$$

Now we are in the same situation as in step 2a) of the proof of Theorem 10.5. The same arguments therefore show that $\Delta_{p,w,1}$ and $\Delta_{p,w,2}$ are resolvent commuting. Exactly as in step 2b), Proposition 4.19a) now proves the claim for $N = 2$.

Step 3: Finally, we prove the assertion for $N > 2$.

We first assume $\partial V_i \neq \emptyset$ for all $i = 1, \ldots, N$. W.l.o.g. we can rearrange the different sets V_i such that $\Gamma_0 = \bigcup_{i=1}^l \partial V_i$ and $\Gamma_1 = \bigcup_{i=l+1}^N \partial V_i$. Here $l = N$ corresponds to the pure Dirichlet Laplacian, $l = 0$ to the pure Neumann Laplacian, and $0 < l < N$ to the mixed Dirichlet-Neumann Laplacian. Set $\Omega_0 := \prod_{i=1}^l V_i$ and $\Omega_1 := \prod_{i=l+1}^N V_i$ and let $\Delta_{p,w,0}^D$ and $\Delta_{p,w,1}^N$ denote the extended $L^p(\Omega)$-realizations of the Dirichlet problem on Ω_0 and of the Neumann problem on Ω_1, respectively. Then according to the first step of the proof by iteration we see that $-\Delta_{p,w,0}^D$ and $-\Delta_{p,w,1}^N + \delta$ for every $\delta > 0$ admit each a bounded $\mathcal{RH}^\infty$-calculus with angle less than $\frac{\pi}{2}$. Moreover, the argumentation on commutativity of resolvents performed in step 2 applies to $\Delta^D_{p,w,0}$ and $\Delta^N_{p,w,1}$. As $\Delta_{p,w} = \Delta^D_{p,w,0} + \Delta^N_{p,w,1}$, Proposition 4.19a) gives the desired result. In case $\partial V_i = \emptyset$ for some $i = 1, \ldots, N$, we repeat the former step with the Laplacian on $\Omega_0 \times \Omega_1$ subject to mixed Dirichlet-Neumann boundary conditions and the Laplacian defined on the whole space. The proof is now complete. □

Remark 10.10. There are some situations in which the assertion remains true for $\delta = 0$. In fact, the shift $\delta > 0$ is only required to overcome the lack of injectivity in 'Neumann cross-sections'. Thus, if we assume e.g. pure Dirichlet boundary conditions the assertion remains true for $\delta = 0$ since then every $-\Delta^D_{p,w,i}$ as defined in (10.16) is injective. Furthermore, if at least one V_i, say V_{i_0}, is bounded and the operator on V_{i_0} is the Dirichlet Laplacian $-\Delta^D_{p,w,i_0}$, we obtain $0 \in \rho(-\Delta_{p,w})$ (i.e. in particular $\delta = 0$) for the full Laplacian with mixed Dirichlet-Neumann boundary conditions $-\Delta_{p,w}$ on Ω. This follows thanks to the spectral property $\sigma(-\Delta^D_{p,w,i_0}) \subset (c, \infty)$ for some $c > 0$ which gives $-\Delta^D_{p,w,i_0} - \delta_{i_0} \in \mathcal{H}^\infty(L^p(V_{i_0}))$ for each $\delta_{i_0} < c$ by Cauchy's theorem and step 1 of the proof. Due to the fact that we may choose the shifts $\delta_i > 0$ for the Neumann-Laplacians on other cross-sections arbitrarily small, the sum over all δ_i, $i \neq i_0$, can be absorbed by $-\delta_{i_0}$.

Remark 10.11. We emphasize again that classes of unbounded Lipschitz domains and of mixed boundary conditions are treated simultaneously. Moreover, Theorem 10.9 yields more regularity for $u \in D(\Delta_{p,w})$ than indicated by (10.17) if at least one cross-section V_i is smooth. Also observe that the assertions of Remarks 8.25, 8.26, and 8.27 carry over to the present situation.

Remark 10.12. Finally we note that Remark 10.7 also applies to the situation in Theorem 10.9 if only one non-smooth domain is considered. Recall that a bounded $\mathcal{H}^\infty$-calculus for an E-valued realization of the Laplacian in non-smooth Lipschitz domains was deduced from Fubini's theorem especially for $E := L^p(G)$. Since Fubini's theorem fails to be true for integrability with respect to different values of p, the present proof does not apply in full generality. However, if only V_N is non-smooth, we can apply Theorem 10.5, i.e. it is possible to consider $L^{p_\ell}(V_\ell, L^{p_{\ell+1}}(V_{\ell+1}, \ldots, L^{p_N}(V_N)) \ldots)$-valued realizations of $-\Delta_{p_{\ell-1},\ell-1}$ on standard domains $V_{\ell-1}$, $\ell = 2, \ldots, N$ (cf. Remark 8.24).

Given a bounded Lipschitz domain $V \subset \mathbb{R}^2$, the V-dependent range for p such that $-\Delta^D_p \in \mathcal{RH}^\infty(L^p(V))$ extends to $(4 + \varepsilon)' < p < 4 + \varepsilon$ with some $\varepsilon > 0$. By our technique this range is preserved for higher dimensional Lipschitz cylinders provided the roughness of the boundary is of two dimensional character.

Theorem 10.13. *Let V_i be a C^2 standard domain in $\mathbb{R}^{n_i}$, $n_i \in \mathbb{N}$, or a bounded Lipschitz domain in $\mathbb{R}^2$ and assume that $\Delta_{p,w,i}$ is the Dirichlet Laplacian on Lipschitz cross-sections V_i. Then there exists $\varepsilon > 0$ depending only on the Lipschitz character of the different V_i such that for all $(4 + \varepsilon)' < p < 4 + \varepsilon$ and for every $\delta > 0$ we have $-\Delta_{p,w} + \delta \in \mathcal{RH}^\infty(L^p(\Omega))$ and $\phi^{\mathcal{R},\infty}_{-\Delta_{p,w}+\delta} < \pi/2$, where $\Delta_{p,w}$ is defined as in (10.17).*

Proof. We make use of the better range $(4+\varepsilon)' < p < 4+\varepsilon$ of p in Proposition 8.7. Now we can go on as in the proof of Theorem 10.9.   □

Remark 10.14. Observe that in addition to the better range of p, all observations given in Remarks 10.10, 10.11, and 10.12 apply also here.

The result on non-commuting operators from the previous section can be established particularly in the context of the Dirichlet Laplacian in cylindrical Lipschitz domains.

Theorem 10.15. *Let $1 < p < \infty$. Let V_i be given as in Theorem 10.9 (respectively Theorem 10.13) and let $\Delta_{p,w,i} := \Delta^D_{p,w,i}$ define the weak Dirichlet Laplacian for $i = 1, 2$. Assume that*

(i) $r \in [W^{2,p} + W^{2,\infty}](V_1)$ if $2p > n_1$ and $r \in W^{2,\infty}(V_1)$ else,

(ii) $r(x^1) \in \Sigma_\vartheta$ for all $x^1 \in V_1$ and some $0 \le \vartheta < \pi - \phi^{\mathcal{R}\infty}_{-\Delta^D_{p,w}}$,

(iii) $\frac{\nabla r}{r}, \frac{\Delta r}{r} - 2\frac{|\nabla r|^2}{r^2} \in L^\infty(V_1)$,

(iv) $r^{-1} \in L^\infty(V_1)$ or $0 \in \rho(\Delta_{p,2})$,

and let $\phi > \vartheta$. Then there exists $\varepsilon > 0$ depending only on the Lipschitz character of V_i such that for all $(3 + \varepsilon)' < p < 3 + \varepsilon$ (respectively $(4 + \varepsilon)' < p < 4 + \varepsilon$) there is a $\delta \ge 0$ such that for $-\Delta_{r,p} + \delta := -\Delta_{p,w,1} - M_r\Delta_{p,w,2} + \delta$ defined on $D(\Delta_{r,p}) := D(\Delta_{p,w,1}) \cap D(M_r\Delta_{p,w,2})$ we have that $-\Delta_{r,p} + \delta \in \mathcal{RH}^\infty(L^p(\Omega))$ and $\phi^{\mathcal{R},\infty}_{-\Delta_{r,p}+\delta} \le \phi + \phi^{\mathcal{R}\infty}_{-\Delta^D_{p,w}}$.

Proof. We try to mimic the proof of Theorem 10.8. By the fact that

$$\Delta ru = u\Delta r + \nabla r \cdot \nabla u + r\Delta u$$

we see that also here we have $M_r(D(\Delta_{p,w,1})) \subset D(\Delta_{p,w,1})$. Completely analogous to the preceding we therefore arrive at (10.13) with $A_1 = -\Delta_{p,w,1}$, $A_2 = -\Delta_{p,w,2}$, and

$$Ru = \frac{\nabla r}{r} \cdot \nabla u + \left(\frac{\Delta r}{r} - 2\frac{|\nabla r|^2}{r^2}\right)u.$$

We have to show that R is relatively bounded. Since $D(\Delta_{p,w,1}) \subset W^{2,p}(V_1)$ fails to be true in general, this is not so obvious as above. Recall that by the results obtained in [JK95] we know that $D((-\Delta_{p,w,1})^{1/2}) = W^{1,p}_0(V_1)$. Since $\Delta_{p,w,1} \in \mathcal{H}^\infty(L^p(V_1))$ this yields

$$W^{1,p}_0(V_1) = D((-\Delta_{p,w,1})^{1/2}) = [L^p(V_1), D(\Delta_{p,w,1})]_{1/2}.$$

Hence, the interpolation inequality

$$\|u\|_{W^{1,p}} \le C\|u\|^{1/2}_{D(\Delta_{p,w,1})}\|u\|^{1/2}_p \le C(\varepsilon\|u\|_{D(\Delta_{p,w,1})} + C(\varepsilon)\|u\|_p)$$

holds for all $u \in D(\Delta_{p,w,1})$ and $\varepsilon > 0$. This implies

$$\|Ru\|_p \leq C\|u\|_{W^{1,p}} \leq C(\varepsilon\|\Delta_{p,w,1}u\|_p + C(\varepsilon)\|u\|_p) \quad (u \in D(\Delta_{p,w,1}), \ \varepsilon > 0)$$

from which we deduce that $\mu - \Delta_{p,w,1} - R$ is sectorial for some $\mu \geq 0$. The remaining proof now copies verbatim from Theorem 10.8. $\qquad \square$

Example 10.16. Theorem 10.15 in particular covers the case of heat conduction in a bounded Lipschitz cylinder with either in longitudinal direction or in cross-sections non-constant heat conductivity coefficient (see also Example 8.28).

A Notation and vector-valued function spaces

Throughout this thesis the symbols X, Y and Z stand for Banach spaces. Given a closed operator A, we denote by $D(A)$, $N(A)$, and $R(A)$ the domain of definition, the kernel, and the range of A, respectively. Furthermore, by $\rho(A)$ and $\sigma(A)$ we denote the resolvent set and the A. The symbol $\mathcal{L}(X, Y)$ stands for the Banach space of all bounded linear operators from X to Y equipped with operator norm $\|\cdot\|_{\mathcal{L}(X,Y)}$. As an abbreviation we set $\mathcal{L}(X) := \mathcal{L}(X, X)$.

In the following we collect some definitions and facts on function spaces. This will be done in a vector-valued context, i.e. the functions under consideration are allowed to take values in rather arbitrary Banach spaces. For a more general introduction and further results we refer to the monographs [Ama95] and [Ama09]. Since many vector-valued function spaces will define Banach spaces again, we will find it convenient to employ E and F as further symbols for Banach spaces which merely occur as image spaces.

Let E be an arbitrary Banach space and let $G \subset \mathbb{R}^n$ denote an arbitrary domain. For $1 \leq p \leq \infty$ we denote by $L^p(G, E)$ the E-valued Lebesgue or Bochner spaces. Endowed with the norm

$$\|f\|_p := \|f\|_{p,G} := \|f\|_{p,G,E} := \begin{cases} \left(\int_G \|f(x)\|_E^p dx\right)^{\frac{1}{p}}, & 1 \leq p < \infty, \\ \operatorname{ess\,sup}_{x \in G} \|f(x)\|_E, & p = \infty \end{cases}$$

$L^p(G, E)$ is a Banach space.

For $m \in \mathbb{N}_0 \cup \{\infty\}$ we denote by $C^m(G, E)$ the space of all m-times continuously differentiable functions. In particular $C(G, E) := C^0(G, E)$ denotes the space of continuous functions. Its subspace of bounded and uniformly continuous functions is denoted by $\operatorname{BUC}(G, E)$. Finally for $\alpha \in (0, 1)$ we denote by $C^{m,\alpha}(G, E)$ the space of m-times continuously differentiable functions whose m-th derivative is Hölder continuous with Hölder exponent α. The subindex 0, e.g. $C_0^\infty(G, E)$, will be used to indicate the subspace of in G compactly supported functions of the respective function space. Finally, $C_\infty(\mathbb{R}^n, E)$ denotes the closed subspace of $\operatorname{BUC}(\mathbb{R}^n, E)$ with respect to $\|\cdot\|_\infty$, consisting of all continuous functions vanishing at infinity. We agree to drop the indication E in the notation of all function spaces if $E = \mathbb{C}$.

Let $m \in \mathbb{N}_0$ and $K \subset G$ be bounded. For $f \in C^\infty(G, E)$ define the seminorms

$$p_{m,K}(f) := \max_{|\alpha| \leq m} \|\partial^\alpha f\|_{L^\infty(K,E)}.$$

Then the space of smooth E-valued functions endowed with these seminorms

$$\mathcal{E}(G, E) := \left(C^\infty(G, E), \{p_{m,K};\ m \in \mathbb{N}_0,\ K \subset G \text{ bounded}\}\right)$$

is a Fréchet space.

The space of E-valued test functions, that is, the space $C_0^\infty(G, E)$ equipped with the inductive limit topology as presented e.g. in [Ama03, Section 1.1] is denoted by $\mathcal{D}(G, E)$.

Also recall the Schwartz space of smooth, rapidly decreasing E-valued functions on $\mathbb{R}^n$ denoted by $\mathcal{S}(\mathbb{R}^n, E)$. Setting

$$q_{k,m}(f) := \sup_{x \in \mathbb{R}^n, \, |\alpha| \leq m} (1 + |x|^2)^k \|\partial^\alpha f(x)\|_E,$$

it consists of all $f \in C^\infty(\mathbb{R}^n, E)$ such that $q_{k,m}(f) < \infty$ for each $k, m \in \mathbb{N}_0$. Endowed with the topology induced by $\{q_{k,m}; \, k, m \in \mathbb{N}_0\}$, $\mathcal{S}(\mathbb{R}^n, E)$ becomes a Fréchet space.

As an abbreviation let $\mathfrak{F} \in \{\mathcal{E}, \mathcal{D}, \mathcal{S}\}$. In what follows set $\mathcal{S}(G, E) := \mathcal{S}(\mathbb{R}^n, E)$, i.e. in the context of rapidly decreasing functions G is always assumed to be the whole space. Again we abbreviate $\mathfrak{F}(G) := \mathfrak{F}(G, \mathbb{C})$ and set

$$\mathfrak{F}'(G, E) := \mathcal{L}(\mathfrak{F}(G), E)$$

endowed with the topology of uniform convergence on bounded subsets of $\mathfrak{F}(G, E)$. In particular, $T \in \mathcal{D}'(G, E)$ if and only if for each compact set $K \subset G$ there exist $m \in \mathbb{N}_0$ and $C > 0$ such that

(A.1) $\|T(\varphi)\|_E \leq C p_{m,K}(\varphi)$

holds for all $\varphi \in C_0^\infty(G, E)$ with $\operatorname{supp} \varphi \subset K$. A sequence $(T_j)_{j \in \mathbb{N}} \subset \mathcal{D}'(G, E)$ is said to converge to $T \in \mathcal{D}'(G, E)$, i.e. $T_j \to T$ if and only if $T_j(\varphi) \to T(\varphi)$ for all $\varphi \in C_0^\infty(G)$. As usual, we call $\mathcal{D}'(G, E)$ the space of E-valued distributions on G, $\mathcal{E}'(G, E)$ the subspace of E-valued distributions with compact support and $\mathcal{S}'(\mathbb{R}^n, E)$ the space of E-valued tempered distributions. It is well-known that

$$\mathcal{D}(G, E) \hookrightarrow \mathcal{E}(G, E) \quad \text{and} \quad \mathcal{D}(\mathbb{R}^n, E) \hookrightarrow \mathcal{S}(\mathbb{R}^n, E) \hookrightarrow \mathcal{E}(\mathbb{R}^n, E)$$

are dense embeddings. Since $C_0^\infty(\mathbb{R}^n, E)$ is dense in $L^p(\mathbb{R}^n, E)$ for $1 \leq p < \infty$ the same is true for $\mathcal{S}(\mathbb{R}^n, E)$. Furthermore, the relations between $\mathcal{D}, \mathcal{S}$ and $\mathcal{E}$ immediately yield

$$\mathcal{E}'(G, E) \hookrightarrow \mathcal{D}'(G, E) \quad \text{and} \quad \mathcal{E}'(\mathbb{R}^n, E) \hookrightarrow \mathcal{S}'(\mathbb{R}^n, E) \hookrightarrow \mathcal{D}'(\mathbb{R}^n, E).$$

We write $f \in L^p_{loc}(G, E)$ if $\varphi f \in L^p(G, E)$ for all $\varphi \in C_0^\infty(\mathbb{R}^n, E)$. Given any function $f \in L^1_{loc}(G, E)$ we set

$$T_f(\varphi) := [f](\varphi) := (f, \varphi) := \int_G f\varphi \, dx, \quad (\varphi \in C_0^\infty(G)).$$

Then $T_f \colon \mathcal{D}(G) \to E$ is linear and for each compact set $K \subset G$ it holds that

$$\|T_f(\varphi)\|_E \leq \|f\|_{1,K,E} \, p_{0,K}(\varphi) \quad (\varphi \in C_0^\infty(G), \, \operatorname{supp} \varphi \subset K).$$

Taking $C = \|f\|_{1,K,E}$ and $m = 0$ in (A.1) shows $T_f \in \mathcal{D}'(G, E)$. Moreover, the mapping

$$L^1_{loc}(G, E) \to \mathcal{D}'(G, E); \quad f \mapsto T_f$$

is linear and injective so that $L^1_{loc}(G, E) \hookrightarrow \mathcal{D}'(G, E)$. For $f \in L^1_{loc}(G, E)$ the distributions T_f are called regular distributions. Restricted to smaller function spaces sometimes more can be said about the according regular distributions. For example, it is easily seen that $C_0(G, E) \hookrightarrow \mathcal{E}'(\mathbb{R}^n, E)$. In view of Fourier transform we will make use of the important fact that $L^p(\mathbb{R}^n, E) \hookrightarrow \mathcal{S}'(\mathbb{R}^n, E)$ for $1 \le p \le \infty$. For the sake of simplicity we will frequently write $f \in \mathfrak{F}'(G, E)$ instead of $T_f \in \mathfrak{F}'(G, E)$ if no confusion seems likely.

For $\alpha \in \mathbb{N}_0^n$ the distributional derivative $\partial^\alpha T$ of $T \in \mathcal{D}'(G, E)$ is defined by

$$(\partial^\alpha T)(\varphi) := (-1)^{|\alpha|} T(\partial^\alpha \varphi) \quad (\varphi \in \mathcal{D}(G)).$$

In view of (A.1) it is easily verified that $T_j \to T$ implies $\partial^\alpha T_j \to \partial^\alpha T$. Along the same lines derivation of tempered distributions defines a continuous mapping in $\mathcal{S}'(\mathbb{R}^n, E)$.

For each $m \in \mathbb{N}_0$ and $1 \le p \le \infty$ the E-valued Sobolev space $W^{m,p}(G, E)$ is defined as the space of all $f \in L^p(G, E)$ such that $\partial^\alpha f \in L^p(G, E)$ for all $|\alpha| \le m$. Endowed with the norm

$$\|f\|_{m,p} := \|f\|_{m,p,G} := \|f\|_{m,p,G,E} := \begin{cases} \left(\sum_{|\alpha| \le m} \|\partial^\alpha f\|_p^p \right)^{\frac{1}{p}}, & 1 \le p < \infty, \\ \max_{|\alpha| \le m} \|\partial^\alpha f\|_\infty, & p = \infty \end{cases}$$

$W^{m,p}(G, E)$ becomes a Banach space. By this definition $W^{0,p}(G, E) = L^p(G, E)$. For $1 \le p < \infty$ we denote by $W_0^{m,p}(G, E)$ the closure of $C_0^\infty(G, E)$ in $W^{m,p}(G, E)$. We further define the Fréchet spaces $W_{loc}^{m,p}(G, E)$ for $\le p \le \infty$ and $m \in \mathbb{N}$ which consist of all $u \in L^p_{loc}(G, E)$ such that $u\varphi \in W^{m,p}(G, E)$ for all $\varphi \in C_0^\infty(G)$.

In the subsequent lines, we briefly focus on the task to find a primitive with respect to x_j of a distribution $T \in \mathcal{D}'(\mathbb{R}^n, E)$, that is, a distribution $F \in \mathcal{D}'(\mathbb{R}^n, E)$ such that $\partial_j F = T$. For that purpose, given $n \in \mathbb{N}$, $n \ge 2$, and $x \in \mathbb{R}^n$ for each $j \in \{1, \ldots, n\}$ we define

$$(x_j, x') := (x', x_j) := (x_1, \ldots, x_{j-1}, x_j, x_{j+1}, \ldots, x_n) = x.$$

Then for $\varphi \in C_0^\infty(\mathbb{R}^n)$ there exists $\psi \in C_0^\infty(\mathbb{R}^n)$ such that $\varphi = \partial_j \psi$ if and only if the function $\Phi(x') := \int_\mathbb{R} \varphi(x', s) ds$ fulfills $\Phi \equiv 0$. Let $\mathcal{D}_j(\mathbb{R}^n)$ denote the subspace of functions $\varphi \in \mathcal{D}(\mathbb{R}^n)$ for which one of these equivalent conditions holds true. Let $\kappa \in C_0^\infty(\mathbb{R})$ be such that $\int_\mathbb{R} \kappa(s) ds = 1$. Then every $\varphi \in \mathcal{D}(\mathbb{R}^n)$ admits a unique decomposition of the form

$$\varphi(x', x_j) = \varphi_0(x', x_j) + \kappa(x_j)\Phi(x')$$

such that $\varphi_0 \in \mathcal{D}_j(\mathbb{R}^n)$. This allows to define a projection onto $\mathcal{D}_j(\mathbb{R}^n)$ given by $P_j(\varphi) := \varphi_0 = \varphi - \kappa \otimes \Phi$, where $\kappa \otimes \Phi(x) := \kappa(x_j)\Phi(x')$ denotes the tensor product of functions. Hence, $\psi_\varphi(x) := \int_{-\infty}^{x_j} P_j\varphi(x', s)ds$ defines a test function $\psi \in C_0^\infty(\mathbb{R}^n)$ and

$$(A.2) \qquad\qquad F(\varphi) := -T(\psi_\varphi)$$

fulfills $\partial_j F = T$. Moreover, $\partial_j F = 0$ if and only if F is independent of x_j which means

$$(A.3) \qquad\qquad F(\varphi) = F(\varphi(\cdot + ce_j)) \quad (\varphi \in C_0^\infty(\mathbb{R}^n),\ c \in \mathbb{R}).$$

For $T \in \mathcal{D}'(\mathbb{R}^n, E)$ this implies

$$(A.4) \qquad\qquad \nabla T = 0 \quad \Longleftrightarrow \quad T = T_\eta = \eta \quad \text{with} \quad \eta \in E$$

(see [Wal94, Chapter 6]). It is easily seen that a regular distribution T_f is independent of x_j if and only if $f(x', x_j) = f(x', x_j + c)$ for all $c \in \mathbb{R}$, i.e. if $f(x) = f(x')$ x-almost everywhere.

Given $T \in \mathfrak{F}'(\mathbb{R}^n, E)$ and $\varphi \in \mathfrak{F}(\mathbb{R}^n)$, for $\mathfrak{F} \in \{\mathcal{D}, \mathcal{E}\}$ the convolution $T * \varphi$ is a well-defined element of $\mathcal{E}(\mathbb{R}^n, E)$ (see [Ama95, Section III 4.2]). To be more precise, let $\check{\varphi}(x) := \varphi(-x)$ and $\tau_a\varphi(x) := \varphi(x - a)$ with fixed $a \in \mathbb{R}^n$ for $x \in \mathbb{R}^n$. Then

$$(A.5) \qquad\qquad T * \varphi(x) := T(\tau_x\check{\varphi}) \quad (x \in \mathbb{R}^n)$$

simultaneously defines the bilinear and separately continuous mappings

$$\mathfrak{F}'(\mathbb{R}^n, E) \times \mathfrak{F}(\mathbb{R}^n) \to \mathcal{E}(\mathbb{R}^n, E) \quad \text{and} \quad \mathcal{E}'(\mathbb{R}^n, E) \times \mathcal{D}(\mathbb{R}^n) \to \mathcal{D}(\mathbb{R}^n, E).$$

This allows to define the convolution of an E-valued and a scalar-valued distribution T and F, if at least one of them is compactly supported, as for $\varphi \in \mathcal{D}(\mathbb{R}^n)$

$$(A.6) \qquad\qquad (T * F) * \varphi = T * (F * \varphi)$$

is a well-defined element in $\mathcal{E}(\mathbb{R}^n, E) \hookrightarrow \mathcal{D}'(\mathbb{R}^n, E)$. Again this defines the bilinear and separately continuous mappings

$$\mathcal{D}'(\mathbb{R}^n, E) \times \mathcal{E}'(\mathbb{R}^n) \to \mathcal{D}'(\mathbb{R}^n, E) \quad \text{and} \quad \mathcal{E}'(\mathbb{R}^n, E) \times \mathcal{D}'(\mathbb{R}^n) \to \mathcal{D}'(\mathbb{R}^n, E).$$

A proof of this result and of hypocontinuity of the above mappings can be found in [Ama03, Proposition 1.2.3]. We close this chapter with a well-known and useful example which immediately follows from (A.5) and (A.6).

Lemma A.1. Let $\delta \in \mathcal{E}'(\mathbb{R}^n)$ denote the Dirac distribution, i.e. $\delta(\varphi) := \varphi(0)$ for $\varphi \in \mathcal{E}(\mathbb{R}^n)$. Then $T * \delta = T$ for $T \in \mathcal{D}'(\mathbb{R}^n, E)$.

B Related topics in the literature

In the subsequent lines we comment on topics in the literature that are related to some parts of this thesis.

As frequently mentioned throughout this thesis, E-valued parabolic boundary value problems in standard domains were extensively studied in [DHP03] and [DDH$^+$04]. For classical papers on scalar-valued boundary value problems we refer to [DN55], [ADN59], [ADN64], and [Sol66] in the elliptic case and to [AV64], [ADF97], and [LSU68] in the parameter-elliptic and parabolic case. For a more comprehensive list see also [DHP03]. An approach to a class of elliptic differential operators with Dirichlet boundary conditions in uniform C^2-domains can be found in [Kun03]. All cited results above are based on standard localization procedures for the domain contrary to the approach presented in this thesis.

In [AB02] solution isomorphisms for the first order equation

$$(B.1) \qquad u_t + Au = f$$

subject to pure periodic boundary conditions in Lebesgue and Hölder spaces are established. A generalization of these results on equation (B.1) to pure periodic first order integro-differential equations in Lebesgue, Besov, and Hölder spaces is given in [KL04]. Here the concept of 1-regularity in the context of sequences is introduced (cf. Remark 7.10). In [KL06] one finds a comprehensive treatment of periodic second order equations of type

$$(B.2) \qquad u_{tt} + aAu_t + \alpha Au = f$$

in Lebesgue and Hölder spaces. The special case $a = 0$ and $\alpha = 1$ subject to pure periodic boundary conditions, pure Dirichlet conditions, and pure Neumann conditions in Lebesgue spaces is also treated in [AB02]. In [KLP09] more general equations are investigated in the mentioned spaces as well as in Triebel-Lizorkin spaces. Moreover, applications to nonlinear equations are presented. Note that all articles mentioned so far consider equations depending on a single variable related to time. In contrast to that, in this thesis we apply multiplier techniques with respect to multiple space variables. In particular, the solution isomorphisms for the one-dimensional equation (B.1) subject to pure periodic boundary conditions from [AB02] are extended to solution isomorphisms for the multi-dimensional equation (1.6) subject to ν-periodic boundary conditions in Theorem 7.15.

In [AR09] various properties as e.g. Fredholmness of the operator $\partial_t - A(\cdot)$ associated with the non-autonomous periodic first order Cauchy problem in the L^q-context are investigated. Results on this operator based on Floquet theory are obtained in the PhD-thesis [Gau01]. We remark that in Floquet theory ν-periodic instead of pure periodic boundary conditions appear in a natural way.

Maximal regularity of second order initial value problems of the type

$$u_{tt}(t) + Bu_t(t) + Au(t) = f(t) \quad (t \in [0, T)),$$
$$u(0) = u_t(0) = 0$$

is treated in [CS05] and [CS08]. In particular, q-independence of maximal regularity for second order problems of this type is shown. The same equation involving dynamic boundary conditions is studied in [XL04]. The non-autonomous second order problem involving t-dependent operators $B(t)$ and $A(t)$ appears in [BCS08]. We also refer to [XL98] for the investigation of higher order Cauchy problems.

For results in boundary value problems in $(0, 1)$ with operator-valued coefficients subject to numerous types of homogeneous and inhomogeneous boundary conditions we refer to [FLM$^+$08], [FSY09], [FY10], and the references therein. Their approaches mainly rely on semigroup theory and do not allow for an easy generalization to $(0, 1)^n$. In [FSY09], however, applications to boundary value problems in the cylindrical space domain $(0, 1) \times V$ can be found.

As far as the author of this thesis knows, the usage of operator-valued multipliers to treat cylindrical boundary value problems was first carried out in [Gui04] and [Gui05] in a Besov space setting. In these papers the author constructs semiclassical fundamental solutions for a class of elliptic operators on infinite cylindrical domains $\mathbb{R}^n \times V$. This proves to be a strong tool for the treatment of related elliptic and parabolic (see [Gui04] and [Gui05]) as well as of hyperbolic (see [Gui05]) problems. In particular, this approach leads to semiclassical representation formulas for solutions of related elliptic and parabolic boundary value problems. Based on these formulas and on a multiplier result of Amann [Ama97], the author derives a couple of interesting results for these problems in a Besov space setting. In particular, the given applications include asymptotic behavior in the large, singular perturbations, exact boundary conditions on artificial boundaries, and the validity of maximum principles. With a remark on possible generalizations, in [Gui04] and [Gui05] the author assumes $A_1 = -\Delta$ which avoids the investigation of Dore-Venni-type angle conditions. Very recently in [DCGL10] the wellposedness of a class of parabolic boundary value problems in a vector-valued Hölder space setting is proved. In this article $\Omega = [0, L] \times V$, the first part is given by $A_1 = a(x_n)\partial_n^{2m}$, $x_n \in [0, L]$, and A_2 is uniformly elliptic. In contrast to [Gui04], [Gui05], and [DCGL10] here we present the L^p-approach to cylindrical boundary value problems. Therefore, the notion of $\mathcal{R}$-boundedness comes into play which is not required in the framework of Besov or Hölder spaces.

Results on sums of closed operators were already applied to the second order problem (B.2) with $a = 0$ and $\alpha = 1$ subject to Dirichlet boundary conditions in [HP98]. In the same article the method is applied to the Dirichlet problem

$$(B.3) \qquad\qquad \Delta u = f \quad \text{in } \Omega, \quad u|_{\partial\Omega} = 0,$$

where Ω defines a cone in $\mathbb{R}^2$. In virtue of a change of variables, Ω is transferred to a strip-type domain. This allows for an application of the Dore-Venni-Theorem to deduce unique solvability by means of suitable weights. Note that in contrast to strip-type domains in this thesis cylindrical domains are treated where more than one cross-section has a non-vanishing boundary. The parabolic problem associated with (B.3) where the space domain is given by a cone in $\mathbb{R}^n$ or more general a wedge in $\mathbb{R}^{m+n}$ is investigated recently in [PS07]. A change of variables leads to the space domain $\mathbb{R}^m \times M$ where M defines an open subset of the n-sphere with smooth boundary.

For literature on the Laplacian in general bounded Lipschitz domains we refer to [JK95] ans [Woo07] for the Dirichlet problem and to [FMM98] and [Woo07] for the Neumann problem and to the references therein. For a treatment of the Laplacian with mixed boundary conditions in more general Lipschitz domains we refer to [HDKR08], [HDR09], and to the literature cited therein.

For more classical results on the Helmholtz projection as well as on Stokes and Navier-Stokes equations we refer to the textbook [Gal94]. Out of the wide range of results on the Stokes operator on bounded domains we only name [NS03] where a bounded $\mathcal{H}^\infty$-calculus in L^p-spaces on bounded and exterior C^3-domains subject to non-slip boundary conditions is proved and refer to the references therein. It is based on the related half space result from [DHP01], a localization procedure, and perturbation arguments for the $\mathcal{H}^\infty$-calculus. For a recent result on Stokes and Navier-Stokes equations with partial-slip boundary conditions in a half space see [Saa06]. The Helmholtz projection in layers and infinite cylinders Ω with a standard cross-section is constructed by Farwig in [Far03] by means of continuous Fourier multiplier results. Here the $\mathcal{R}$-boundedness assumptions on the symbol are inferred from an equivalent condition involving arbitrary Muckenhoupt weights. Together with Ri, this author serializes results on resolvent estimates, maximal regularity and boundedness of the $\mathcal{H}^\infty$-calculus for the Stokes operator in $L^p_\sigma(\Omega)$ in [FR07c], [FR07a], and [FR08]. In [FR07b] these results and results on bounded domains are merged to according results on domains with finite exits to infinity. For general unbounded domains of class C^1 the existence of the Helmholtz projection in $L^2 \cap L^p$ for $p > 2$ and in $L^2 + L^p$ for $1 < p < 2$ instead of in L^p is proved in [FKS07]. Resolvent estimates for the Stokes operator in infinite layers by means of continuous Fourier multipliers were also proved in [AW05a]. In the series [Abe05a]/[Abe05b], these estimates are extended to various boundary conditions and to layer-like domains. Moreover, a representation of the Helmholtz projection in L^p-spaces on this type of domains by means of singular Green operators is deduced. Furthermore, a bounded $\mathcal{H}^\infty$-calculus of the Stokes operator subject to Dirichlet boundary conditions is proved.

Bibliography

[AB02]   W. Arendt and S. Bu, *The operator-valued Marcinkiewicz multiplier theorem and maximal regularity*, Math. Z. **240** (2002), no. 2, 311–343.

[Abe05a] H. Abels, *Reduced and generalized Stokes resolvent equations in asymptotically flat layers. I. Unique solvability*, J. Math. Fluid Mech. **7** (2005), no. 2, 201–222.

[Abe05b] _____, *Reduced and generalized Stokes resolvent equations in asymptotically flat layers. II. H_∞-calculus*, J. Math. Fluid Mech. **7** (2005), no. 2, 223–260.

[ABF+08] W. Arendt, M. Beil, F. Fleischer, S. Lück, S. Portet, and V. Schmidt, *The Laplacian in a stochastic model for spatiotemporal reaction systems*, Ulmer Seminare **13** (2008), 133–144.

[ABHN01] W. Arendt, C. J. K. Batty, M. Hieber, and F. Neubrander, *Vector-valued Laplace Transforms and Cauchy Problems*, Monographs in Mathematics, vol. 96, Birkhäuser Verlag, Basel, 2001.

[ADF97]   M. S. Agranovich, R. Denk, and M. Faierman, *Weakly smooth nonselfadjoint spectral elliptic boundary problems*, Spectral Theory, Microlocal Analysis, Singular Manifolds, Math. Top., vol. 14, Akademie Verlag, Berlin, 1997, pp. 138–199.

[ADN59]   S. Agmon, A. Douglis, and L. Nirenberg, *Estimates near the boundary for solutions of elliptic partial differential equations satisfying general boundary conditions. I*, Comm. Pure Appl. Math. **12** (1959), 623–727.

[ADN64]   _____, *Estimates near the boundary for solutions of elliptic partial differential equations satisfying general boundary conditions. II*, Comm. Pure Appl. Math. **17** (1964), 35–92.

[Ama95]   H. Amann, *Linear and Quasilinear Parabolic Problems. Vol. I*, Monographs in Mathematics, vol. 89, Birkhäuser Boston Inc., Boston, MA, 1995.

[Ama97]   _____, *Operator-valued Fourier multipliers, vector-valued Besov spaces, and applications*, Math. Nachr. **186** (1997), 5–56.

[Ama03]   _____, *Vector-valued distributions and Fourier multipliers*, unpublished manuscript, 2003.

[Ama09]   _____, *Anisotropic Function Spaces and Maximal Regularity for Parabolic Problems. Part 1: Function Spaces*, Lecture Notes, vol. 6, Jindrich Necas Center for Mathematical Modeling, 2009.

[AR09]   W. Arendt and P. J. Rabier, *Linear evolution operators on spaces of periodic functions*, Commun. Pure Appl. Anal. **8** (2009), no. 1, 5–36.

[AV64]   M. S. Agranovič and M. I. Visik, *Elliptic problems with a parameter and parabolic problems of general type*, Russian Math. Surveys **19** (1964), no. 3, 53–157.

[AW05a]   H. Abels and M. Wiegner, *Resolvent estimates for the Stokes operator on an infinite layer*, Differential Integral Equations **18** (2005), no. 10, 1081–1110.

[AW05b]   H. Amann and C. Walker, *Local and global strong solutions to continuous coagulation-fragmentation equations with diffusion*, J. Differential Equations **218** (2005), no. 1, 159–186.

[BCS08]   C. J. K. Batty, R. Chill, and S. Srivastava, *Maximal regularity for second order non-autonomous Cauchy problems*, Studia Math. **189** (2008), no. 3, 205–223.

[BK04]   S. Bu and J.-M. Kim, *Operator-valued Fourier multiplier theorems on L_p-spaces on $\mathbb{T}^d$*, Arch. Math. **82** (2004), no. 5, 404–414.

[Bu06]   S. Bu, *On operator-valued Fourier multipliers*, Sci. China Ser. A **49** (2006), no. 4, 574–576.

[CDMY96]   M. Cowling, I. Doust, A. McIntosh, and A. Yagi, *Banach space operators with a bounded H^∞ functional calculus*, J. Austral. Math. Soc. Ser. A **60** (1996), no. 1, 51–89.

[CP01]   P. Clément and J. Prüss, *An operator-valued transference principle and maximal regularity on vector-valued L_p-spaces*, Evolution Equations and Their Applications in Physical and Life Sciences (Bad Herrenalb, 1998), Lecture Notes in Pure and Appl. Math., vol. 215, Dekker, New York, 2001, pp. 67–87.

[CS05]   R. Chill and S. Srivastava, *L^p-maximal regularity for second order Cauchy problems*, Math. Z. **251** (2005), no. 4, 751–781.

[CS08]   _____, *L^p maximal regularity for second order Cauchy problems is independent of p*, Boll. Unione Mat. Ital. (9) **1** (2008), no. 1, 147–157.

[CW77]   R. R. Coifman and G. Weiss, *Transference Methods in Analysis*, American Mathematical Society, Providence, R.I., 1977, Conference Board of the Mathematical Sciences Regional Conference Series in Mathematics, No. 31.

[DCGL10]   M. Di Cristo, D. Guidetti, and A. Lorenzi, *Abstract parabolic equations with applications to problems in cylindrical space domains*, Adv. Differential Equations **15** (2010), no. 1-2, 1–42.

[DDH⁺04]   R. Denk, G. Dore, M. Hieber, J. Prüss, and A. Venni, *New thoughts on old results of R. T. Seeley*, Math. Ann. **328** (2004), no. 4, 545–583.

[DHP01]   W. Desch, M. Hieber, and J. Prüss, L^p-*theory of the Stokes equation in a half space*, J. Evol. Equ. **1** (2001), no. 1, 115–142.

[DHP03]   R. Denk, M. Hieber, and J. Prüss, $\mathcal{R}$-*boundedness, Fourier multipliers and problems of elliptic and parabolic type*, Mem. Amer. Math. Soc. **166** (2003), no. 788, viii+114.

[DMT02]   R. Denk, M. Möller, and C. Tretter, *The spectrum of a parametrized partial differential operator occurring in hydrodynamics*, J. London Math. Soc. (2) **65** (2002), no. 2, 483–492.

[DN55]   A. Douglis and L. Nirenberg, *Interior estimates for elliptic systems of partial differential equations*, Comm. Pure Appl. Math. **8** (1955), 503–538.

[DN11]   R. Denk and T. Nau, *Discrete Fourier multipliers and cylindrical boundary value problems*, preprint, 2011.

[Dor93]   G. Dore, L^p *regularity for abstract differential equations*, Functional Analysis and Related Topics, 1991 (Kyoto), Lecture Notes in Math., vol. 1540, Springer, Berlin, 1993, pp. 25–38.

[Duo90]   X. T. Duong, H_∞ *functional calculus of second order elliptic partial differential operators on L^p spaces*, Miniconference on Operators in Analysis (Sydney, 1989), Proc. Centre Math. Anal. Austral. Nat. Univ., vol. 24, Austral. Nat. Univ., Canberra, 1990, pp. 91–102.

[DV87]   G. Dore and A. Venni, *On the closedness of the sum of two closed operators*, Math. Z. **196** (1987), no. 2, 189–201.

[DV05]   ———, H^∞ *functional calculus for sectorial and bisectorial operators*, Studia Math. **166** (2005), no. 3, 221–241.

[Far03]   R. Farwig, *Weighted L^q-Helmholtz decompositions in infinite cylinders and in infinite layers*, Adv. Differential Equations **8** (2003), no. 3, 357–384.

[FKS07]   R. Farwig, H. Kozono, and H. Sohr, *On the Helmholtz decomposition in general unbounded domains*, Arch. Math. **88** (2007), no. 3, 239–248.

[FLM⁺08]   A. Favini, R. Labbas, S. Maingot, H. Tanabe, and A. Yagi, *A simplified approach in the study of elliptic differential equations in UMD spaces and new applications*, Funkcial. Ekvac. **51** (2008), no. 2, 165–187.

[FMM98]   E. Fabes, O. Mendez, and M. Mitrea, *Boundary layers on Sobolev-Besov spaces and Poisson's equation for the Laplacian in Lipschitz domains*, J. Funct. Anal. **159** (1998), no. 2, 323–368.

[FR07a]   R. Farwig and M.-H. Ri, *An $L^q(L^2)$-theory of the generalized Stokes resolvent system in infinite cylinders*, Studia Math. **178** (2007), no. 3, 197–216.

[FR07b]   _____, *The resolvent problem and H^∞-calculus of the Stokes operator in unbounded cylinders with several exits to infinity*, J. Evol. Equ. **7** (2007), no. 3, 497–528.

[FR07c]   _____, *Stokes resolvent systems in an infinite cylinder*, Math. Nachr. **280** (2007), no. 9-10, 1061–1082.

[FR08]   _____, *Resolvent estimates and maximal regularity in weighted L^q-spaces of the Stokes operator in an infinite cylinder*, J. Math. Fluid Mech. **10** (2008), no. 3, 352–387.

[FSY09]   A. Favini, V. Shakhmurov, and Y. Yakubov, *Regular boundary value problems for complete second order elliptic differential-operator equations in UMD Banach spaces*, Semigroup Forum **79** (2009), no. 1, 22–54.

[FY10]   A. Favini and Y. Yakubov, *Irregular boundary value problems for second order elliptic differential-operator equations in UMD Banach spaces*, Math. Ann. **348** (2010), no. 3, 601–632.

[Gal94]   G. P. Galdi, *An Introduction to the Mathematical Theory of the Navier-Stokes Equations. Vol. I*, Springer Tracts in Natural Philosophy, vol. 38, Springer-Verlag, New York, 1994.

[Gau01]   T. Gauss, *Floquet Theory for a Class of Periodic Evolution Equations in an L^p-Setting*, Dissertation, KIT Scientific Publishing, Karlsruhe, 2001.

[Gri85]   P. Grisvard, *Elliptic Problems in Nonsmooth Domains*, Monographs
           and Studies in Mathematics, vol. 24, Pitman, Boston, MA, 1985.

[Gui04]   P. Guidotti, *Elliptic and parabolic problems in unbounded domains*,
           Math. Nachr. **272** (2004), 32–45.

[Gui05]   ———, *Semiclassical fundamental solutions*, Abstr. Appl. Anal. **1**
           (2005), 45–57.

[GW03]   M. Girardi and L. Weis, *Criteria for R-boundedness of operator fam-
           ilies*, Evolution Equations, Lecture Notes in Pure and Appl. Math.,
           vol. 234, Dekker, New York, 2003, pp. 203–221.

[Haa06]   M. Haase, *The Functional Calculus for Sectorial Operators*, Opera-
           tor Theory: Advances and Applications, vol. 169, Birkhäuser Verlag,
           Basel, 2006.

[HD03]   R. Haller-Dintelmann, *Methoden der Banachraum-wertigen Analy-
           sis und Anwendungen auf parabolische Probleme*, Dissertation, Wis-
           senschaftlicher Verlag, Berlin, 2003.

[HDKR08]   R. Haller-Dintelmann, H.-C. Kaiser, and J. Rehberg, *Elliptic model
           problems including mixed boundary conditions and material hetero-
           geneities*, J. Math. Pures Appl. (9) **89** (2008), no. 1, 25–48.

[HDR09]   R. Haller-Dintelmann and J. Rehberg, *Maximal parabolic regularity
           for divergence operators including mixed boundary conditions*, J. Dif-
           ferential Equations **247** (2009), no. 5, 1354–1396.

[HHN02]   R. Haller, H. Heck, and A. Noll, *Mikhlin's theorem for operator-valued
           Fourier multipliers in n variables*, Math. Nachr. **244** (2002), 110–130.

[HP98]   M. Hieber and J. Prüss, *Functional calculi for linear operators in
           vector-valued L^p-spaces via the transference principle*, Adv. Differen-
           tial Equations **3** (1998), no. 6, 847–872.

[Jan71]   L. Jantscher, *Distributionen*, Walter de Gruyter, Berlin-New York,
           1971.

[JK95]   D. Jerison and C. E. Kenig, *The inhomogeneous Dirichlet problem in
           Lipschitz domains*, J. Funct. Anal. **130** (1995), no. 1, 161–219.

[KKW06]   N. Kalton, P. C. Kunstmann, and L. Weis, *Perturbation and interpo-
           lation theorems for the H^∞-calculus with applications to differential
           operators*, Math. Ann. **336** (2006), no. 4, 747–801.

[KL04]   V. Keyantuo and C. Lizama, *Fourier multipliers and integro-differen-tial equations in Banach spaces*, J. London Math. Soc. (2) **69** (2004), no. 3, 737–750.

[KL06]   ———, *Periodic solutions of second order differential equations in Banach spaces*, Math. Z. **253** (2006), no. 3, 489–514.

[KLP09]   V. Keyantuo, C. Lizama, and V. Poblete, *Periodic solutions of integro-differential equations in vector-valued function spaces*, J. Differential Equations **246** (2009), no. 3, 1007–1037.

[Kun03]   P. C. Kunstmann, *Maximal L_p-regularity for second order elliptic op-erators with uniformly continuous coefficients on domains*, Evolution Equations: Applications to Physics, Industry, Life Sciences and Sco-nomics (Levico Terme, 2000), Progr. Nonlinear Differential Equations Appl., vol. 55, Birkhäuser, Basel, 2003, pp. 293–305.

[KW01]   N. J. Kalton and L. Weis, *The H^∞-calculus and sums of closed oper-ators*, Math. Ann. **321** (2001), no. 2, 319–345.

[KW04]   P. C. Kunstmann and L. Weis, *Maximal L_p-regularity for parabolic equations, Fourier multiplier theorems and H^∞-functional calculus*, Functional Analytic Methods for Evolution Equations, Lecture Notes in Mathematics, vol. 1855, Springer, Berlin, 2004, pp. 65–311.

[LSU68]   O. A. Ladyzhenskaya, V. A. Solonnikov, and N. N. Ural'tseva, *Lin-ear and Quasi-Linear Equations of Parabolic Type*, Translations of Mathematical Monographs, vol. 23, American Mathematical Society (AMS), Providence, RI, 1968.

[MM08]   M. Mitrea and S. Monniaux, *The regularity of the Stokes operator and the Fujita-Kato approach to the Navier-Stokes initial value problem in Lipschitz domains*, J. Funct. Anal. **254** (2008), no. 6, 1522–1574.

[NS03]   A. Noll and J. Saal, *H^∞-calculus for the Stokes operator on L_q-spaces*, Math. Z. **244** (2003), no. 3, 651–688.

[NS11a]   T. Nau and J. Saal, *$\mathcal{H}^\infty$-calculus for cylindrical boundary value prob-lems*, preprint, 2011.

[NS11b]   ———, *$\mathcal{R}$-sectoriality of cylindrical boundary value problems*, Parabo-lic Problems. The Herbert Amann Festschrift, Progr. Nonlinear Differ-ential Equations Appl., vol. 80, Birkhäuser, Basel, 2011, pp. 479–505.

[Prü93]   J. Prüss, *Evolutionary Integral Equations and Applications*, Mono-graphs in Mathematics, vol. 87, Birkhäuser Verlag, Basel, 1993.

[PS90]   J. Prüss and H. Sohr, *On operators with bounded imaginary powers in Banach spaces*, Math. Z. **203** (1990), no. 3, 429–452.

[PS93]   ———, *Imaginary powers of elliptic second order differential operators in L^p-spaces*, Hiroshima Math. J. **23** (1993), no. 1, 161–192.

[PS07]   J. Prüss and G. Simonett, *H^∞-calculus for the sum of non-commuting operators*, Trans. Amer. Math. Soc. **359** (2007), no. 8, 3549–3565 (electronic).

[Ran86] R. M. Range, *Holomorphic Functions and Integral Representations in Several Complex Variables*, Graduate Texts in Mathematics, vol. 108, Springer-Verlag, New York, 1986.

[RdF86] J. L. Rubio de Francia, *Martingale and integral transforms of Banach space valued functions*, Probability and Banach Spaces (Zaragoza, 1985), Lecture Notes in Math., vol. 1221, Springer, Berlin, 1986, pp. 195–222.

[Saa06] J. Saal, *Stokes and Navier-Stokes equations with Robin boundary conditions in a half-space*, J. Math. Fluid Mech. **8** (2006), no. 2, 211–241.

[Sob64] P. E. Sobolevskiĭ, *Coerciveness inequalities for abstract parabolic equations*, Dokl. Akad. Nauk SSSR **157** (1964), 52–55, translated in: Soviet. Math. (Doklady) **5** (1964), 894-897.

[Sol66] V. A. Solonnikov, *On general boundary problems for systems which are elliptic in the sense of A. Douglis and L. Nirenberg*, Amer. Math. Soc. Transl. Ser. **56** (1966), 193–232.

[SS05]   H.-J. Schmeißer and W. Sickel, *Vector-valued Sobolev spaces and Gagliardo-Nirenberg inequalities*, Nonlinear Elliptic and Parabolic Problems, Progr. Nonlinear Differential Equations Appl., vol. 64, Birkhäuser, Basel, 2005, pp. 463–472.

[ŠW07]   Ž. Štrkalj and L. Weis, *On operator-valued Fourier multiplier theorems*, Trans. Amer. Math. Soc. **359** (2007), no. 8, 3529–3547 (electronic).

[Tri78] H. Triebel, *Interpolation Theory, Function Spaces, Differential Operators*, North-Holland Mathematical Library, vol. 18, North-Holland Publishing Co., Amsterdam, 1978.

[Wal94] W. Walter, *Einführung in die Theorie der Distributionen*, Bibliographisches Institut, Mannheim, 1994.

[Wei01a]   L. Weis, *A new approach to maximal L_p-regularity*, Evolution Equations and Their Applications in Physical and Life Sciences (Bad Herrenalb, 1998), Lecture Notes in Pure and Appl. Math., vol. 215, Dekker, New York, 2001, pp. 195–214.

[Wei01b]   _____, *Operator-valued Fourier multiplier theorems and maximal L_p-regularity*, Math. Ann. **319** (2001), no. 4, 735–758.

[Wlo87]   J. Wloka, *Partial Differential Equations*, Cambridge University Press, Cambridge, 1987.

[Woo07]   I. Wood, *Maximal L^p-regularity for the Laplacian on Lipschitz domains*, Math. Z. **255** (2007), no. 4, 855–875.

[XL98]   T.-J. Xiao and J. Liang, *The Cauchy Problem for Higher-Order Abstract Differential Equations*, Lecture Notes in Mathematics, vol. 1701, Springer, Berlin, 1998.

[XL04]   _____, *Second order parabolic equations in Banach spaces with dynamic boundary conditions*, Trans. Amer. Math. Soc. **356** (2004), no. 12, 4787–4809 (electronic).

Index